Computer-Controlled Systems

PRENTICE HALL INFORMATION AND SYSTEM SCIENCES SERIES
Thomas Kailath, Editor

Computer-Controlled Systems

Theory and Design

THIRD EDITION

Karl J. Åström

Björn Wittenmark

Prentice Hall, Upper Saddle River, New Jersey 07458

Library of Congress Cataloging-in-Publication Data

Åström, Karl J. (Karl Johan)
   Computer-Controlled systems : theory and design / Karl J. Åström
Björn Wittenmark. -- 3rd ed.
    p. cm.
   Includes bibliographical references and index.
   ISBN 0-13-314899-8
   1. Automatic control--Data processing. I. Wittenmark, Björn.
  II. Title.
  TJ213.A78 1997   96-36745
  629.8'9--dc20   CIP

Publisher: Tom Robbins
Associate editor: Alice Dworkin
Editorial production supervision: Joseph Scordato
Editor-in-chief: Marcia Horton
Managing editor: Bayani Mendoza DeLeon
Copyeditor: Peter J. Zurita
Cover designer: Bruce Kenselaar
Director of production and manufacturing: David W. Riccardi
Manufacturing buyer: Donna Sullivan
Editorial assistant: Nancy Garcia

©1997 by Prentice-Hall, Inc.
Simon & Schuster/A Viacom Company
Upper Saddle River, NJ 07458

The author and publisher of this book have used their best efforts in preparing this book. These efforts
include the development, research, and testing of the theories and programs to determine their effectiveness.
The author and publisher make no warranty of any kind, expressed or implied, with regard to these programs
or the documentation contained in this book. The author and publisher shall not be liable in any event for
incidental or consequential damages in connection with, or arising out of, the furnishing, performance, or use
of these programs.

Printed in the United States of America

10 9 8 7 6 5 4 3 2 1

ISBN 0-13-314899-8

Prentice-Hall International (UK) Limited, London
Prentice-Hall of Australia Pty. Limited, Sydney
Prentice-Hall Canada Inc., Toronto
Prentice-Hall Hispanoamericana, S.A., Mexico
Prentice-Hall of India Private Limited, New Delhi
Prentice-Hall of Japan, Inc., Tokyo
Simon & Schuster Asia Pte. Ltd., Singapore
Editora Prentice-Hall do Brasil, Ltda., Rio de Janeiro

To Bia and Karin

Preface

A consequence of the revolutionary advances in microelectronics is that practically all control systems constructed today are based on microprocessors and sophisticated microcontrollers. By using computer-controlled systems it is possible to obtain higher performance than with analog systems, as well as new functionality. New software tools have also drastically improved the engineering efficiency in analysis and design of control systems.

Goal of the book   This book provides the necessary insight, knowledge, and understanding required to effectively analyze and design computer-controlled systems.

The new edition   This third edition is a major revision based on the advances in technology and the experiences from teaching to academic and industrial audiences. The material has been drastically reorganized with more than half the text rewritten. The advances in theory and practice of computer-controlled systems and a desire to put more focus on design issues have provided the motivation for the changes in the third edition. Many new results have been incorporated. By ruthless trimming and rewriting we are now able to include new material without increasing the size of the book. Experiences of teaching from a draft version have shown the advantages of the changes. We have been very pleased to note that students can indeed deal with design at a much earlier stage. This has also made it possible to go much more deeply into design and implementation.

Another major change in the third edition is that the computational tools MATLAB® and SIMULINK® have been used extensively. This changes the pedagogy in teaching substantially. All major results are formulated in such a way that the computational tools can be applied directly. This makes it easy to deal with complicated problems. It is thus possible to deal with many realistic design issues in the courses. The use of computational tools has been balanced by a strong emphasis of principles and ideas. Most key results have also been illustrated by simple pencil and paper calculations so that the students understand the workings of the computational tools.

Outline of the Book

Background Material   A broad outline of computer-controlled systems is presented in the first chapter. This gives a historical perspective on the development of computers, control systems, and relevant theory. Some key points of the theory and the behavior of computer-control systems are also given, together with many examples.

Analysis and Design of Discrete-Time Systems   It is possible to make drastic simplifications in analysis and design by considering only the behavior of the system at the sampling instants. We call this the computer-oriented view. It is the view of the system obtained by observing its behavior through the numbers in the computer. The reason for the simplicity is that the system can be described by linear difference equations with constant coefficients. This approach is covered in Chapters 2, 3, 4 and 5. Chapter 2 describes how the discrete-time systems are obtained by sampling continuous-time systems. Both state-space models and input-output models are given. Basic properties of the models are also given together with mathematical tools such as the z-transform. Tools for analysis are presented in Chapter 3.

Chapter 4 deals with the traditional problem of state feedback and observers, but it goes much further than what is normally covered in similar textbooks. In particular, the chapter shows how to deal with load disturbances, feedforward, and command-signal following. Taken together, these features give the controller a structure that can cope with many of the cases typically found in applications. An educational advantage is that students are equipped with tools to deal with real design issues after a very short time.

Chapter 5 deals with the problems of Chapter 4 from the input-output point of view, thereby giving an alternative view on the design problem. All issues discussed in Chapter 4 are also treated in Chapter 5. This affords an excellent way to ensure a good understanding of similarities and differences between state-space and polynomial approaches. The polynomial approach also makes it possible to deal with the problems of modeling errors and robustness, which cannot be conveniently handled by state-space techniques.

Having dealt with specific design methods, we present general aspects of the design of control systems in Chapter 6. This covers structuring of large systems as well as bottom-up and top-down techniques.

Broadening the View   Although many issues in computer-controlled systems can be dealt with using the computer-oriented view, there are some questions that require a detailed study of the behavior of the system between the sampling instants. Such problems arise naturally if a computer-controlled system is investigated through the analog signals that appear in the process. We call this the process-oriented view. It typically leads to linear systems with periodic coefficients. This gives rise to phenomena such as aliasing, which may lead to very undesirable effects unless special precautions are taken. It is very important to understand both this and the design of anti-aliasing filters when investigating computer-controlled systems. Tools for this are developed in Chapter 7.

When upgrading older control equipment, sometimes analog designs of controllers may be available already. In such cases it may be cost effective to have methods to translate analog designs to digital control directly. Methods for this are given in Chapter 8.

Implementation   It is not enough to know about methods of analysis and design. A control engineer should also be aware of implementation issues. These are treated in Chapter 9, which covers matters such as prefiltering and computational delays, numerics, programming, and operational aspects. At this stage the reader is well prepared for all steps in design, from concepts to computer implementation.

More Advanced Design Methods   To make more effective designs of control systems it is necessary to better characterize disturbances. This is done in Chapter 10. Having such descriptions it is then possible to design for optimal performance. This is done using state-space methods in Chapter 11 and by using polynomial techniques in Chapter 12. So far it has been assumed that models of the processes and their disturbances are available. Experimental methods to obtain such models are described in Chapter 13.

Prerequisites

The book is intended for a final-year undergraduate or a first-year graduate course for engineering majors. It is assumed that the reader has had an introductory course in automatic control. The book should be useful for an industrial audience.

Course Configurations

The book has been organized so that it can be used in different ways. An introductory course in computer-controlled systems could cover Chapters 1, 2, 3, 4, 5, and 9. A more advanced course might include all chapters in the book. A course for an industrial audience could contain Chapters 1, parts of Chapters 2, 3, 4, and 5, and Chapters 6, 7, 8, and 9. To get the full benefit of a course, it is important to supplement lectures with problem-solving sessions, simulation exercises, and laboratory experiments.

Computational Tools

Computer tools for analysis, design, and simulation are indispensable tools when working with computer-controlled systems. The methods for analysis and design presented in this book can be performed very conveniently using MATLAB®. Many of the exercises also cover this. Simulation of the system can similarly be done with Simnon® or SIMULINK®. There are 30 figures that illustrate various aspects of analysis and design that have been performed using MATLAB®, and 73 figures from simulations using SIMULINK®. Macros and m-files are available from anonymous FTP from `ftp.control.lth.se`, directory `/pub/books/ccs`. Other tools such as Simnon® and Xmath® can be used also.

Supplements

Complete solutions are available from the publisher for instructors who have adopted our book. Simulation macros, transparencies, and examples of examinations are available on the World Wide Web at `http://www.control.lth.se`; see Education/Computer-Controlled Systems.

Wanted: Feedback

As teachers and researchers in automatic control, we know the importance of feedback. Therefore, we encourage all readers to write to us about errors, potential miscommunications, suggestions for improvement, and also about what may be of special valuable in the material we have presented.

Acknowledgments

During the years that we have done research in computer-controlled systems and that we have written the book, we have had the pleasure and privilege of interacting with many colleagues in academia and industry throughout the world. Consciously and subconsciously, we have picked up material from the knowledge base called computer control. It is impossible to mention everyone who has contributed ideas, suggestions, concepts, and examples, but we owe each one our deepest thanks. The long-term support of our research by the Swedish Board of Industrial and Technical Development (NUTEK) and by the Swedish Research Council for Engineering Sciences (TFR) are gratefully acknowledged.

Finally, we want to thank some people who, more than others, have made it possible for us to write this book. We wish to thank Leif Andersson, who has been our TEXpert. He and Eva Dagnegård have been invaluable in solving many of our TEX problems. Eva Dagnegård and Agneta Tuszynski have done an excellent job of typing many versions of the manuscript. Most of the illustrations have been done by Britt-Marie Mårtensson. Without all their patience and understanding of our whims, never would there have been a final book. We also want to thank the staff at Prentice Hall for their support and professionalism in textbook production.

KARL J. ÅSTRÖM
BJÖRN WITTENMARK

Department of Automatic Control
Lund Institute of Technology
Box 118, S-221 00 Lund, Sweden

`karl_johan.astrom@control.lth.se`
`bjorn.wittenmark@control.lth.se`

Contents

Computer-Controlled Systems

1

Computer Control

1.1 Introduction

Practically all control systems that are implemented today are based on computer control. It is therefore important to understand computer-controlled systems well. Such systems can be viewed as approximations of analog-control systems, but this is a poor approach because the full potential of computer control is not used. At best the results are only as good as those obtained with analog control. It is much better to master computer-controlled systems, so that the full potential of computer control can be used. There are also phenomena that occur in computer-controlled systems that have no correspondence in analog systems. It is important for an engineer to understand this. The main goal of this book is to provide a solid background for understanding, analyzing, and designing computer-controlled systems.

A computer-controlled system can be described schematically as in Fig. 1.1. The output from the process $y(t)$ is a continuous-time signal. The output is converted into digital form by the analog-to-digital (A-D) converter. The A-D converter can be included in the computer or regarded as a separate unit, according to one's preference. The conversion is done at the sampling times, t_k. The computer interprets the converted signal, $\{y(t_k)\}$, as a sequence of numbers, processes the measurements using an algorithm, and gives a new sequence of numbers, $\{u(t_k)\}$. This sequence is converted to an analog signal by a digital-to-analog (D-A) converter. The events are synchronized by the real-time clock in the computer. The digital computer operates sequentially in time and each operation takes some time. The D-A converter must, however, produce a continuous-time signal. This is normally done by keeping the control signal constant between the conversions. In this case the system runs open loop in the time interval between the sampling instants because the control signal is constant irrespective of the value of the output.

The computer-controlled system contains both continuous-time signals and *sampled*, or *discrete-time*, signals. Such systems have traditionally been called

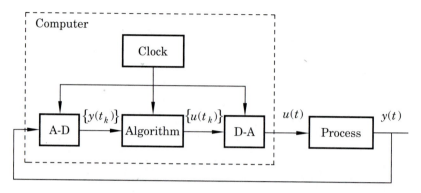

Figure 1.1 Schematic diagram of a computer-controlled system.

sampled-data systems, and this term will be used here as a synonym for *computer-controlled systems*.

The mixture of different types of signals sometimes causes difficulties. In most cases it is, however, sufficient to describe the behavior of the system at the sampling instants. The signals are then of interest only at discrete times. Such systems will be called *discrete-time systems*. Discrete-time systems deal with sequences of numbers, so a natural way to represent these systems is to use difference equations.

The purpose of the book is to present the control theory that is relevant to the analysis and design of computer-controlled systems. This chapter provides some background. A brief overview of the development of computer-control technology is given in Sec. 1.2. The need for a suitable theory is discussed in Sec. 1.3. Examples are used to demonstrate that computer-controlled systems cannot be fully understood by the theory of linear time-invariant continuous-time systems. An example shows not only that computer-controlled systems can be designed using continuous-time theory and approximations, but also that substantial improvements can be obtained by other techniques that use the full potential of computer control. Section 1.4 gives some examples of inherently sampled systems. The development of the theory of sampled-data systems is outlined in Sec. 1.5.

1.2 Computer Technology

The idea of using digital computers as components in control systems emerged around 1950. Applications in missile and aircraft control were investigated first. Studies showed that there was no potential for using the general-purpose digital computers that were available at that time. The computers were too big, they consumed too much power, and they were not sufficiently reliable. For this reason special-purpose computers—digital differential analyzers (DDAs)—were developed for the early aerospace applications.

The idea of using digital computers for process control emerged in the mid-1950s. Serious work started in March 1956 when the aerospace company Thomson Ramo Woodridge (TRW) contacted Texaco to set up a feasibility study. After preliminary discussions it was decided to investigate a polymerization unit at the Port Arthur, Texas, refinery. A group of engineers from TRW and Texaco made a thorough feasibility study, which required about 30 people-years. A computer-controlled system for the polymerization unit was designed based on the RW-300 computer. The control system went on-line March 12, 1959. The system controlled 26 flows, 72 temperatures, 3 pressures, and 3 compositions. The essential functions were to minimize the reactor pressure, to determine an optimal distribution among the feeds of 5 reactors, to control the hot-water inflow based on measurement of catalyst activity, and to determine the optimal recirculation.

The pioneering work done by TRW was noticed by many computer manufacturers, who saw a large potential market for their products. Many different feasibility studies were initiated and vigorous development was started. To discuss the dramatic developments, it is useful to introduce six periods:

Pioneering period ≈ 1955

Direct-digital-control period ≈ 1962

Minicomputer period ≈ 1967

Microcomputer period ≈ 1972

General use of digital control ≈ 1980

Distributed control ≈ 1990

It is difficult to give precise dates, because the development was highly diversified. There was a wide difference between different application areas and different industries; there was also considerable overlap. The dates given refer to the emergence of new approaches.

Pioneering Period

The work done by TRW and Texaco evoked substantial interest in process industries, among computer manufacturers, and in research organizations. The industries saw a potential tool for increased automation, the computer industries saw new markets, and universities saw a new research field. Many feasibility studies were initiated by the computer manufacturers because they were eager to learn the new technology and were very interested in knowing what a proper process-control computer should look like. Feasibility studies continued throughout the sixties.

The computer systems that were used were slow, expensive, and unreliable. The earlier systems used vacuum tubes. Typical data for a computer around 1958 were an addition time of 1 ms, a multiplication time of 20 ms, and a mean time between failures (MTBF) for a central processing unit of 50–100 h. To make full use of the expensive computers, it was necessary to have them perform many tasks. Because the computers were so unreliable, they controlled the process by printing instructions to the process operator or by changing the set points of analog regulators. These supervisory modes of operation were referred to as an *operator guide* and a *set-point control*.

The major tasks of the computer were to find the optimal operating conditions, to perform scheduling and production planning, and to give reports about production and raw-material consumption. The problem of finding the best operating conditions was viewed as a static optimization problem. Mathematical models of the processes were necessary in order to perform the optimization. The models used—which were quite complicated—were derived from physical models and from regression analysis of process data. Attempts were also made to carry out on-line optimization.

Progress was often hampered by lack of process knowledge. It also became clear that it was not sufficient to view the problems simply as static optimization problems; dynamic models were needed. A significant proportion of the effort in many of the feasibility studies was devoted to modeling, which was quite time-consuming because there was a lack of good modeling methodology. This stimulated research into system-identification methods.

A lot of experience was gained during the feasibility studies. It became clear that process control puts special demands on computers. The need to respond quickly to demands from the process led to development of the *interrupt feature*, which is a special hardware device that allows an external event to interrupt the computer in its current work so that it can respond to more urgent process tasks. Many sensors that were needed were not available. There were also several difficulties in trying to introduce a new technology into old industries.

The progress made was closely monitored at conferences and meetings and in journals. A series of articles describing the use of computers in process control was published in the journal *Control Engineering*. By March 1961, 37 systems had been installed. A year later the number of systems had grown to 159. The applications involved control of steel mills and chemical industries and generation of electric power. The development progressed at different rates in different industries. Feasibility studies continued through the 1960s and the 1970s.

Direct-Digital-Control Period

The early installations of control computers operated in a supervisory mode, either as an operator guide or as a set-point control. The ordinary analog-control equipment was needed in both cases. A drastic departure from this approach was made by Imperial Chemical Industries (ICI) in England in 1962. A complete analog instrumentation for process control was replaced by one computer, a Ferranti Argus. The computer measured 224 variables and controlled 129 valves directly. This was the beginning of a new era in process control: Analog technology was simply replaced by digital technology; the function of the system was the same. The name *direct digital control* (DDC) was coined to emphasize that the computer-controlled the process directly. In 1962 a typical process-control computer could add two numbers in 100 μs and multiply them in 1 ms. The MTBF was around 1000 h.

Cost was the major argument for changing the technology. The cost of an analog system increased linearly with the number of control loops; the initial cost of a digital system was large, but the cost of adding an additional loop was small. The digital system was thus cheaper for large installations. Another advantage was that operator communication could be changed drastically; an operator communication panel could replace a large wall of analog instruments. The panel used in the ICI system was very simple—a digital display and a few buttons.

Flexibility was another advantage of the DDC systems. Analog systems were changed by rewiring; computer-controlled systems were changed by reprogramming. Digital technology also offered other advantages. It was easy to have interaction among several control loops. The parameters of a control loop could be made functions of operating conditions. The programming was simplified by introducing special DDC languages. A user of such a language did not need to know anything about programming, but simply introduced inputs, outputs, regulator types, scale factors, and regulator parameters into tables. To the user the systems thus looked like a connection of ordinary regulators. A drawback of the systems was that it was difficult to do unconventional control strategies. This certainly hampered development of control for many years.

DDC was a major change of direction in the development of computer-controlled systems. Interest was focused on the basic control functions instead of the supervisory functions of the earlier systems. Considerable progress was made in the years 1963–1965. Specifications for DDC systems were worked out jointly between users and vendors. Problems related to choice of sampling period and control algorithms, as well as the key problem of reliability, were discussed extensively. The DDC concept was quickly accepted although DDC systems often turned out to be more expensive than corresponding analog systems.

Minicomputer Period

There was substantial development of digital computer technology in the 1960s. The requirements on a process-control computer were neatly matched with progress in integrated-circuit technology. The computers became smaller, faster, more reliable, and cheaper. The term *minicomputer* was coined for the new computers that emerged. It was possible to design efficient process-control systems by using minicomputers.

The development of minicomputer technology combined with the increasing knowledge gained about process control with computers during the pioneering and DDC periods caused a rapid increase in applications of computer control. Special process-control computers were announced by several manufacturers. A typical process computer of the period had a word length of 16 bits. The primary memory was 8–124 k words. A disk drive was commonly used as a secondary memory. The CDC 1700 was a typical computer of this period, with an addition time of 2 μs and a multiplication time of 7 μs. The MTBF for a central processing unit was about 20,000 h.

An important factor in the rapid increase of computer control in this period was that digital computer control now came in a smaller "unit." It was thus possible to use computer control for smaller projects and for smaller problems. Because of minicomputers, the number of process computers grew from about 5000 in 1970 to about 50,000 in 1975.

Microcomputer Period and General Use of Computer Control

The early use of computer control was restricted to large industrial systems because digital computing was only available in expensive, large, slow, and unreliable machines. The minicomputer was still a fairly large system. Even as performance continued to increase and prices to decrease, the price of a minicomputer mainframe in 1975 was still about $10,000. This meant that a small system rarely cost less than $100,000. Computer control was still out of reach for a large number of control problems. But with the development of the microcomputer in 1972, the price of a card computer with the performance of a 1975 minicomputer dropped to $500 in 1980. Another consequence was that digital computing power in 1980 came in quanta as small as $50. The development of microelectronics has continued with advances in very large-scale integration (VLSI) technology; in the 1990s microprocessors became available for a few dollars. This has had a profound impact on the use of computer control. As a result practically all controllers are now computer-based. Mass markets such as automotive electronics has also led to the development of special-purpose computers, called microcontrollers, in which a standard computer chip has been augmented with A-D and D-A converters, registers, and other features that make it easy to interface with physical equipment.

Practically all control systems developed today are based on computer control. Applications span all areas of control, generation, and distribution of electricity; process control; manufacturing; transportation; and entertainment. Mass-market applications such as automotive electronics, CD players, and videos are particularly interesting because they have motivated computer manufacturers to make chips that can be used in a wide variety of applications.

As an illustration Fig. 1.2 shows an example of a single-loop controller for process control. Such systems were traditionally implemented using pneumatic or electronic techniques, but they are now always computer-based. The controller has the traditional proportional, integral, and derivative actions (PID), which are implemented in a microprocessor. With digital control it is also possible to obtain added functionality. In this particular case, the regulator is provided with automatic tuning, gain scheduling, and continuous adaptation of feedforward and feedback gains. These functions are difficult to implement with analog techniques. The system is a typical case that shows how the functionality of a traditional product can be improved substantially by use of computer control.

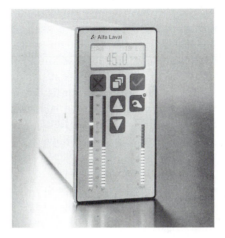

Figure 1.2 A standard single-loop controller for process control. (By courtesy of Alfa Laval Automation, Stockholm, Sweden.)

Logic, Sequencing, and Control

Industrial automation systems have traditionally had two components, controllers and relay logic. Relays were used to sequence operations such as startup and shutdown and they were also used to ensure safety of the operations by providing interlocks. Relays and controllers were handled by different categories of personnel at the plant. Instrument engineers were responsible for the controllers and electricians were responsible for the relay systems. We have already discussed how the controllers were influenced by microcomputers. The relay systems went through a similar change with the advent of microelectronics. The so-called *programmable logic controller (PLC)* emerged in the beginning of the 1970s as replacements for relays. They could be programmed by electricians and in familiar notations, that is, as rungs of relay contact logic or as logic (AND/OR) statements. Americans were the first to bring this novelty to the market, relying primarily on relay contact logic, but the Europeans were hard on their heels, preferring logic statements. The technology became a big success, primarily in the discrete parts manufacturing industry (for obvious reasons). However, in time, it evolved to include regulatory control and data-handling capabilities as well, a development that has broadened the range of applications for it. The attraction was, and is, the ease with which controls, including intraloop dependencies, can be implemented and changed, without any impact on hardware.

Distributed Control

The microprocessor has also had a profound impact on the way computers were applied to control entire production plants. It became economically feasible to develop systems consisting of several interacting microcomputers sharing the overall workload. Such systems generally consist of process stations, controlling the process; operator stations, where process operators monitor activities; and various auxiliary stations, for example, for system configuration and programming, data storage, and so on, all interacting by means of some kind of communications network. The allure was to boost performance by facilitating parallel multitasking, to improve overall availability by not putting "all the eggs in one basket," to further expandability and to reduce the amount of control cabling. The first system of this kind to see the light of day was Honeywell's TDC 2000 (the year was 1975), but it was soon followed by others. The term "distributed control" was coined. The first systems were oriented toward regulatory control, but over the years distributed control systems have adopted more and more of the capabilities of programmable (logic) controllers, making today's distributed control systems able to control all aspects of production and enabling operators to monitor and control activities from a single computer console.

Plantwide Supervision and Control

The next development phase in industrial process-control systems was facilitated by the emergence of common standards in computing, making it possible to integrate virtually all computers and computer systems in industrial plants into a monolithic whole to achieve real-time exchange of data across what used to be closed system borders. Such interaction enables

- top managers to investigate all aspects of operations

- production managers to plan and schedule production on the basis of current information

- order handlers and liaison officers to provide instant and current information to inquiring customers

- process operators to look up the cost accounts and the quality records of the previous production run to do better next time

all from the computer screens in front of them, all in real time. An example of such a system is shown in Fig. 1.3. ABB's Advant OCS (open control system) seems to be a good exponent of this phase. It consists of process controllers with local and/or remote I/O, operator stations, information management stations, and engineering stations that are interconnected by high-speed communications buses at the field, process-sectional, and plantwide levels. By supporting industry standards in computing such as Unix, Windows, and SQL, it makes interfacing with the surrounding world of computers easy. The system features a real-time process database that is distributed among the process controllers of the system to avoid redundancy in data storage, data inconsistency, and to

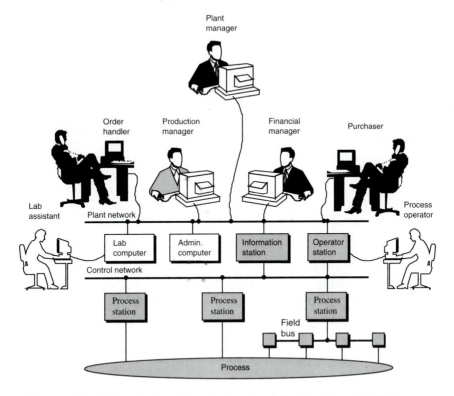

Figure 1.3 Modern industrial control systems like the Advant OCS tie computers together and help create a common uniform computer environment supporting all industrial activities, from input to output, from top to bottom. (By courtesy of ABB Industrial System, Västerås, Sweden.)

improve system availability. All database instances are addressed by their symbolic names for ease of access. The functional relationships among database instances are defined by application programs that are developed graphically by laying out the required function blocks on diagram sheets, then connecting these blocks to other blocks so as to process inputs and outputs. There are function blocks as simple as logic AND and OR gates and as complex as adaptive controllers. Interconnections can be given symbolic names for easy reference and the programming tool assigns page references automatically. The result is in the form of graphic program diagrams to IEC's standards to the applicable extent that diagrams can be easily understood by all concerned. Process operators then monitor and control the process from a variety of live displays, from process schematics and trend graphs to alarm lists and reports. Live measured values and status indications reveal the current situation. Information from process controllers, information management stations, and external computers can be mixed and matched at will. The interaction style is every bit as friendly and intuitive as that of today's personal computers.

Information-Handling Capabilities

Advant OCS offers basic ready-to-use information management functions such as historical data storage and playback, a versatile report generator, and a supplementary calculation package. It also offers open interfaces to third-party applications and to other computers in the plant. The historical data-storage and -retrieval service enables users to collect data from any system station at specified intervals, on command or on occurrence of specified events, performs a wide range of calculations on this data, and stores the results in so-called logs. Such logs can be accessed for presentation on any operator station or be used by applications on information stations or on external stations for a wide range of purposes. A report generator makes it possible to collect data for reports from the process database, from other reports, or the historical database. Output can be generated at specified times, on occurrence of specified events, or on request by an operator or software application. Unix- or Windows-based application programming interfaces offer a wide range of system services that give programmers a head start and safeguard engineering quality. Applications developed on this basis can be installed on the information management stations of the system, that is, close enough to the process to offer real-time performance.

The Future

Based on the dramatic developments in the past, it is tempting to speculate about the future. There are four areas that are important for the development of computer process control.

- Process knowledge

- Measurement technology

- Computer technology

- Control theory

Knowledge about process control and process dynamics is increasing slowly but steadily. The possibilities of learning about process characteristics are increasing substantially with the installation of process-control systems because it is then easy to collect data, perform experiments, and analyze the results. Progress in system identification and data analysis has also provided valuable information.

Progress in measurement technology is hard to predict. Many things can be done using existing techniques. The possibility of combining outputs of several different sensors with mathematical models is interesting. It is also possible to obtain automatic calibration with a computer. The advent of new sensors will, however, always offer new possibilities.

Spectacular developments are expected in computer technology with the introduction of VLSI. The ratio of price to performance will continue to drop substantially. The future microcomputers are expected to have computing power greater than the large mainframes of today. Substantial improvements are also expected in display techniques and in communications.

Programming has so far been one of the bottlenecks. There were only marginal improvements in productivity in programming from 1950 to 1970. At the end of the 1970s, many computer-controlled systems were still programmed in assembler code. In the computer-control field, it has been customary to overcome some of the programming problems by providing table-driven software. A user of a DDC, system is thus provided with a so-called DDC package that allows the user to generate a DDC system simply by filling in a table, so very little effort is needed to generate a system. The widespread use of packages hampers development, however, because it is very easy to use DDC, but it is a major effort to do something else. So only the well-proven methods are tried.

Control theory has made substantial progress since 1955. Only some of this theory, however, has made its way into existing computer-controlled systems, even though feasibility studies have indicated that significant improvements can be made. Model predictive control and adaptive control are some of the theoretical areas that are being applied in the industry today. To use these theories, it is necessary to fully understand the basic concepts of computer control. One reason for not using more complex digital controllers is the cost of programming. As already mentioned, it requires little effort to use a package provided by a vendor. It is, however, a major effort to try to do something else. Several signs show that this situation can be expected to change. Personal computers with interactive high-level languages are starting to be used for process control. With an interactive language, it is very easy to try new things. It is, however, unfortunately very difficult to write *safe* real-time control systems. This will change as better interactive systems become available.

Thus, there are many signs that point to interesting developments in the field of computer-controlled systems. A good way to be prepared is to learn the theory presented in this book.

1.3 Computer-Control Theory

Using computers to implement controllers has substantial advantages. Many of the difficulties with analog implementation can be avoided. For example, there are no problems with accuracy or drift of the components. It is very easy to have sophisticated calculations in the control law, and it is easy to include logic and nonlinear functions. Tables can be used to store data in order to accumulate knowledge about the properties of the system. It is also possible to have effective user interfaces.

A schematic diagram of a computer-controlled system is shown in Fig. 1.1. The system contains essentially five parts: the process, the A-D and D-A converters, the control algorithm, and the clock. Its operation is controlled by the clock. The times when the measured signals are converted to digital form are called the *sampling instants*; the time between successive samplings is called the *sampling period* and is denoted by h. Periodic sampling is normally used, but there are, of course, many other possibilities. For example, it is possible to sample when the output signals have changed by a certain amount. It is also

possible to use different sampling periods for different loops in a system. This
is called *multirate sampling*.

In this section we will give examples that illustrate the differences and the
similarities of analog and computer-controlled systems. It will be shown that
essential new phenomena that require theoretical attention do indeed occur.

Time Dependence

The presence of the the clock in Fig. 1.1 makes computer-controlled systems
time-varying. Such systems can exhibit behavior that does not occur in linear
time-invariant systems.

Example 1.1   Time dependence in digital filtering

A digital filter is a simple example of a computer-controlled system. Suppose that
we want to implement a compensator that is simply a first-order lag. Such a compensator can be implemented using A-D conversion, a digital computer, and D-A

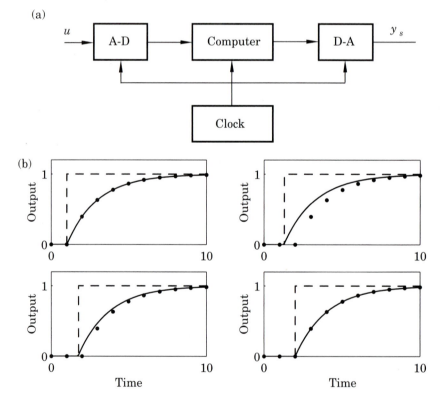

Figure 1.4   (a) Block diagram of a digital filter. (b) Step responses (dots)
of a digital computer implementation of a first-order lag for different delays
in the input step (dashed) compared with the first sampling instant. For
comparison the response of the corresponding continuous-time system (solid)
is also shown.

conversion. The first-order differential equation is approximated by a first-order difference equation. The step response of such a system is shown in Fig. 1.4. The figure clearly shows that the sampled system is not time-invariant because the response depends on the time when the step occurs. If the input is delayed, then the output is delayed by the same amount only if the delay is a multiple of the sampling period.   ∎

The phenomenon illustrated in Fig. 1.4 depends on the fact that the system is controlled by a clock (compare with Fig. 1.1). The response of the system to an external stimulus will then depend on how the external event is synchronized with the internal clock of the computer system.

Because sampling is often periodic, computer-controlled systems will often result in closed-loop systems that are linear *periodic systems*. The phenomenon shown in Fig. 1.4 is typical for such systems. Later we will illustrate other consequences of periodic sampling.

A Naive Approach to Computer-Controlled Systems

We may expect that a computer-controlled system behaves as a continuous-time system if the sampling period is sufficiently small. This is true under very reasonable assumptions. We will illustrate this with an example.

Example 1.2   Controlling the arm of a disk drive

A schematic diagram of a disk-drive assembly is shown in Fig. 1.5. Let J be the moment of inertia of the arm assembly. The dynamics relating the position y of the arm to the voltage u of the drive amplifier is approximately described by the transfer function

$$G(s) = \frac{k}{Js^2} \tag{1.1}$$

where k is a constant. The purpose of the control system is to control the position of the arm so that the head follows a given track and that it can be rapidly moved to a different track. It is easy to find the benefits of improved control. Better trackkeeping allows narrower tracks and higher packing density. A faster control system reduces the search time. In this example we will focus on the search problem, which is a typical servo problem. Let u_c be the command signal and denote Laplace transforms with capital letters. A simple servo controller can be described by

$$U(s) = \frac{bK}{a} U_c(s) - K\frac{s+b}{s+a} Y(s) \tag{1.2}$$

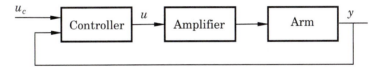

Figure 1.5   A system for controlling the position of the arm of a disk drive.

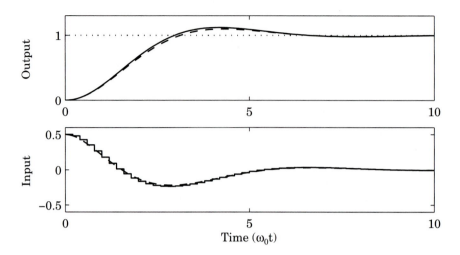

Figure 1.6 Simulation of the disk arm servo with analog (dashed) and computer control (solid). The sampling period is $h = 0.2/\omega_0$.

This controller is a two-degree-of-freedom controller where the feedback from the measured signal is simply a lead-lag filter. If the controller parameters are chosen as

$$a = 2\omega_0$$
$$b = \omega_0/2$$
$$K = 2\,\frac{J\omega_0^2}{k}$$

a closed system with the characteristic polynomial

$$P(s) = s^3 + 2\omega_0 s^2 + 2\omega_0^2 s + \omega_0^3$$

is obtained. This system has a reasonable behavior with a settling time to 5% of $5.52/\omega_0$. See Fig. 1.6. To obtain an algorithm for a computer-controlled system, the control law given by (1.2) is first written as

$$U(s) = \frac{bK}{a} U_c(s) - KY(s) + K\,\frac{a-b}{s+a}\,Y(s) = K\left(\frac{b}{a} U_c(s) - Y(s) + X(s)\right)$$

This control law can be written as

$$u(t) = K\left(\frac{b}{a} u_c(t) - y(t) + x(t)\right)$$

$$\frac{dx}{dt} = -ax + (a-b)y \tag{1.3}$$

To obtain an algorithm for a control computer, the derivative dx/dt is approximated with a difference. This gives

$$\frac{x(t+h) - x(t)}{h} = -ax(t) + (a-b)y(t)$$

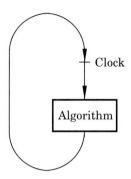

Figure 1.7 Scheduling a computer program.

The following approximation of the continuous algorithm (1.3) is then obtained:

$$u(t_k) = K\left(\frac{b}{a}u_c(t_k) - y(t_k) + x(t_k)\right)$$

$$x(t_k + h) = x(t_k) + h\Big((a - b)y(t_k) - ax(t_k)\Big)$$

(1.4)

This control law should be executed at each sampling instant. This can be accomplished with the following computer program.

```
y: = adin(in2)            {read process value}
u:=K*(a/b*uc-y+x)
dout(u)                   {output control signal}
newx:=x+h*((b-a)*y-a*x)
```

Arm position y is read from an analog input. Its desired value u_c is assumed to be given digitally. The algorithm has one state, variable x, which is updated at each sampling instant. The control law is computed and the value is converted to an analog signal. The program is executed periodically with period h by a scheduling program, as illustrated in Fig. 1.7. Because the approximation of the derivative by a difference is good if the interval h is small, we can expect the behavior of the computer-controlled system to be close to the continuous-time system. This is illustrated in Fig. 1.6, which shows the arm positions and the control signals for the systems with $h = 0.2/\omega_0$. Notice that the control signal for the computer-controlled system is constant between the sampling instants. Also notice that the difference between the outputs of the systems is very small. The computer-controlled system has slightly higher overshoot and the settling time to 5% is a little longer, $5.7/\omega_0$ instead of $5.5/\omega_0$. The difference between the systems decreases when the sampling period decreases. When the sampling period increases the computer-controlled system will, however, deteriorate. This is illustrated in Fig. 1.8, which shows the behavior of the system for the sampling periods $h = 0.5/\omega_0$ and $h = 1.08/\omega_0$. The response is quite reasonable for short sampling periods, but the system becomes unstable for long sampling periods. ∎

We have thus shown that it is straightforward to obtain an algorithm for computer control simply by writing the continuous-time control law as a differential equation and approximating the derivatives by differences. The example indi-

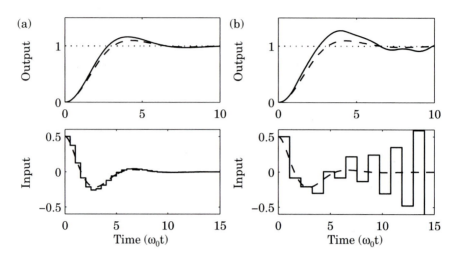

Figure 1.8 Simulation of the disk arm servo with computer control having sampling rates (a) $h = 0.5/\omega_0$ and (b) $h = 1.08/\omega_0$. For comparison, the signals for analog control are shown with dashed lines.

cated that the procedure seemed to work well if the sampling period was sufficiently small. The overshoot and the settling time are, however, a little larger for the computer-controlled system. This approach to design of computer-controlled systems will be discussed fully in the following chapters.

Deadbeat Control

Example 1.2 seems to indicate that a computer-controlled system will be inferior to a continuous-time example. We will now show that this is not necessarily the case. The periodic nature of the control actions can be actually used to obtain control strategies with superior performance.

Example 1.3   Disk drive with deadbeat control

Consider the disk drive in the previous example. Figure 1.9 shows the behavior of a computer-controlled system with a very long sampling interval $h = 1.4/\omega_0$. For comparison we have also shown the arm position, its velocity, and the control signal for the continuous controller used in Example 1.2. Notice the excellent behavior of the computer-controlled system. It settles much quicker than the continuous-time system even if control signals of the same magnitude are used. The 5% settling time is $2.34/\omega_0$, which is much shorter than the settling time $5.5/\omega_0$ of the continuous system. The output also reaches the desired position without overshoot and it remains constant when it has achieved its desired value, which happens in finite time. This behavior cannot be obtained with continuous-time systems because the solutions to such systems are sums of functions that are products of polynomials and exponential functions. The behavior obtained can be also described in the following way: The arm accelerates with constant acceleration until is is halfway to the desired position and it then decelerates with constant retardation. The control

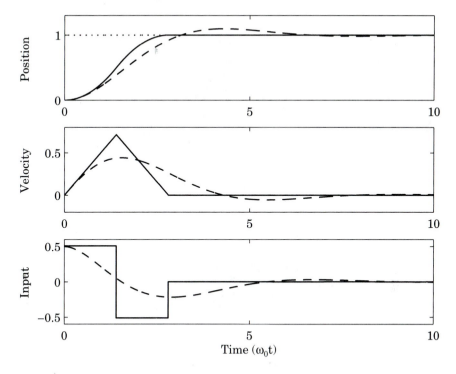

Figure 1.9   Simulation of the disk arm servo with deadbeat control (solid). The sampling period is $h = 1.4/\omega_0$. The analog controller from Example 1.2 is also shown (dashed).

strategy used has the same form as the control strategy in Example 1.2, that is,

$$u(t_k) = t_0 u_c(t_k) + t_1 u_c(t_{k-1}) - s_0 y(t_k) - s_1 y(t_{k-}) - r_1 u(t_{k-1}) \tag{1.5}$$

The parameter values are different. When controlling the disk drive, the system can be implemented in such a way that sampling is initiated when the command signal is changed. In this way it is possible to avoid the extra time delay that occurs due to the lack of synchronization of sampling and command signal changes illustrated in Fig. 1.4.   ∎

The example shows that control strategies with different behavior can be obtained with computer control. In the particular example the response time can be reduced by a factor of 2. The control strategy in Example 1.3 is called *deadbeat control* because the system is at rest when the desired position is reached. Such a control scheme cannot be obtained with a continuous-time controller.

Aliasing

One property of the time-varying nature of computer-controlled systems was illustrated in Fig. 1.4. We will now illustrate another property that has far-reaching consequences. Stable linear time-invariant systems have the property

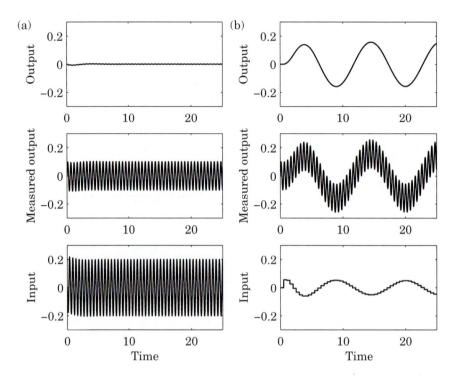

Figure 1.10 Simulation of the disk arm servo with analog and computer control. The frequency ω_0 is 1, the sampling period is $h = 0.5$, and there is a measurement noise $n = 0.1 \sin 12t$. (a) Continuous-time system; (b) sampled-data system.

that the steady-state response to sinusoidal excitations is sinusoidal with the frequency of the excitation signal. It will be shown that computer-controlled systems behave in a much more complicated way because sampling will create signals with new frequencies. This can drastically deteriorate performance if proper precautions are not taken.

Example 1.4   Sampling creates new frequencies

Consider the systems for control of the disk drive arm discussed in Example 1.2. Assume that the frequency ω_0 is 1 rad/s, let the sampling period be $h = 0.5/\omega_0$, and assume that there is a sinusoidal measurement noise with amplitude 0.1 and frequency 12 rad/s. Figure 1.10 shows interesting variables for the continuous-time system and the computer-controlled system. There is clearly a drastic difference between the systems. For the continuous-time system, the measurement noise has very little influence on the arm position. It does, however, create substantial control action with the frequency of the measurement noise. The high-frequency measurement noise is not noticeable in the control signal for the computer-controlled system, but there is also a substantial low-frequency component.

To understand what happens, we can consider Fig. 1.11, which shows the control signal and the measured signal on an expanded scale. The figure shows

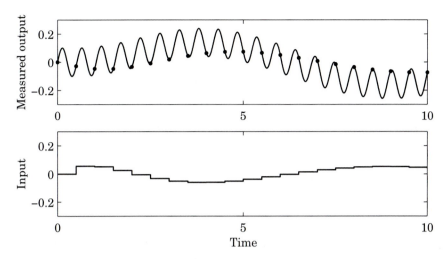

Figure 1.11 Simulation of the disk arm servo with computer control. The frequency ω_0 is 1, the sampling period is $h = 0.5$, and there is a measurement noise $n = 0.1 \sin 12t$.

that there is a considerable variation of the measured signal over the sampling period and the low-frequency variation is obtained by sampling the high-frequency signal at a slow rate.   ∎

We have thus made the striking observation that sampling creates signals with new frequencies. This is clearly a phenomenon that we must understand in order to deal with computer-controlled systems. At this stage we do not wish to go into the details of the theory; let it suffice to mention that sampling of a signal with frequency ω creates signal components with frequencies

$$\omega_{\text{sampled}} = n\omega_s \pm \omega \tag{1.6}$$

where $\omega_s = 2\pi/h$ is the sampling frequency, and n is an arbitrary integer. Sampling thus creates new frequencies. This is further discussed in Sec. 7.4.

In the particular example we have $\omega_s = 4\pi = 12.57$, and the measurement signal has the frequency 12 rad/s. In this case we find that sampling creates a signal component with the frequency 0.57 rad/s. The period of this signal is thus 11 s. This is the low-frequency component that is clearly visible in Fig. 1.11.

Example 1.4 illustrated that lower frequencies can be created by sampling. It follows from (1.6) that sampling also can give frequencies that are higher than the excitation frequency. This is illustrated in the following example.

Example 1.5 Creation of higher frequencies by sampling

Figure 1.12 shows what can happen when a sinusoidal signal of frequency 4.9 Hz is applied to the system in Example 1.1, which has a sampling period of 10 Hz. It follows from Eq. (1.6) that a signal component with frequency 5.1 Hz is created by sampling. This signal interacts with the original signal with frequency 4.9 Hz to give the beating of 0.1 Hz shown in the figure.   ∎

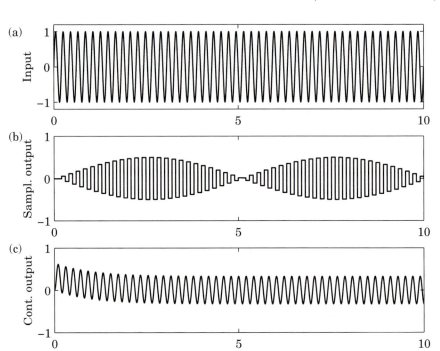

Figure 1.12 Sinusoidal excitation of the sampled system in Example 1.5. (a) Input sinusoidal with frequency 4.9 Hz. (b) Sampled-system output. The sampling period is 0.1 s. (c) Output of the corresponding continuous-time system.

There are many aspects of sampled systems that indeed can be understood by linear time-invariant theory. The examples given indicate, however, that the sampled systems cannot be fully understood within that framework. It is thus useful to have other tools for analysis.

The phenomenon that the sampling process creates new frequency components is called *aliasing*. A consequence of Eq. (1.6) is that there will be low-frequency components created whenever the sampled signal contains frequencies that are larger than half the sampling frequency. The frequency $\omega_N = \omega_s/2$ is called the *Nyquist frequency* and is an important parameter of a sampled system.

Presampling Filters or Antialiasing Filters

To avoid the difficulties illustrated in Fig. 1.10, it is essential that all signal components with frequencies higher than the Nyquist frequency are removed before a signal is sampled. By doing this the signals sampled will not change much over a sampling interval and the difficulties illustrated in the previous examples are avoided. The filters that reduce the high-frequency components of the signals are called *antialiasing filters*. These filters are an important component of computer-controlled systems. The proper selection of sampling periods and antialiasing filters are important aspects of the design of computer-controlled systems.

Difference Equations

Although a computer-controlled system may have a quite complex behavior, it is very easy to describe the behavior of the system at the sampling instants. We will illustrate this by analyzing the disk drive with a deadbeat controller.

Example 1.6   Difference equations

The input-output properties of the process Eq. (1.1) can be described by

$$y(t_k) - 2y(t_{k-1}) + y(t_{k-2}) = \frac{kh^2}{2J}\left(u(t_{k-1}) + u(t_{k-2})\right) \tag{1.7}$$

This equation is exact if the control signal is constant over the sampling intervals. The deadbeat control strategy is given by Eq. (1.5) and the closed-loop system thus can be described by the equations.

$$y(t_k) - 2y(t_{k-1}) + y(t_{k-2}) = \alpha\left(u(t_{k-1}) + u(t_{k-2})\right)$$
$$u(t_{k-1}) + r_1 u(t_{k-2}) = t_0 u_c(t_{k-1}) - s_0 y(t_{k-1}) - s_1 y(t_{k-2}) \tag{1.8}$$

where $\alpha = kh^2/2J$. Elimination of the control signal u between these equations gives

$$y(t_k) + (r_1 - 2 + \alpha s_0)y(t_{k-1}) + \left(1 - 2r_1 + \alpha(s_0 + s_1)\right)y(t_{k-2}) + (r_1 + \alpha s_1)y(t_{k-3})$$
$$= \frac{\alpha t_0}{2}\left(u_c(t_{k-1}) + u_c(t_{k-2})\right)$$

The parameters of the deadbeat controller are given by

$$r_1 = 0.75$$
$$s_0 = \frac{1.25}{\alpha} = \frac{2.5J}{kh^2}$$
$$s_1 = -\frac{0.75}{\alpha} = -\frac{1.5J}{kh^2}$$
$$t_0 = \frac{1}{4\alpha} = \frac{1}{2}$$

With these parameters the closed-loop system becomes

$$y(t_k) = \frac{1}{2}\left(u_c(t_{k-1}) + u_c(t_{k-2})\right)$$

It follows from this equation that the output is the average value of the past two values of the command signal. Compare with Fig. 1.9.   ∎

The example illustrates that the behavior of the computer-controlled system at the sampling instants is described by a linear difference equation. This observation is true for general linear systems. Difference equations, therefore, will be a key element of the theory of computer-controlled systems, they play the same role as differential equations for continuous systems, and they will give the values of the important system variables at the sampling instants. If we are satisfied by this knowledge, it is possible to develop a simple theory for analysis and design of sampled systems. To have a more complete knowledge of the behavior of the systems, we must also analyse the behavior between the sampling instants and make sure that the system variables do not change too much over a sampling period.

Is There a Need for a Theory for Computer-Controlled Systems?

The examples in this section have demonstrated that computer-controlled systems can be designed simply by using continuous-time theory and approximating the differential equations describing the controllers by difference equations. The examples also have shown that computer-controlled systems have the potential of giving control schemes, such as the deadbeat strategy, with behavior that cannot be obtained by continuous-time systems. It also has been demonstrated that sampling can create phenomena that are not found in linear time-invariant systems. It also has been demonstrated that the selection of the sampling period is important and that it is necessary to use antialiasing filters. These issues clearly indicate the need for a theory for computer-controlled systems.

1.4 Inherently Sampled Systems

Sampled models are natural descriptions for many phenomena. The theory of sampled-data systems, therefore, has many applications outside the field of computer control.

Sampling due to the Measurement System

In many cases, sampling will occur naturally in connection with the measurement procedure. A few examples follow.

Example 1.7   Radar

When a radar antenna rotates, information about range and direction is naturally obtained once per revolution of the antenna. A sampled model is thus the natural way to describe a radar system. Attempts to describe radar systems were, in fact, one of the starting points of the theory of sampled systems.   ∎

Example 1.8   Analytical instruments

In process-control systems, there are many variables that cannot be measured on-line, so a sample of the product is analyzed off-line in an analytical instrument such as a mass spectrograph or a chromatograph.   ∎

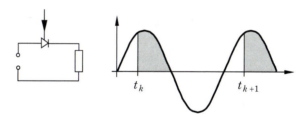

Figure 1.13   Thyristor control circuit.

Example 1.9   Economic systems

Accounting procedures in economic systems are often tied to the calendar. Although transactions may occur at any time, information about important variables is accumulated only at certain times—for example, daily, weekly, monthly, quarterly, or yearly.   ∎

Example 1.10   Magnetic flow meters

A magnetic flow meter is based on the principle that current that moves in a magnetic field generates a voltage. In a typical meter a magnetic field is generated across the pipe and the voltage is measured in a direction orthogonal to the field. To compensate for eletrolytic voltages that often are present, it is common to use a pulsed operation in which the field is switched on and off periodically. This switching causes an inherent sampling.   ∎

Sampling due to Pulsed Operation

Many systems are inherently sampled because information is transmitted using pulsed information. Electronic circuits are a prototype example. They were also one source of inspiration for the development of sampled-data theory. Other examples follow.

Example 1.11   Thyristor control

Power electronics using thyristors are sampled systems. Consider the circuit in Fig. 1.13. The current can be switched on only when the voltage is positive. When the current is switched on, it remains on until the current has a zero crossing. The current is thus synchronized to the periodicity of the power supply. The variation of the ingition time will cause the sampling period to vary, which must be taken care of when making models for thyristor circuits.   ∎

Example 1.12   Biological systems

Biological systems are fundamentally sampled because the signal transmission in the nervous system is in the form of pulses.   ∎

Example 1.13   Internal-combustion engines

An internal-combustion engine is a sampled system. The ignition can be viewed as a clock that synchronizes the operation of the engine. A torque pulse is generated at each ignition.   ∎

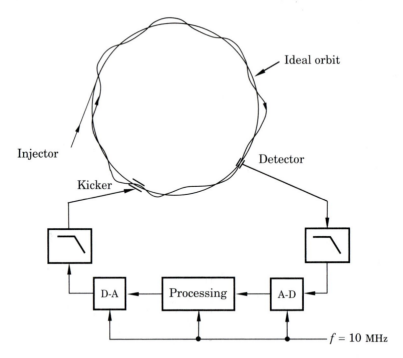

Figure 1.14   Particle accelerator with stochastic cooling.

Example 1.14   Particle accelerators

Particle accelerators are the key experimental tool in particle physics. The Dutch engineer Simnon van der Meer made a major improvement in accelerators by introducing feedback to control particle paths, which made it possible to increase the beam intensity and to improve the beam quality substantially. The method, which is called stochastic cooling, was a key factor in the successful experiments at CERN. As a result van der Meer shared the 1984 Nobel Prize in Physics with Carlo Rubbia.

     A schematic diagram of the system is shown in Fig. 1.14. The particles enter into a circular orbit via the injector. The particles are picked up by a detector at a fixed position and the energy of the particles is increased by the kicker, which is located at a fixed position. The system is inherently sampled because the particles are only observed when they pass the detector and control only acts when they pass the kicker.

     From the point of view of sampled systems, it is interesting to observe that there is inherent sampling both in sensing and actuation.   ∎

The systems in these examples are periodic because of their pulsed operation. Periodic systems are quite difficult to handle, but they can be considerably simplified by studying the systems at instants synchronized with the pulses— that is, by using sampled-data models. The processes then can be described as time-invariant discrete-time systems at the sampling instants. Examples 1.11 and 1.13 are of this type.

1.5 How Theory Developed

Although the major applications of the theory of sampled systems are currently in computer control, many of the problems were encountered earlier. In this section some of the main ideas in the development of the theory are discussed. Many of the ideas are extensions of the ideas for continuous-time systems.

The Sampling Theorem

Because all computer-controlled systems operate on values of the process variables at discrete times only, it is very important to know the conditions under which a signal can be recovered from its values in discrete points only. The key issue was explored by Nyquist, who showed that to recover a sinusoidal signal from its samples, it is necessary to sample at least twice per period. A complete solution was given in an important work by Shannon in 1949. This is very fundamental for the understanding of some of the phenomena occuring in discrete-time systems.

Difference Equations

The first germs of a theory for sampled systems appeared in connection with analyses of specific control systems. The behavior of the chopper-bar galvanometer, investigated in Oldenburg and Sartorius (1948), was one of the earliest contributions to the theory. It was shown that many properties could be understood by analyzing a linear time-invariant difference equation. The difference equation replaced the differential equations in continuous-time theory. For example, stability could be investigated by the Schur-Cohn method, which is equivalent to the Routh-Hurwitz criterion.

Numerical Analysis

The theory of sampled-data analysis is closely related to numerical analysis. Integrals are evaluated numerically by approximating them with sums. Many optimization problems can be described in terms of difference equations. Ordinary differential equations are integrated by approximating them by difference equations. For instance, step-length adjustment in integration routines can be regarded as a sampled-data control problem. A large body of theory is available that is related to computer-controlled systems. Difference equations are an important element of this theory, too.

Transform Methods

During and after World War II, a lot of activity was devoted to analysis of radar systems. These systems are naturally sampled because a position measurement is obtained once per antenna revolution. One problem was to find ways to describe these new systems. Because transform theory had been so useful for continuous-time systems, it was natural to try to develop a similar theory for sampled systems. The first steps in this direction were taken by Hurewicz (1947). He introduced the transform of a sequence $f(kh)$, defined by

$$\mathcal{Z}\{f(kh)\} = \sum_{k=0}^{\infty} z^{-k} f(kh)$$

This transform is similar to the *generating function*, which had been used so successfully in many branches of applied mathematics. The transform was later defined as the *z-transform* by Ragazzini and Zadeh (1952). Transform theory was developed independently in the Soviet Union, in the United States, and in Great Britain. Tsypkin (1949) and Tsypkin (1950) called the transform the *discrete Laplace transform* and developed a systematic theory for pulse-controlled systems based on the transform. The transform method was also independently developed by Barker (1952) in England.

In the United States the transform was further developed in a Ph.D. dissertation by Jury at Columbia University. Jury developed tools both for analysis and design. He also showed that sampled systems could be better than their continuous-time equivalents. (See Example 1.3 in Sec. 1.3.) Jury also emphasized that it was possible to obtain a closed-loop system that exactly achieved steady state in finite time. In later works he also showed that sampling can cause cancellation of poles and zeros. A closer investigation of this property later gave rise to the notions of observability and reachability.

The *z*-transform theory leads to comparatively simple results. A limitation of the theory, however, is that it tells what happens to the system only at the sampling instants. The behavior between the sampling instants is not just an academic question, because it was found that systems could exhibit *hidden oscillations*. These oscillations are zero at the sampling instants, but very noticeable in between.

Another approach to the theory of sampled system was taken by Linvill (1951). Following ideas due to MacColl (1945), he viewed the sampling as an amplitude modulation. Using a describing-function approach, Linvill effectively described intersample behavior. Yet another approach to the analysis of the problem was the *delayed z-transform*, which was developed by Tsypkin in 1950, Barker in 1951, and Jury in 1956. It is also known as the *modified z-transform*.

Much of the development of the theory was done by a group at Columbia University led by John Ragazzini. Jury, Kalman, Bertram, Zadeh, Franklin, Friedland, Kranc, Freeman, Sarachik, and Sklansky all did their Ph.D. work for Ragazzini.

Toward the end of the 1950s, the *z*-transform approach to sampled systems had matured, and several textbooks appeared almost simultaneously: Jury (1958), Ragazzini and Franklin (1958), Tsypkin (1958), and Tou (1959). This theory, which was patterned after the theory of linear time-invariant continuous-time systems, gave good tools for analysis and synthesis of sampled systems. A few modifications had to be made because of the time-varying nature of sampled systems. For example, all operations in a block-diagram representation do not commute!

State-Space Theory

A very important event in the late 1950s was the development of state-space theory. The major inspiration came from mathematics and the theory of ordinary differential equations and from mathematicians such as Lefschetz, Pontryagin, and Bellman. Kalman deserves major credit for the state-space approach to control theory. He formulated many of the basic concepts and solved many of the important problems.

Several of the fundamental concepts grew out of an analysis of the problem of whether it would be possible to get systems in which the variables achieved steady state in finite time. The analysis of this problem led to the notions of reachability and observability. Kalman's work also led to a much simpler formulation of the analysis of sampled systems: The basic equations could be derived simply by starting with the differential equations and integrating them under the assumption that the control signal is constant over the sampling period. The discrete-time representation is then obtained by only considering the system at the sampling points. This leads to a very simple state-space representation of sampled-data systems.

Optimal and Stochastic Control

There were also several other important developments in the late 1950s. Bellman (1957) and Pontryagin et al. (1962) showed that many design problems could be formulated as optimization problems. For nonlinear systems this led to nonclassical calculus of variations. An explicit solution was given for linear systems with quadratic loss functions by Bellman, Glicksberg, and Gross (1958). Kalman (1960a) showed in a celebrated paper that the linear quadratic problem could be reduced to a solution of a Riccati equation. Kalman also showed that the classical Wiener filtering problem could be reformulated in the state-space framework. This permitted a "solution" in terms of recursive equations, which were very well suited to computer calculation.

In the beginning of the 1960s, a stochastic variational problem was formulated by assuming that disturbances were random processes. The optimal control problem for linear systems could be formulated and solved for the case of quadratic loss functions. This led to the development of *stochastic control theory*. The work resulted in the so-called *Linear Quadratic Gaussian (LQG)* theory. This is now a major design tool for multivariable linear systems.

Algebraic System Theory

The fundamental problems of linear system theory were reconsidered at the end of the 1960s and the beginning of the 1970s. The algebraic character of the problems was reestablished, which resulted in a better understanding of the foundations of linear system theory. Techniques to solve specific problems using polynomial methods were another result [see Kalman, Falb, and Arbib (1969), Rosenbrock (1970), Wonham (1974), Kučera (1979, 1991), and Blomberg and Ylinen (1983)].

System Identification

All techniques for analysis and design of control systems are based on the availability of appropriate models for process dynamics. The success of classical control theory that almost exclusively builds on Laplace transforms was largely due to the fact that the transfer function of a process can be determined experimentally using frequency response. The development of digital control was accompanied by a similar development of system identification methods. These allow experimental determination of the pulse-transfer function or the difference equations that are the starting point of analysis and design of digital control systems. Good sources of information on these techniques are Åström and Eykhoff (1971), Norton (1986), Ljung (1987), Söderström and Stoica (1989), and Johansson (1993).

Adaptive Control

When digital computers are used to implement a controller, it is possible to implement more complicated control algorithms. A natural step is to include both parameter estimation methods and control design algorithms. In this way it is possible to obtain adaptive control algorithms that determine the mathematical models and perform control system design on-line. Research on adaptive control began in the mid-1950s. Significant progress was made in the 1970s when feasibility was demonstrated in industrial applications. The advent of the microprocessor made the algorithms cost-effective, and commercial adaptive regulators appeared in the early 1980s. This has stimulated vigorous research on theoretical issues and significant product development. See, for instance, Åström and Wittenmark (1973, 1980, 1995), Åström (1983b, 1987), and Goodwin and Sin (1984).

Automatic Tuning

Controller parameters are often tuned manually. Experience has shown that it is difficult to adjust more than two parameters manually. From the user point of view it is therefore helpful to have tuning tools built into the controllers. Such systems are similar to adaptive controllers. They are, however, easier to design and use. With computer-based controllers it is easy to incorporate tuning tools. Such systems also started to appear industrially in the mid-1980s. See Åström and Hägglund (1995).

1.6 Notes and References

To acquire mature knowledge about a field it is useful to know its history and to read some of the original papers. Jury and Tsypkin (1971), and Jury (1980), written by two of the originators of sampled-data theory, give a useful perspective. Early work on sampled systems is found in MacColl (1945), Hurewicz

(1947), and Oldenburg and Sartorius (1948). The sampling theorem was given in Kotelnikov (1933) and Shannon (1949).

Major contributions to the early theory of sampled-data systems were obtained in England by Lawden (1951) and Barker (1952); in the United States by Linvill (1951), Ragazzini and Zadeh (1952), and Jury (1956); and in the Soviet Union by Tsypkin (1949) and Tsypkin (1950).The first textbooks on sampled-data theory appeared toward the end of the 1950s. They were Jury (1958), Ragazzini and Franklin (1958), Tsypkin (1958), and Tou (1959). A large number of textbooks have appeared since then. Among the more common ones we can mention Ackermann (1972, 1996), Kuo (1980), Franklin and Powell (1989), and Isermann (1989, 1991).

The idea of formulating control problems in the state space also resulted in a reformulation of sampled-data theory. Kalman (1961) is seminal.

Some fundamental references on optimal and stochastic control are Bellman (1957), Bellman, Glicksberg, and Gross (1958), Kalman (1960a), Pontryagin et al. (1962), and Åström (1970). The algebraic system approach is discussed in Kalman, Falb, and Arbib (1969), Rosenbrock (1970), Wonham (1974), Kučera (1979, 1991, 1993), and Blomberg and Ylinen (1983).

System identification is surveyed in Åström and Eykhoff (1971), Ljung and Söderström (1983), Norton (1986), Ljung (1987), Söderström and Stoica (1989), and Johansson (1993). Adaptive control is discussed in Bellman (1961), Åström and Wittenmark (1973, 1980, 1995), Åström (1983b, 1987), Goodwin and Sin (1984), Gupta (1986), and Åström and Hägglund (1995).

A survey of distributed computer systems is found in Lucas (1986). In Gustafsson, Lundh, and Söderlind (1988), it is shown how step-length control in numerical integration can be regarded as a control problem. This is also discussed in Hairer and Wanner (1991).

Many additional references are given in the following sections. We also recommend the proceedings of the IFAC Symposia on Digital Computer Applications to Process Control and on Identification and System Parameter Estimation, which are published by Pergamon Press.

2

Discrete-Time Systems

2.1 Introduction

Mathematical models for computer-controlled systems are introduced in this chapter. A key idea is to show how a continuous-time system can be transformed into a discrete-time system by considering the behavior of the signals at the sampling instants.

In this chapter the system is studied as seen from the computer. The computer receives measurements from the process at discrete times and transmits new control signals at discrete times. The goal then is to describe the change in the signals from sample to sample and disregard the behavior between the samples. The use of difference equations then becomes a natural tool. It should be emphasized that computer-oriented mathematical models only give the behavior at the sampling points—the physical process is still a continuous-time system. Looking at the problem this way, however, will greatly simplify the treatment. We will give formulas that allow a computation of intersample behavior, but a full treatment of process-oriented models, which takes continuous-time behavior into account, is given in Chapter 7.

One point that must be treated with some caution is that the sampled-data system is time-varying (see Example 1.1). This problem is also discussed in Chapter 7. In this chapter the problem of time variation is avoided by studying the signals at time instances that are synchronized with the clock in the computer. This gives models described by difference equations in state-space and input-output forms. Section 2.2 gives a description of the sampling mechanism. Section 2.3 treats the problem of finding the discrete-time representation of a continuous-time state-space model by using zero-order-hold devices. The inverse problem of finding the continuous-time system that corresponds to a given discrete-time system is also treated in Sec. 2.3. The general solution of forced difference equations is given in Sec. 2.4. Sections 2.5 and 2.6 deal with transformation of state-space models and the connection between state-space and input-output models. Shift operators are used to describe input-output models. Shift-operator calculus is equivalent to the use of differential operators for continuous-time systems. The discrete-time equivalent of the Laplace transform is the z-transform, which is covered in Sec. 2.7.

The treatment of state-space models in Sec. 2.3 covers the multivariable case. The discussion of input-output models is, however, restricted to single-input–single-output systems. Extensions to the multivariable case are possible, but are not used in this book because they require the mathematics of polynomial matrices.

In order to design computer-controlled systems, it is important to understand how poles and zeros of continuous-time and discrete-time models are related. This is treated in Sec. 2.8. The selection of sampling period is discussed in Sec. 2.9. Rules of thumb based on the appearances of transient responses are given in terms of samples per rise time.

2.2 Sampling Continuous-Time Signals

According to dictionaries, *sampling* means "the act or process of taking a small part or quantity of something as a sample for testing or analysis." In the context of control and communication, sampling means that *a continuous-time signal is replaced by a sequence of numbers*, which represents the values of the signal at certain times.

Sampling is a fundamental property of computer-controlled systems because of the discrete-time nature of the digital computer. Consider, for example, the system shown in Fig. 1.1. The process variables are sampled in connection with the analog conversion and then converted to digital representation for processing. The continuous-time signal that represents the process variables is thus converted to a sequence of numbers, which is processed by the digital computer. The processing gives a new sequence of numbers, which is converted to a continuous-time signal and applied to the process. In the system shown in Fig. 1.1, this is handled by the D-A converter. The process of converting a sequence of numbers into a continuous-time signal is called *signal reconstruction*.

For the purpose of analysis, it is useful to have a mathematical description of sampling. Sampling a continuous-time signal simply means to replace the signal by its values in a discrete set of points. Let Z be the positive and negative integers $Z = \{\dots, -1, 0, 1, \dots\}$ and let $\{t_k : k \in Z\}$ be a subset of the real numbers called the sampling instants. The sampled version of the signal f is then the sequence $\{f(t_k) : k \in Z\}$. Sampling is a linear operation. The sampling instants are often equally spaced in time, that is, $t_k = k \cdot h$. This case is called *periodic sampling* and h is called the *sampling period*, or the *sampling time*. The corresponding frequency $f_s = 1/h$ (Hz) or $\omega_s = 2\pi/h$ (rad/s) is called the *sampling frequency*. It is also convenient to introduce a notation for half the sampling frequency $f_N = 1/(2h)$ (Hz) or $\omega_N = \pi/h$ (rad/s), which is called the *Nyquist frequency*.

More complicated sampling schemes can also be used. For instance, different sampling periods can be used for different control loops. This is called multirate sampling and can be considered to be the superposition of several periodic sampling schemes.

The case of periodic sampling is well understood. Most theory is devoted to this case, but systems with multirate sampling are becoming more important because of the increased use of multiprocessor systems. With modern software for concurrent processes, it is also possible to design a system as if it were composed of many different processes running asynchronously. There are also technical advantages in using different sampling rates for different variables.

2.3 Sampling a Continuous-Time State-Space System

A fundamental problem is how to describe a continuous-time system connected to a computer via A-D and D-A converters. Consider the system shown in Fig. 2.1. The signals in the computer are the sequences $\{u(t_k)\}$ and $\{y(t_k)\}$. The key problem is to find the relationship between these sequences. To find the discrete-time equivalent of a continuous-time system is called *sampling a continuous-time system*. The model obtained is also called a *stroboscopic model* because it gives a relationship between the system variables at the sampling instants only. To obtain the desired descriptions, it is necessary to describe the converters and the system. Assume that the continuous-time system is given in the following state-space form:

$$\frac{dx}{dt} = Ax(t) + Bu(t)$$
$$y(t) = Cx(t) + Du(t)$$
(2.1)

The system has r inputs, p outputs, and is of order n.

Zero-Order-Hold Sampling of a System

A common situation in computer control is that the D-A converter is so constructed that it holds the analog signal constant until a new conversion is commanded. This is often called a *zero-order-hold circuit*. It is then natural

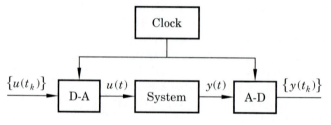

Figure 2.1 Block diagram of a continuous-time system connected to A-D and D-A converters.

to choose the sampling instants, t_k, as the times when the control changes. Because the control signal is discontinuous, it is necessary to specify its behavior at the discontinuities. The convention that the signal is continuous from the right is adopted. The control signal is thus represented by the sampled signal $\{u(t_k) : k = \ldots, -1, 0, 1, \ldots\}$. The relationship between the system variables at the sampling instants will now be determined. Given the state at the sampling time t_k, the state at some future time t is obtained by solving (2.1). The state at time t, where $t_k \le t \le t_{k+1}$, is thus given by

$$
\begin{aligned}
x(t) &= e^{A(t-t_k)}x(t_k) + \int_{t_k}^{t} e^{A(t-s')}Bu(s')\,ds' \\
&= e^{A(t-t_k)}x(t_k) + \int_{t_k}^{t} e^{A(t-s')}\,ds'\,Bu(t_k) \\
&= e^{A(t-t_k)}x(t_k) + \int_{0}^{t-t_k} e^{As}\,ds\,Bu(t_k) \\
&= \Phi(t,t_k)x(t_k) + \Gamma(t,t_k)u(t_k)
\end{aligned}
\tag{2.2}
$$

The second equality follows because u is constant between the sampling instants.

The state vector at time t is thus a linear function of $x(t_k)$ and $u(t_k)$. If the A-D and D-A converters in Fig. 2.1 are perfectly synchronized and if the conversion times are negligible, the input u and the output y can be regarded as being sampled at the same instants. The *system equation* of the sampled system at the sampling instants is then

$$
\begin{aligned}
x(t_{k+1}) &= \Phi(t_{k+1},t_k)x(t_k) + \Gamma(t_{k+1},t_k)u(t_k) \\
y(t_k) &= Cx(t_k) + Du(t_k)
\end{aligned}
\tag{2.3}
$$

where

$$
\begin{aligned}
\Phi(t_{k+1},t_k) &= e^{A(t_{k+1}-t_k)} \\
\Gamma(t_{k+1},t_k) &= \int_{0}^{t_{k+1}-t_k} e^{As}\,ds\,B
\end{aligned}
$$

The relationship between the sampled signals thus can be expressed by the linear difference equation, (2.3). Notice that Equation (2.3) does not involve any approximations. It gives the exact values of the state variables and the output at the sampling instants because the control signal is constant between the sampling instants. The model in (2.3) is therefore called a zero-order-hold sampling of the system in (2.1). The system in (2.3) can also be called the zero-order-hold equivalent of (2.1).

In most cases $D = 0$. One reason for this is because in computer-controlled systems, the output y is first measured and the control signal $u(t_k)$ is then generated as a function of $y(t_k)$. In practice it often happens that there is a significant delay between the A-D and D-A conversions. However, it is easy to make the necessary modifications. The state vector at times between sampling points is given by (2.2). This makes it possible to investigate the intersample behavior of the system. Notice that the responses between the sampling points are parts of step responses, with initial conditions, for the system. This implies that the system is running in open loop between the sampling points.

For periodic sampling with period h, we have $t_k = k \cdot h$ and the model of (2.3) simplifies to the time-invariant system

$$\begin{aligned} x(kh + h) &= \Phi x(kh) + \Gamma u(kh) \\ y(kh) &= C x(kh) + D u(kh) \end{aligned} \tag{2.4}$$

where

$$\begin{aligned} \Phi &= e^{Ah} \\ \Gamma &= \int_0^h e^{As}\, ds\, B \end{aligned} \tag{2.5}$$

It follows from (2.5) that

$$\begin{aligned} \frac{d\Phi(t)}{dt} &= A\Phi(t) = \Phi(t)A \\ \frac{d\Gamma(t)}{dt} &= \Phi(t)B \end{aligned}$$

The matrices Φ and Γ therefore satisfy the equation

$$\frac{d}{dt}\begin{pmatrix} \Phi(t) & \Gamma(t) \\ 0 & I \end{pmatrix} = \begin{pmatrix} \Phi(t) & \Gamma(t) \\ 0 & I \end{pmatrix} \begin{pmatrix} A & B \\ 0 & 0 \end{pmatrix}$$

where I is a unit matrix of the same dimension as the number of inputs. The matrices $\Phi(h)$ and $\Gamma(h)$ for the sampling period h therefore can be obtained from the block matrix

$$\begin{pmatrix} \Phi(h) & \Gamma(h) \\ 0 & I \end{pmatrix} = \exp\left(\begin{pmatrix} A & B \\ 0 & 0 \end{pmatrix} h \right) \tag{2.6}$$

How to Compute Φ and Γ

The calculations required to sample a continuous-time system are the evaluation of a matrix exponential and the integration of a matrix exponential. These can be done in many different ways, for instance, by using the following:

- Numerical calculation in MATLAB® or MATRIX$_X$®

- Series expansion of the matrix exponential

- The Laplace transform—the Laplace transform of $\exp(At)$ is $(sI - A)^{-1}$

- Cayley-Hamilton's theorem (see Appendix B)

- Transformation to diagonal or Jordan forms

- Symbolic computer algebra, using programs such as Maple® and *Mathematica*®.

Calculations by hand are feasible for low-order systems, $n \leq 2$, and for high-order systems with special structures. One way to simplify the computations is to compute

$$\Psi = \int_0^h e^{As}\, ds = Ih + \frac{Ah^2}{2!} + \frac{A^2 h^3}{3!} + \cdots + \frac{A^i h^{i+1}}{(i+1)!} + \cdots$$

The matrices Φ and Γ are given by

$$\Phi = I + A\Psi$$
$$\Gamma = \Psi B$$

Computer evaluation can be done using several different numerical algorithms in MATLAB® or MATRIX$_X$®.

Example 2.1   First-order system

Consider the system

$$\frac{dx}{dt} = \alpha x + \beta u$$

Applying Eqs. (2.5) we get

$$\Phi = e^{\alpha h}$$
$$\Gamma = \int_0^h e^{\alpha s}\, ds\, \beta = \frac{\beta}{\alpha}\,(e^{\alpha h} - 1)$$

The sampled system thus becomes

$$x(kh + h) = e^{\alpha h} x(kh) + \frac{\beta}{\alpha}\,(e^{\alpha h} - 1)u(kh)$$

∎

Example 2.2   Double integrator

The double integrator (see Example A.1 in Appendix A) is described by

$$\frac{dx}{dt} = \begin{pmatrix} 0 & 1 \\ 0 & 0 \end{pmatrix} x + \begin{pmatrix} 0 \\ 1 \end{pmatrix} u$$
$$y = \begin{pmatrix} 1 & 0 \end{pmatrix} x$$

Hence

$$\Phi = e^{Ah} = I + Ah + A^2 h^2/2 + \cdots = \begin{pmatrix} 1 & 0 \\ 0 & 1 \end{pmatrix} + \begin{pmatrix} 0 & h \\ 0 & 0 \end{pmatrix} = \begin{pmatrix} 1 & h \\ 0 & 1 \end{pmatrix}$$

$$\Gamma = \int_0^h \begin{pmatrix} s \\ 1 \end{pmatrix} ds = \begin{pmatrix} \dfrac{h^2}{2} \\ h \end{pmatrix}$$

The discrete-time model of the double integrator is

$$x(kh + h) = \begin{pmatrix} 1 & h \\ 0 & 1 \end{pmatrix} x(kh) + \begin{pmatrix} \dfrac{h^2}{2} \\ h \end{pmatrix}$$

$$y(kh) = \begin{pmatrix} 1 & 0 \end{pmatrix} x(kh)$$

(2.7)

∎

Example 2.3   Motor

A simple normalized model of an electrical DC motor (see Example A.2 in Appendix A) is given by

$$\frac{dx}{dt} = \begin{pmatrix} -1 & 0 \\ 1 & 0 \end{pmatrix} x + \begin{pmatrix} 1 \\ 0 \end{pmatrix} u$$

$$y = \begin{pmatrix} 0 & 1 \end{pmatrix} x$$

The Laplace transform method gives

$$(sI - A)^{-1} = \begin{pmatrix} s+1 & 0 \\ -1 & s \end{pmatrix}^{-1} = \frac{1}{s(s+1)} \begin{pmatrix} s & 0 \\ 1 & s+1 \end{pmatrix} = \begin{pmatrix} \dfrac{1}{s+1} & 0 \\ \dfrac{1}{s(s+1)} & \dfrac{1}{s} \end{pmatrix}$$

Hence

$$\Phi = e^{Ah} = \mathcal{L}^{-1}(sI - A)^{-1} = \begin{pmatrix} e^{-h} & 0 \\ 1 - e^{-h} & 1 \end{pmatrix}$$

and

$$\Gamma = \int_0^h \begin{pmatrix} e^{-v} \\ 1 - e^{-v} \end{pmatrix} dv = \begin{pmatrix} 1 - e^{-h} \\ h - 1 + e^{-h} \end{pmatrix}$$

where $\mathcal{L}^{-1}$ is the inverse of the Laplace transform. ∎

The Inverse of Sampling

Sampling a system defines a map from continuous-time systems, as in (2.1), to discrete-time systems, as in (2.4). A natural question is if and when it is possible to get the corresponding continuous-time system from a discrete-time description.

Example 2.4   Inverse sampling

Consider the first-order difference equation

$$x(kh + h) = ax(kh) + bu(kh)$$

From Example 2.1 we find that the corresponding continuous-time system is obtained from

$$e^{\alpha h} = a$$

$$\frac{\beta}{\alpha} \left(e^{\alpha h} - 1 \right) = b$$

This gives

$$\alpha = \frac{1}{h} \ln a$$

$$\beta = \frac{1}{h} \ln a \cdot \frac{b}{a - 1}$$

This example shows that a continuous-time system with real coefficients is obtained only when a is positive. ∎

To invest the process of sampling in the general case we note that it follows from (2.6) that

$$\begin{pmatrix} A & B \\ 0 & 0 \end{pmatrix} = \frac{1}{h} \ln \begin{pmatrix} \Phi & \Gamma \\ 0 & I \end{pmatrix}$$

where ln $(\cdot)$ is the matrix logarithmic function. The continuous-time system is thus obtained by taking the matrix logarithm function of a block matrix. Computation of matrix logarithm is discussed in Appendix B. From the Cayley-Hamilton theorem it must be assumed that the logarithm exists only when the matrix Φ does not have any eigenvalues on the negative real axis. There is also a nonuniqueness in the matrix logarithmic function for complex arguments, which is illustrated by the following example.

Example 2.5   Harmonic oscillator

The discrete-time system

$$x(kh + h) = \begin{pmatrix} \cos \alpha h & \sin \alpha h \\ -\sin \alpha h & \cos \alpha h \end{pmatrix} x(kh) + \begin{pmatrix} 1 - \cos \alpha h \\ \sin \alpha h \end{pmatrix} u(kh)$$

can be obtained by sampling a continuous-time system with

$$A = \begin{pmatrix} 0 & \omega \\ -\omega & 0 \end{pmatrix} \quad \text{and} \quad B = \begin{pmatrix} 0 \\ \omega \end{pmatrix}$$

where

$$\omega = \alpha + \frac{2\pi}{h} \cdot n \qquad n = 0, 1, \ldots$$

In this case the inverse problem has many solutions (compare Examples A.3 and B.1). This is generally the case if the matrix Φ has complex eigenvalues. Notice that there always exists a unique ω in the interval $-\omega_N \le \omega \le \omega_N$, where $\omega_N = \pi/h$ is the Nyquist frequency associated with the sampling period h. ∎

Sampling a System with Time Delay

Time delays are common in mathematical models of industrial processes. The theory of continuous-time systems with time delays is complicated because the systems are infinite-dimensional.

It is, however, easy to sample systems with time delays because the control signal is constant between sampling instants, which makes the sampled-data system finite-dimensional. Let the system be described by

$$\frac{dx(t)}{dt} = Ax(t) + Bu(t - \tau) \tag{2.8}$$

It is assumed initially that the time delay τ is less than or equal to the sampling period. The zero-order-hold sampling of the system (2.8) will now be calculated. Integration of (2.8) over one sampling period gives

$$x(kh + h) = e^{Ah}x(kh) + \int_{kh}^{kh+h} e^{A(kh+h-s')}Bu(s' - \tau)\,ds' \tag{2.9}$$

Because the signal $u(t)$ is piecewise constant over the sampling interval, the delayed signal $u(t-\tau)$ is also piecewise constant. The delayed signal will, however, change between the sampling instants (see Fig. 2.2). To evaluate the integral of (2.9), it is then convenient to split the integration interval into two parts so

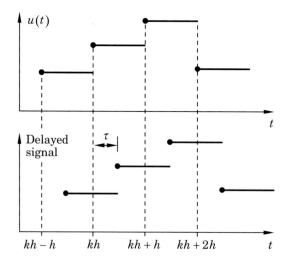

Figure 2.2 The relationship among $u(t)$, the delayed signal $u(t - \tau)$, and the sampling instants.

that $u(t - \tau)$ is constant in each part. Hence

$$\int_{kh}^{kh+h} e^{A(kh+h-s')} B u(s' - \tau) \, ds'$$

$$= \int_{kh}^{kh+\tau} e^{A(kh+h-s')} B \, ds' \, u(kh - h) + \int_{kh+\tau}^{kh+h} e^{A(kh+h-s')} B \, ds' \, u(kh)$$

$$= \Gamma_1 u(kh - h) + \Gamma_0 u(kh)$$

Sampling the continuous-time system (2.8) thus gives

$$x(kh + h) = \Phi x(kh) + \Gamma_0 u(kh) + \Gamma_1 u(kh - h) \qquad (2.10)$$

where

$$\Phi = e^{Ah}$$

$$\Gamma_0 = \int_0^{h-\tau} e^{As} \, ds \, B \qquad (2.11)$$

$$\Gamma_1 = e^{A(h-\tau)} \int_0^{\tau} e^{As} \, ds \, B$$

A state-space model of (2.10) is given by

$$\begin{pmatrix} x(kh + h) \\ u(kh) \end{pmatrix} = \begin{pmatrix} \Phi & \Gamma_1 \\ 0 & 0 \end{pmatrix} \begin{pmatrix} x(kh) \\ u(kh - h) \end{pmatrix} + \begin{pmatrix} \Gamma_0 \\ I \end{pmatrix} u(kh)$$

Notice that r extra state variables $u(kh - h)$, which represent the past values of the control signal, are introduced. The continuous-time system of (2.8) is infinite dimensional; the corresponding sampled system, however, is a finite-dimensional system. Thus time delays are considerably simpler to handle if the system is sampled, for the following reason: To specify the state of the system, it is necessary to store the input over a time interval equal to the time delay. With zero-order-hold reconstruction, the input signal can be represented always by a finite number of values.

Example 2.6   First-order system with time delay

Consider the system

$$\frac{dx(t)}{dt} = \alpha x(t) + \beta u(t - \tau)$$

Assume that the system is sampled with period h, where $0 \leq \tau \leq h$. Equation (2.11) gives

$$\Phi = a = e^{\alpha h}$$

$$\Gamma_0 = b_0 = \int_0^{h-\tau} e^{\alpha s} \beta \, ds = \frac{\beta}{\alpha} \left(e^{\alpha(h-\tau)} - 1 \right)$$

$$\Gamma_1 = b_1 = e^{\alpha(h-\tau)} \int_0^{\tau} e^{\alpha s} \beta \, ds = \frac{\beta}{\alpha} \left(e^{\alpha h} - e^{\alpha(h-\tau)} \right)$$

The sampled system is thus

$$x(kh + h) = ax(kh) + b_0 u(kh) + b_1 u(kh - h)$$

∎

Example 2.7 Double integrator with delay

The double integrator in Example 2.2 with a time delay $0 \le \tau \le h$ gives

$$\Phi = e^{Ah} = \begin{pmatrix} 1 & h \\ 0 & 1 \end{pmatrix}$$

$$\Gamma_1 = e^{A(h-\tau)} \int_0^\tau e^{As}\, ds\, B = \begin{pmatrix} 1 & h - \tau \\ 0 & 1 \end{pmatrix} \begin{pmatrix} \dfrac{\tau^2}{2} \\ \tau \end{pmatrix} = \begin{pmatrix} \tau\left(h - \dfrac{\tau}{2}\right) \\ \tau \end{pmatrix}$$

$$\Gamma_0 = \int_0^{h-\tau} e^{As}\, ds\, B = \begin{pmatrix} \dfrac{(h-\tau)^2}{2} \\ h - \tau \end{pmatrix}$$

∎

Longer Time Delays

If the time delay is longer than h, then the previous analysis has to be modified a little. If

$$\tau = (d - 1)h + \tau' \qquad 0 < \tau' \le h$$

where d is an integer, the following equation is obtained:

$$x(kh + h) = \Phi x(kh) + \Gamma_0 u(kh - (d - 1)h) + \Gamma_1 u(kh - dh)$$

where Γ_0 and Γ_1 are given by (2.11) with τ replaced by τ'. The corresponding state-space description is

$$\begin{pmatrix} x(kh + h) \\ u(kh - (d-1)h) \\ \vdots \\ u(kh - h) \\ u(kh) \end{pmatrix} = \begin{pmatrix} \Phi & \Gamma_1 & \Gamma_0 & \cdots & 0 \\ 0 & 0 & I & \cdots & 0 \\ \vdots & \vdots & \vdots & \ddots & \vdots \\ 0 & 0 & 0 & \cdots & I \\ 0 & 0 & 0 & \cdots & 0 \end{pmatrix} \begin{pmatrix} x(kh) \\ u(kh - dh) \\ \vdots \\ u(kh - 2h) \\ u(kh - h) \end{pmatrix} + \begin{pmatrix} 0 \\ 0 \\ \vdots \\ 0 \\ I \end{pmatrix} u(kh)$$

$$(2.12)$$

Notice that if $\tau > 0$, then $d \cdot r$ extra state variables are used to describe the delay, where r is the number of inputs. The characteristics polynomial of the state-space description is $\lambda^{dr} A(\lambda)$, where $A(\lambda)$ is the characteristic polynomial of Φ.

An example illustrates use of the general formula.

Example 2.8 Simple paper-machine model

Determine the zero-order-hold sampling of the system (see Example A.4).

$$\frac{dx(t)}{dt} = -x(t) + u(t - 2.6)$$

with sampling interval $h = 1$. In this case $d = 3$ and $\tau' = 0.6$, and (2.12) becomes

$$x(k + 1) = \Phi x(k) + \Gamma_0 u(k - 2) + \Gamma_1 u(k - 3)$$

where

$$\Phi = e^{-1} = 0.3679$$

$$\Gamma_0 = \int_0^{0.4} e^{-s}\, ds = 1 - e^{-0.4} = 0.3297$$

$$\Gamma_1 = e^{-0.4} \int_0^{0.6} e^{-s}\, ds = e^{-0.4} - e^{-1} = 0.3024$$

∎

System with Internal Time Delay

In the previous derivation it is assumed that the time delay of the system is at the input (or the output) of the system. Many physical systems have the structure shown in Fig. 2.3, that is, the time delay is internal. Let the system be described by the equations

$$S_1: \frac{dx_1(t)}{dt} = A_1 x_1(t) + B_1 u(t)$$

$$y_1(t) = C_1 x_1(t) + D_1 u(t)$$

$$S_2: \frac{dx_2(t)}{dt} = A_2 x_2(t) + B_2 u_2(t) \tag{2.13}$$

$$u_2(t) = y_1(t - \tau)$$

It is assumed that $u(t)$ is piecewise constant over the sampling interval h. We now want to find the recursive equations for $x_1(kh)$ and $x_2(kh)$.

Sampling (2.13) when $\tau = 0$ using the sampling period h gives the partitioned system

$$\begin{pmatrix} x_1(kh + h) \\ x_2(kh + h) \end{pmatrix} = \begin{pmatrix} \Phi_1(h) & 0 \\ \Phi_{21}(h) & \Phi_2(h) \end{pmatrix} \begin{pmatrix} x_1(kh) \\ x_2(kh) \end{pmatrix} + \begin{pmatrix} \Gamma_1(h) \\ \Gamma_2(h) \end{pmatrix} u(kh)$$

We have the following theorem.

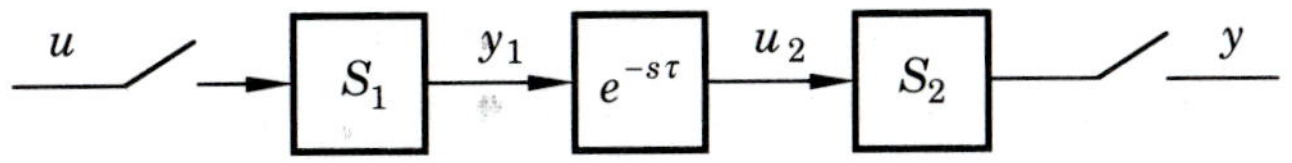

Figure 2.3 System with inner time delay.

THEOREM 2.1   INNER TIME DELAY   Periodic sampling of the system (2.13) with the sampling interval h and with $0 < \tau \leq h$ gives the sampled-data representation

$$
\begin{aligned}
x_1(kh + h) &= \Phi_1(h)x_1(kh) + \Gamma_1(h)u(kh) \\
x_2(kh + h) &= \Phi_{21}^{-}x_1(kh - h) + \Phi_2(h)x_2(kh) \\
&\quad + \Gamma_2^{-}u(kh - h) + \Gamma_2(h - \tau)u(kh)
\end{aligned}
\tag{2.14}
$$

where

$$
\begin{aligned}
\Phi_i(t) &= e^{A_i t} \qquad i = 1,\, 2 \\
\Phi_{21}(t) &= \int_0^t e^{A_2 s} B_2 C_1 e^{A_1(t-s)}\, ds \\
\Gamma_1(t) &= \int_0^t e^{A_1 s} B_1\, ds \\
\Gamma_2(t) &= \int_0^t e^{A_2 s} B_2 C_1 \Gamma_1(t - s)\, ds \\
\Phi_{21}^{-} &= \Phi_{21}(h)\Phi_1(h - \tau) \\
\Gamma_2^{-} &= \Phi_{21}(h)\Gamma_1(h - \tau) + \Phi_{21}(h - \tau)\Gamma_1(\tau) + \Phi_2(h - \tau)\Gamma_2(\tau)
\end{aligned}
\tag{2.15}
$$

Reference to proof of the theorem is given at the end of the chapter.   ∎

Remark.   The sampled-data system (2.14) for the time delay τ is obtained by sampling (2.13) without any time delay for the sampling intervals h, $h - \tau$, and τ. This gives Φ_1, Φ_2, Φ_{21}, Γ_1, and Γ_2 for the needed sampling intervals. This implies that standard software for sampling systems can be used to obtain (2.14).

Intersample Behavior

The discrete-time models (2.3) and (2.4) give the values of the state variables and the outputs at the sampling instants t_k. The values of the variables between the sampling points are also of interest. These values are given by (2.2). Other ways of obtaining the intersample behavior are discussed in Sec. 2.7 and Chapter 7.

2.4 Discrete-Time Systems

The previous section showed how to transform a continuous-time system into discrete-time form. In most of the remaining part of this chapter we will disregard how the difference equation representing the discrete-time system has been obtained. Instead we will concentrate on the properties of difference equations. Time-invariant discrete-time systems can be described by the difference equation

$$
\begin{aligned}
x(k + 1) &= \Phi x(k) + \Gamma u(k) \\
y(k) &= C x(k) + D u(k)
\end{aligned}
\tag{2.16}
$$

For simplicity the sampling time is used as the time unit, $h = 1$.

Solution of the System Equation

To analyze discrete-time systems it is necessary to solve the system equation (2.16). Assume that the initial condition $x(k_0)$ and the input signals $u(k_0), u(k_0 + 1), \ldots$ are given. How is the state then evolving? It is possible to solve (2.16) by simple iterations.

$$
\begin{aligned}
x(k_0 + 1) &= \Phi x(k_0) + \Gamma u(k_0) \\
x(k_0 + 2) &= \Phi x(k_0 + 1) + \Gamma u(k_0 + 1) \\
&= \Phi^2 x(k_0) + \Phi \Gamma u(k_0) + \Gamma u(k_0 + 1) \\
&\;\;\vdots \\
x(k) &= \Phi^{k-k_0} x(k_0) + \Phi^{k-k_0-1} \Gamma u(k_0) + \cdots + \Gamma u(k - 1) \\
&= \Phi^{k-k_0} x(k_0) + \sum_{j=k_0}^{k-1} \Phi^{k-j-1} \Gamma u(j)
\end{aligned}
\tag{2.17}
$$

The solution consists of two parts: One depends on the initial condition, and the other is a weighted sum of the input signals. Equation (2.17) clearly shows that the eigenvalues of Φ will determine the properties of the solution. The eigenvalues are obtained from the *characteristic equation*

$$
\det(\lambda I - \Phi) = 0
$$

Example 2.9   Solution of the difference equation

Consider the discrete-time system

$$
x(k + 1) = \begin{pmatrix} \lambda_1 & 0 \\ 1 & \lambda_2 \end{pmatrix} x(k)
$$

with $x(0) = \begin{pmatrix} 1 & 1 \end{pmatrix}^T$. It is easily verified that

$$
\Phi^k = \begin{pmatrix} \lambda_1^k & 0 \\ \sum_{j=1}^{k} \lambda_1^{k-j} \lambda_2^{j-1} & \lambda_2^k \end{pmatrix}
$$

and

$$
x(k) = \begin{pmatrix} \lambda_1^k \\ \sum_{j=1}^{k} \lambda_1^{k-j} \lambda_2^{j-1} + \lambda_2^k \end{pmatrix}
$$

If $|\lambda_i| < 1$, $j = 1, 2$, then $x(k)$ will converge to the origin. If one of the eigenvalues of Φ has an absolute value larger than 1, then one or both of the states will diverge.

∎

2.5 Changing Coordinates in State-Space Models

Consider the discrete-time system (2.16). We will now discuss how new coordinates can be introduced. Assume that T is a nonsingular matrix and define a new state vector $z(k) = Tx(k)$. Then

$$
z(k+1) = Tx(k+1) = T\Phi x(k) + T\Gamma u(k) = T\Phi T^{-1}z(k) + T\Gamma u(k)
$$
$$
= \tilde{\Phi}z(k) + \tilde{\Gamma}u(k)
$$

and

$$
y(k) = Cx(k) + Du(k) = CT^{-1}z(k) + Du(k) = \tilde{C}z(k) + \tilde{D}u(k)
$$

The state-space representation thus depends on the coordinate system chosen to represent the state. The invariants under the transformation are of interest.

THEOREM 2.2 INVARIANCE OF THE CHARACTERISTIC EQUATION   The characteristic equation

$$
\det(\lambda I - \Phi) = 0
$$

is invariant when new states are introduced through the nonsingular transformation matrix T.

Proof.

$$
\det(\lambda I - \tilde{\Phi}) = \det(\lambda TT^{-1} - T\Phi T^{-1}) = \det T \det(\lambda I - \Phi) \det T^{-1}
$$
$$
= \det(\lambda I - \Phi)
$$

∎

To find a transformation matrix is the same as solving for the n^2 elements of T from the linear set of equations

$$
T\Phi = \tilde{\Phi}T
$$

Coordinates can be chosen to give simple forms of the system equations.

Diagonal Form

Assume that Φ has distinct eigenvalues. Then there exists a T such that

$$T\Phi T^{-1} = \begin{pmatrix} \lambda_1 & & 0 \\ & \ddots & \\ 0 & & \lambda_n \end{pmatrix}$$

where λ_i are the eigenvalues of Φ. The computation of T is discussed in Sec. 3.4. In this case a set of decoupled first-order difference equations is obtained.

$$z_1(k+1) = \lambda_1 z_1(k) + \beta_1 u(k)$$

$$\vdots$$

$$z_n(k+1) = \lambda_n z_n(k) + \beta_n u(k)$$
$$y(k) = \gamma_1 z_1(k) + \cdots + \gamma_n z_n(k)$$

The solution to the system of equations is now simple. Each mode will have the solution

$$z_i(k) = \lambda_i^k z_i(0) + \sum_{j=0}^{k-1} \lambda_i^{k-j-1} \beta_i u(j) \tag{2.18}$$

Example 2.10   Diagonal form

Consider the motor in Example 2.3 with $h = 1$. Using the transformation

$$T = \begin{pmatrix} 0 & 0.7071 \\ 1.0000 & -0.7071 \end{pmatrix}$$

gives

$$\tilde{\Phi} = \begin{pmatrix} 1.0000 & 0 \\ 0 & 0.3679 \end{pmatrix} \qquad \tilde{\Gamma} = \begin{pmatrix} 0.2601 \\ 0.3720 \end{pmatrix} \qquad \tilde{C} = \begin{pmatrix} 1.4142 & 0 \end{pmatrix}$$

∎

Jordan Form

If Φ has multiple eigenvalues, then it is generally not possible to diagonalize Φ. Let Φ be a $n \times n$ matrix and introduce the notation

$$L_k(\lambda) = \begin{pmatrix} \lambda & 1 & 0 & \cdots & 0 \\ 0 & \lambda & 1 & & 0 \\ \vdots & \ddots & \ddots & \ddots & \\ 0 & \cdots & 0 & \lambda & 1 \\ 0 & \cdots & 0 & 0 & \lambda \end{pmatrix}$$

where L_k is a $k \times k$ matrix. Then there exists a matrix T such that

$$T\Phi T^{-1} = \begin{pmatrix} L_{k_1}(\lambda_1) & & & 0 \\ & L_{k_2}(\lambda_2) & & \\ & & \ddots & \\ 0 & & & L_{k_r}(\lambda_r) \end{pmatrix} \tag{2.19}$$

with $k_1 + k_2 + \cdots + k_r = n$. The λ_i are the eigenvalues of Φ, not necessarily distinct. Equation (2.19) is called the *Jordan form*. See Appendix B. In this form the transformed matrix, $\tilde{\Phi}$, has the eigenvalues in the diagonal and some 1's in the superdiagonal.

2.6 Input-Output Models

A dynamic system can be described using either internal models or external models. Internal models—for instance, the state-space models discussed in Sec. 2.3—describe all internal couplings among the system variables. The external models give only the relationship between the input and the output of the system. In this section, it is first shown that the input-output relationship for a general linear system can be expressed by a pulse-response function. It is then shown that shift-operator calculus can be used to derive input-output relationships directly, which leads to characterization of the input-output behavior in terms of pulse-transfer operators.

The Pulse Response

Consider a discrete-time system with one input and one output. The input and output signals over a finite interval can be represented as finite-dimensional vectors

$$U = \begin{pmatrix} u(0) & \cdots & u(N-1) \end{pmatrix}^T$$

$$Y = \begin{pmatrix} y(0) & \cdots & y(N-1) \end{pmatrix}^T$$

The general linear model that relates Y to U can then be expressed as

$$Y = \overline{H}U + Y_p$$

where $\overline{H}$ is an $N \times N$ matrix. Y_p accounts for the initial conditions. If the relation between U and Y is causal, the matrix $\overline{H}$ must be lower triangular. The element $h(k, m)$ of $\overline{H}$ is thus zero if $m > k$. The input-output relationship for a general linear system can be written as

$$y(k) = \sum_{m=0}^{k} \overline{h}(k, m)u(m) + y_p(k)$$

where the term y_p is introduced to account for initial conditions in the system. The function $\bar{h}(k, m)$ is called the *pulse-response function*, or the *weighting function*, of the system. The pulse-response function is a convenient representation, because it can easily be measured directly by injecting a pulse of unit magnitude and the width of the sampling interval and recording the output. For zero initial conditions, the value $\bar{h}(k, m)$ of the pulse response gives the output at time k for a unit pulse at time m. For systems with many inputs and outputs, the pulse response is simply a matrix-valued function. For time-invariant systems, the pulse response is a function of $k - m$ only, that is,

$$\bar{h}(k, m) = h(k - m)$$

It is easy to compute the pulse response of the system defined by the state-space model in (2.16). It follows from (2.17) that

$$y(k) = C\Phi^{k-k_0}x(k_0) + \sum_{j=k_0}^{k-1} C\Phi^{k-j-1}\Gamma u(j) + Du(k)$$

The pulse-response function for the discrete-time system is thus

$$h(k) = \begin{cases} 0 & k < 0 \\ D & k = 0 \\ C\Phi^{k-1}\Gamma & k \geq 1 \end{cases} \qquad (2.20)$$

The pulse response is a sum of functions of the form

$$\text{Re}\{P(k)\lambda_i^k\}$$

where P is a polynomial in k, and λ_i are the eigenvalues of the matrix Φ.

The pulse response has the following property.

THEOREM 2.3   INVARIANCE OF PULSE RESPONSE   The pulse response (2.20) is invariant with respect to coordinate transformations of the state-space model.

Proof.   Introduce new coordinates $z = Tx$. The pulse response of the transformed system is then, for $k \geq 1$,

$$\tilde{h}(k) = \tilde{C}\tilde{\Phi}^{k-1}\tilde{\Gamma} = CT^{-1}(T\Phi T^{-1})^{k-1}T\Gamma$$
$$= CT^{-1}T\Phi^{k-1}T^{-1}T\Gamma = C\Phi^{k-1}\Gamma = h(k)$$

$\tilde{D} = D$ has to be added for $k = 0$.   ∎

If $h(k) \neq 0$ for only a finite number of k, then the system is called a *finite impulse-response (FIR) system*. This implies that the output only will be influenced by a finite number of inputs.

Shift-Operator Calculus

Differential-operator calculus is a convenient tool for manipulating linear differential equations with constant coefficients. An analogous operator calculus can be developed for systems described by linear difference equations with constant coefficients. In the development of operator calculus, the systems are viewed as operators that map input signals to output signals. To specify an operator it is necessary to give its range—that is, to define the class of input signals and to describe how the operator acts on the signals. In operator calculus, all signals are considered as doubly infinite sequences $\{f(k) : k = \cdots - 1, 0, 1, \dots\}$. For convenience the sampling period is chosen as the time unit.

The *forward-shift operator* is denoted by q. It has the property

$$qf(k) = f(k+1)$$

If the norm of a signal is defined as

$$\| f \| = \sup_k |f(k)|$$

or

$$\| f \|^2 = \sum_{k=-\infty}^{\infty} f^2(k)$$

it follows that the shift operator has unit norm. This means that the calculus of shift operators is simpler than differential-operator calculus, because the differential operator is unbounded. The inverse of the forward-shift operator is called the *backward-shift operator* or the *delay operator* and is denoted by q^{-1}. Hence

$$q^{-1}f(k) = f(k-1)$$

Notice that it is important for the range of the operator to be doubly infinite sequences; otherwise, the inverse of the forward-shift operator may not exist. In discussions of problems related to the characteristic equation of a system, such as stability and system order, it is more convenient to use the forward-shift operator. In discussions of problems related to causality, it is more convenient to use the backward-shift operator. Operator calculus gives compact descriptions of systems and makes it easy to derive relationships among system variables, because manipulation of difference equations is reduced to a purely algebraic problem.

The shift operator is used to simplify the manipulation of higher-order difference equations. Consider the equation

$$
\begin{aligned}
y(k + na) + a_1 y(k + na - 1) + \cdots + a_{na} y(k) \\
= b_0 u(k + nb) + \cdots + b_{nb} u(k)
\end{aligned}
\tag{2.21}
$$

where $na \geq nb$. Use of the shift operator gives

$$(q^{na} + a_1 q^{na-1} + \cdots + a_{na})y(k) = (b_0 q^{nb} + \cdots + b_{nb})u(k)$$

With the introduction of the polynomials

$$A(z) = z^{na} + a_1 z^{na-1} + \cdots + a_{na}$$

and

$$B(z) = b_0 z^{nb} + b_1 z^{nb-1} + \cdots + b_{nb}$$

the difference equation can be written as

$$A(q)y(k) = B(q)u(k) \tag{2.22}$$

When necessary, the degree of a polynomial can be indicated by a subscript, for example, $A_{na}(q)$. Equation (2.22) can be also expressed in terms of the backward-shift operator. Notice that (2.21) can be written as

$$y(k) + a_1 y(k-1) + \cdots + a_{na} y(k-na) = b_0 u(k-d) + \cdots + b_{nb} u(k-d-nb)$$

where $d = na - nb$ is the *pole excess* of the system. The polynomial

$$A^*(z) = 1 + a_1 z + \cdots + a_{na} z^{na} = z^{na} A(z^{-1})$$

which is obtained from the polynomial A by reversing the order of the coefficients, is called the *reciprocal polynomial*. Introduction of the reciprocal polynomials allows the system in (2.22) to be written as

$$A^*(q^{-1})y(k) = B^*(q^{-1})u(k-d)$$

Some care must be exercised when operating with reciprocal polynomials because A^{**} is not necessarily the same as A. The polynomial $A(z) = z$ has the reciprocal $A^*(z) = z \cdot z^{-1} = 1$. The reciprocal of A^* is $A^{**}(z) = 1$, which is different from A. A polynomial $A(z)$ is called *self-reciprocal* if

$$A^*(z) = A(z)$$

A Difficulty

The goal of algebraic system theory is to convert manipulations of difference equations to purely algebraic problems. It follows from the definition of the shift operator that the difference equation of (2.22) can be multiplied by powers of q, which simply means a forward shift of time. Equations for shifted times can also be multiplied by real numbers and added, which corresponds to multiplying Eq. (2.22) by a polynomial in q. If (2.22) holds, it is thus also true that

$$C(q)A(q)y(k) = C(q)B(q)u(k)$$

To obtain a convenient algebra, it is also useful to be able to divide an equation like (2.22) with a polynomial in q. For example, if

$$A(q)y(k) = 0$$

it would then possible to conclude that

$$y(k) = 0$$

If division is possible, an equation like (2.22) can be solved with respect to $y(k)$. A simple example shows that it is not possible to divide by a polynomial in q unless special assumptions are made.

Example 2.11   Role of initial conditions

Consider the difference equation

$$y(k + 1) - ay(k) = u(k)$$

where $|a| < 1$. In operator notation the equation can be written as

$$(q - a)y(k) = u(k) \tag{2.23}$$

If $y(k_0) = y_0$ it follows from (2.17) that the solution can be written as

$$
\begin{aligned}
y(k) &= a^{k-k_0}y_0 + \sum_{j=k_0}^{k-1} a^{k-j-1}u(j) \\
&= a^{k-k_0}y_0 + \sum_{i=1}^{k-k_0} a^{i-1}u(k-i)
\end{aligned}
\tag{2.24}
$$

A formal solution of the operator equation (2.23) can be obtained as follows:

$$y(k) = \frac{1}{q - a} u(k) = \frac{q^{-1}}{1 - aq^{-1}} u(k)$$

Because q^{-1} has unit norm, the right-hand side can be expressed as a convergent series.

$$
\begin{aligned}
y(k) &= q^{-1}(1 + aq^{-1} + a^2q^{-2} + \cdots)u(k) \\
&= \sum_{i=1}^{\infty} a^{i-1}u(k-i)
\end{aligned}
\tag{2.25}
$$

It is clear that solutions in (2.24) and (2.25) are the same only if it is assumed that $y_0 = 0$ or that $k - k_0 \to \infty$.   ∎

It is possible to develop an operator algebra that allows division by an arbitrary polynomial in q if it is assumed that there is some k_0 such that all sequences are zero for $k \leq k_0$. This algebra then allows the normal manipulations of multiplication and division of equations by polynomials in the shift operator as well as addition and subtraction of equations. However, the assumption does imply that all initial conditions for the difference equation are zero, which is the convention used in this book. (Compare with Example 2.11.)

If no assumptions on the input sequences are made, it is possible to develop a slightly different shift-operator algebra that allows division only by polynomials with zeros inside the unit disc. This corresponds to the fact that effects of initial conditions on stable modes will eventually vanish. This algebra is slightly more complicated because it does not allow normal division.

The Pulse-Transfer Operator

Use of operator calculus allows the input-output relationship to be conveniently expressed as a rational function in either the forward- or the backward-shift operator. This function is called the *pulse-transfer operator* and is easily obtained from any system description by eliminating internal variables using purely algebraic manipulations.

Consider, for example, the state-space model of (2.16). To obtain the input-output relationship, the state vector must be eliminated. It follows from (2.16) that

$$x(k + 1) = qx(k) = \Phi x(k) + \Gamma u(k)$$

Hence

$$(qI - \Phi)x(k) = \Gamma u(k)$$

This gives

$$y(k) = Cx(k) + Du(k) = \left(C(qI - \Phi)^{-1}\Gamma + D \right) u(k)$$

The pulse-transfer operator for the system (2.16) is thus given by

$$H(q) = C(qI - \Phi)^{-1}\Gamma + D$$

The pulse-transfer operator can be also expressed in terms of the backward-shift operator.

$$H^*(q^{-1}) = C(I - q^{-1}\Phi)^{-1}q^{-1}\Gamma + D = H(q)$$

The pulse-transfer operator for the system of (2.16) is thus a matrix whose elements are rational functions in q. For a system with one input and one output,

$$H(q) = C(qI - \Phi)^{-1}\Gamma + D = \frac{B(q)}{A(q)} \tag{2.26}$$

If the state vector is of dimension n and if the polynomials $A(q)$ and $B(q)$ do not have common factors, then the polynomial A is of degree n. It follows from (2.26) that the polynomial A is also the characteristic polynomial of the matrix Φ, which means that the input-output model can be written as

$$y(k) + a_1y(k-1) + \cdots + a_ny(k-n) = b_0u(k) + \cdots + b_nu(k-n)$$

where a_i are the coefficients of the characteristic polynomial of Φ. The most common case in computer-control systems is that $b_0 = 0$, that is, there is no direct term in the discrete-time model. Usually $y(k)$ is measured first, and then $u(k)$ is determined. Then $y(k)$ cannot be influenced by $u(k)$ even if the continuous-time system has a direct term.

Example 2.12 Double integrator

Consider the double integrator in Example 2.2 when $h = 1$. From (2.26)

$$H(q) = \begin{pmatrix} 1 & 0 \end{pmatrix} \begin{pmatrix} q-1 & -1 \\ 0 & q-1 \end{pmatrix}^{-1} \begin{pmatrix} 0.5 \\ 1 \end{pmatrix} = \frac{0.5(q+1)}{(q-1)^2} = \frac{0.5(q^{-1}+q^{-2})}{1-2q^{-1}+q^{-2}}$$

∎

Example 2.13 Double integrator with time delay

Use $h = 1$ for the double integrator and introduce a time delay of 0.5 s. Then from (2.10) and Example 2.7,

$$H(q) = C(qI - \Phi)^{-1}(\Gamma_0 + \Gamma_1 q^{-1})$$

$$= \begin{pmatrix} 1 & 0 \end{pmatrix} \frac{\begin{pmatrix} q-1 & -1 \\ 0 & q-1 \end{pmatrix}}{(q-1)^2} \begin{pmatrix} 0.125 + 0.375q^{-1} \\ 0.5 + 0.5q^{-1} \end{pmatrix}$$

$$= \frac{0.125(q^2 + 6q + 1)}{q(q^2 - 2q + 1)} = \frac{0.125(q^{-1} + 6q^{-2} + q^{-3})}{1 - 2q^{-1} + q^{-2}}$$

∎

Section 2.5 shows that different state-space representations can be used. Of course, this does not change the input-output model.

THEOREM 2.4 INVARIANCE OF THE PULSE-TRANSFER OPERATOR The pulse-transfer operator $H(q)$ for the state-space model (2.16) is independent of the state-space representation.

Proof. Given the pulse-transfer operator

$$H(q) = C(qI - \Phi)^{-1}\Gamma + D$$

and a transformation matrix T. In the new coordinates

$$\tilde{H}(q) = \tilde{C}(qI - \tilde{\Phi})^{-1}\tilde{\Gamma} + \tilde{D} = CT^{-1}(qTT^{-1} - T\Phi T^{-1})^{-1}T\Gamma + D$$

$$= CT^{-1}\left(T(qI - \Phi)T^{-1}\right)^{-1}T\Gamma + D = CT^{-1}T(qI - \Phi)^{-1}T^{-1}T\Gamma + D$$

$$= C(qI - \Phi)^{-1}\Gamma + D = H(q)$$

∎

The input-output models of a system with a zero-order hold can be obtained by using (2.5) and (2.26). In order to simplify the computation of the pulse-transfer operator $H(q)$, it is convenient to use Table 2.1, which gives $H(q)$ for some standard systems.

Programs for computer algebra such as Maple® and *Mathematica*® are very convenient for performing sampling because the result is obtained in algebraic form and it can easily be converted to computer code. This approach makes tables obsolete and it also reduces the potential sources of mistakes in manual calculations.

Poles and Zeros

The *poles* of a system are the zeros of the denominator of $H(q)$, the characteristic polynomial $A(q)$. The *zeros* are obtained from $B(q) = 0$, the poles of the inverse system. For instance, the system in Example 2.12 has one zero in -1; the system has two poles in 1.

Time delay in a system gives rise to poles at the origin. The system in Example 2.13 has three poles: two in 1, and one at the origin. There are two zeros: $-3 \pm \sqrt{8}$.

The interpretations of poles and zeros are discussed in Sec. 2.8.

The Order of a System

The *order* of a system is the same as the dimension of a state-space representation or, equivalently, the number of poles of the system. Notice that it is important to use the forward-shift form to determine the order because of the time delays. The determination of the poles, zeros, and order of a system are occasions when it is important to use the forward-shift form.

2.7 The z-Transform

In the analysis of continuous-time systems the Laplace transform plays an important role. The transformation makes it possible to introduce the transfer function and the frequency interpretation of a system. The combination of time-domain and frequency-domain aspects gives an increasing understanding of systems. The discrete-time analogy of the Laplace transform is the z-transform—a convenient tool to study linear difference equations with or without initial conditions.

The z-transform maps a *semi-infinite time sequence* into a function of a complex variable. Notice the difference in range for the z-transform and the operator calculus. In the operator calculus we consider double-infinite time sequences. The main difference is because the z-transform also takes the initial values into consideration. The variable z is a complex variable and should be distinguished from the operator q.

Table 2.1 Zero-order hold sampling of a continuous-time system, $G(s)$. The table gives the zero-order-hold equivalent of the continuous-time system, $G(s)$, preceded by a zero-order hold. The sampled system is described by its pulse-transfer operator. The pulse-transfer operator is given in terms of the coefficients of

$$H(q) = \frac{b_1 q^{n-1} + b_2 q^{n-2} + \cdots + b_n}{q^n + a_1 q^{n-1} + \cdots + a_n}$$

$G(s)$	$H(q)$ or the coefficients in $H(q)$
$\dfrac{1}{s}$	$\dfrac{h}{q-1}$
$\dfrac{1}{s^2}$	$\dfrac{h^2(q+1)}{2(q-1)^2}$
$\dfrac{1}{s^m}$	$\dfrac{q-1}{q} \lim_{a \to 0} \dfrac{(-1)^m}{m!} \dfrac{\partial^m}{\partial a^m} \left(\dfrac{q}{q - e^{-ah}} \right)$
e^{-sh}	q^{-1}
$\dfrac{a}{s+a}$	$\dfrac{1 - \exp(-ah)}{q - \exp(-ah)}$

For $\dfrac{a}{s(s+a)}$:

$$b_1 = \frac{1}{a}(ah - 1 + e^{-ah}) \qquad b_2 = \frac{1}{a}\left(1 - e^{-ah} - ahe^{-ah}\right)$$
$$a_1 = -(1 + e^{-ah}) \qquad a_2 = e^{-ah}$$

For $\dfrac{a^2}{(s+a)^2}$:

$$b_1 = 1 - e^{-ah}(1 + ah) \qquad b_2 = e^{-ah}(e^{-ah} + ah - 1)$$
$$a_1 = -2e^{-ah} \qquad a_2 = e^{-2ah}$$

For $\dfrac{s}{(s+a)^2}$:

$$\frac{(q-1)he^{-ah}}{(q - e^{-ah})^2}$$

For $\dfrac{ab}{(s+a)(s+b)}$, $a \neq b$:

$$b_1 = \frac{b(1 - e^{-ah}) - a(1 - e^{-bh})}{b - a}$$
$$b_2 = \frac{a(1 - e^{-bh})e^{-ah} - b(1 - e^{-ah})e^{-bh}}{b - a}$$
$$a_1 = -(e^{-ah} + e^{-bh})$$
$$a_2 = e^{-(a+b)h}$$

Table 2.1 *continued*

$G(s)$	$H(q)$ or the coefficients in $H(q)$
$\dfrac{(s+c)}{(s+a)(s+b)}$ $a \neq b$	$b_1 = \dfrac{e^{-bh} - e^{-ah} + (1 - e^{-bh})c/b - (1 - e^{-ah})c/a}{a - b}$ $b_2 = \dfrac{c}{ab}e^{-(a+b)h} + \dfrac{b-c}{b(a-b)}e^{-ah} + \dfrac{c-a}{a(a-b)}e^{-bh}$ $a_1 = -e^{-ah} - e^{-bh} \qquad a_2 = e^{-(a+b)h}$
$\dfrac{\omega_0^2}{s^2 + 2\zeta\omega_0 s + \omega_0^2}$	$b_1 = 1 - \alpha\left(\beta + \dfrac{\zeta\omega_0}{\omega}\gamma\right) \qquad \omega = \omega_0\sqrt{1-\zeta^2} \qquad \zeta < 1$ $b_2 = \alpha^2 + \alpha\left(\dfrac{\zeta\omega_0}{\omega}\gamma - \beta\right) \qquad \alpha = e^{-\zeta\omega h}$ $a_1 = -2\alpha\beta \qquad\qquad\qquad \beta = \cos(\omega h)$ $a_2 = \alpha^2 \qquad\qquad\qquad\quad \gamma = \sin(\omega h)$
$\dfrac{s}{s^2 + 2\zeta\omega_0 s + \omega_0^2}$	$b_1 = \dfrac{1}{\omega}e^{-\zeta\omega_0 h}\sin(\omega h) \qquad b_2 = -b_1$ $a_1 = -2e^{-\zeta\omega_0 h}\cos(\omega h) \qquad a_2 = e^{-\zeta\omega_0 h}$ $\omega = \omega_0\sqrt{1-\zeta^2}$
$\dfrac{a^2}{s^2 + a^2}$	$b_1 = 1 - \cos ah \qquad b_2 = 1 - \cos ah$ $a_1 = -2\cos ah \qquad b_2 = 1 - \cos ah$
$\dfrac{s}{s^2 + a^2}$	$b_1 = \dfrac{1}{a}\sin ah \qquad b_2 = \dfrac{1}{a}\sin ah$ $a_1 = -2\cos ah \qquad a_2 = 1$
$\dfrac{a}{s^2(s+a)}$	$b_1 = \dfrac{1-a}{a^2} + h\left(\dfrac{h}{2} - \dfrac{1}{a}\right) \qquad \alpha = e^{-ah}$ $b_2 = (1-\alpha)\left(\dfrac{h^2}{2} - \dfrac{2}{a^2}\right) + \dfrac{h}{a}(1+\alpha)$ $b_3 = (1-\alpha)\left[\dfrac{1}{a^2}(\alpha - 1) + \alpha h\left(\dfrac{h}{2} + \dfrac{1}{a}\right)\right]$ $a_1 = -(\alpha + 2) \qquad a_2 = 2\alpha + 1 \qquad a_3 = -\alpha$

DEFINITION 2.1   z-TRANSFORM   Consider the *discrete-time signal* $\{f(kh): k = 0, 1, \ldots\}$. The z-transform of $f(kh)$ is defined as

$$\mathcal{Z}\{f(kh)\} = F(z) = \sum_{k=0}^{\infty} f(kh)z^{-k} \qquad (2.27)$$

where z is a complex variable. The z-transform of f is denoted by $\mathcal{Z}f$ or F. The inverse transform is given by

$$f(kh) = \frac{1}{2\pi i} \oint F(z)z^{k-1}\, dz \qquad (2.28)$$

where the contour of integration encloses all singularities of $F(z)$.   ∎

Example 2.14   Transform of a ramp

Consider a ramp signal defined by $y(kh) = kh$ for $k \geq 0$. Then

$$Y(z) = 0 + hz^{-1} + 2hz^{-2} + \cdots = h(z^{-1} + 2z^{-2} + \cdots) = \frac{hz}{(z-1)^2}$$

∎

Some properties of the z-transform are collected in Table 2.2. Notice that the formulas for forward and backward time shifts are not the same. This is a consequence of the assumption that the time sequences are semi-infinite.

The z-transform can be used to solve difference equations; for instance,

$$x(k+1) = \Phi x(k) + \Gamma u(k)$$
$$y(k) = C x(k) + D u(k)$$

If the z-transform of both sides is taken,

$$\sum_{k=0}^{\infty} z^{-k}x(k+1) = z\left(\sum_{k=0}^{\infty} z^{-k}x(k) - x(0)\right) = \sum_{k=0}^{\infty} \Phi z^{-k}x(k) + \sum_{k=0}^{\infty} \Gamma z^{-k}u(k)$$

Hence

$$z\Big(X(z) - x(0)\Big) = \Phi X(z) + \Gamma U(z)$$
$$X(z) = (zI - \Phi)^{-1}\Big(zx(0) + \Gamma U(z)\Big)$$

and

$$Y(z) = C(zI - \Phi)^{-1}zx(0) + \Big(C(zI - \Phi)^{-1}\Gamma + D\Big)U(z)$$

The *pulse-transfer function* can now be introduced.

$$H(z) = C(zI - \Phi)^{-1}\Gamma + D \qquad (2.29)$$

which is the same as (2.26) with q replaced by z. The time sequence $y(k)$ can now be obtained using the inverse transform. The following theorem is analogous to that of continuous-time systems.

Table 2.2   Some properties of the z-transform.

---

1. Definition.

$$F(z) = \sum_{k=0}^{\infty} f(kh)z^{-k}$$

2. Inversion.

$$f(kh) = \frac{1}{2\pi i} \oint F(z)z^{k-1} dz$$

3. Linearity.

$$\mathcal{Z}\{\alpha f + \beta g\} = \alpha \mathcal{Z}f + \beta \mathcal{Z}g$$

4. Time shift.

$$\mathcal{Z}\{q^{-n}f\} = z^{-n}F$$

$$\mathcal{Z}\{q^{n}f\} = z^{n}(F - F_1) \text{ where } F_1(z) = \sum_{j=0}^{n-1} f(jh)z^{-j}$$

5. Initial-value theorem.

$$f(0) = \lim_{z \to \infty} F(z)$$

6. Final-value theorem.

If $(1 - z^{-1})F(z)$ does not have any poles on or outside the unit circle, then $\lim_{k \to \infty} f(kh) = \lim_{z \to 1}(1 - z^{-1})F(z)$.

7. Convolution.

$$\mathcal{Z}\{f * g\} = \mathcal{Z}\left\{\sum_{n=0}^{k} f(n)g(k-n)\right\} = (\mathcal{Z}f)(\mathcal{Z}g)$$

THEOREM 2.5   The pulse response of (2.20) and the pulse-transfer function (2.29) are a z-transform pair, that is, $\mathcal{Z}\{h(k)\} = H(z)$.   ∎

Computation of the Pulse-Transfer Function

The pulse-transfer function can be determined directly from the continuous-time transfer function. Let the system be described by the transfer function $G(s)$ preceded by a zero-order hold (see Fig. 2.4). The pulse-transfer function is

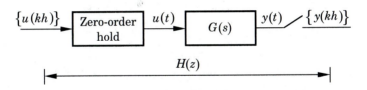

Figure 2.4   Sampling a continuous-time system.

uniquely determined by the response to a given signal. Consider, for instance, a unit-step input. The sequence $\{u(kh)\}$ is then a sequence of ones and the signal $u(t)$ is then also a unit step. Let $Y(s)$ denote the Laplace transform of $y(t)$, that is,

$$Y(s) = \frac{G(s)}{s}$$

Let the sampled output $\{y(kh)\}$ have the z-transform $\tilde{Y} = \mathcal{Z}\{\mathcal{L}^{-1}Y\}$. Division of $\tilde{Y}$ by the pulse-transfer function of the input, which is $z/(z-1)$, gives

$$H(z) = (1 - z^{-1})\tilde{Y}(z)$$

The pulse-transfer function is now obtained as follows:

1. Determine the step response of the system with the transfer function $G(s)$.

2. Determine the corresponding z-transform of the step response.

3. Divide by the z-transform of the step function.

By using this procedure the following formula can be derived:

$$H(z) = \frac{z-1}{z} \frac{1}{2\pi i} \int_{\gamma - i\infty}^{\gamma + i\infty} \frac{e^{sh}}{z - e^{sh}} \frac{G(s)}{s} \, ds \qquad (2.30)$$

If the transfer function $G(s)$ goes to zero at least as fast as $|s|^{-1}$ for a large s and has distinct poles, none of which are at the origin, we get

$$H(z) = \sum_{s=s_i} \frac{1}{z - e^{sh}} \operatorname{Res} \left\{ \frac{e^{sh} - 1}{s} \right\} G(s) \qquad (2.31)$$

where s_i are the poles of $G(s)$ and Res denotes the residue. A proof of this formula is given in Sec. 7.8. If $G(s)$ has multiple poles or a pole in the origin, (2.31) must be modified to take multiple poles into consideration when calculating the residues. Table 2.3 shows some time functions and the corresponding Laplace and z-transforms. The table can thus be used to combine steps 1 and 2. Tables in textbooks are usually found in this form.

   Warning. Notice that $\mathcal{Z}f$ in Table 2.3 does not give the zero-order-hold sampling of a system with the transfer function $\mathcal{L}f$. Examine Table 2.1. It is a very common mistake to believe that it does. The desired pulse-transfer function is obtained through the procedure given.

Shift-Operator Calculus and z-transforms

There are strong formal relations between shift-operator calculus and calculations with z-transforms. When manipulating difference equations we can use either. The expressions obtained look formally very similar. In many textbooks the same notation is in fact used for both. The situation is very similar to the

Table 2.3 Some time functions and corresponding Laplace and z-transforms. Warning: Use the table only as prescribed!

f	$\mathcal{L}f$	$\mathcal{Z}f$
$\delta(k)$ (pulse)	$-$	1
$1 \quad k \geq 0$ (step)	$\dfrac{1}{s}$	$\dfrac{z}{z-1}$
kh	$\dfrac{1}{s^2}$	$\dfrac{hz}{(z-1)^2}$
$\dfrac{1}{2}(kh)^2$	$\dfrac{1}{s^3}$	$\dfrac{h^2 z(z+1)}{2(z-1)^3}$
$e^{-kh/T}$	$\dfrac{T}{1+sT}$	$\dfrac{z}{z-e^{-h/T}}$
$1 - e^{-kh/T}$	$\dfrac{1}{s(1+sT)}$	$\dfrac{z(1-e^{-h/T})}{(z-1)(z-e^{-h/T})}$
$\sin \omega k h$	$\dfrac{\omega}{s^2+\omega^2}$	$\dfrac{z \sin \omega h}{z^2 - 2z \cos \omega h + 1}$

difference between the differential operator $p = d/dt$ and the Laplace transform s for continuous-time systems. First, we may notice that q is an operator that acts on sequences and z is a complex variable. From a purely mathematical point of view, it clearly makes sense to make a distinction between such different objects. There is, however, also a good system-theoretic reason for making a distinction. We illustrate this by an example.

Example 2.15   Pole-zero cancellations

Consider the difference equation

$$y(k+1) + ay(k) = u(k+1) + au(k) \tag{2.32}$$

If (2.32) is considered as a dynamical system its pulse-transfer function is obtained as

$$H(z) = \frac{z+a}{z+a} = 1$$

The last equality is obtained because z is a complex variable. We may be thus misled to believe that the system (2.32) is identical to

$$y(k) = u(k) \tag{2.33}$$

This is clearly not true because the difference equation (2.32) has the solution

$$y(k) = (-a)^k y(0) + u(k) \qquad k \geq 1$$

which is identical to (2.33) *only if the initial condition $y(0)$ is zero*. It may be reasonable to neglect the initial conditions if $|a| < 1$, but not reasonable if $|a| \geq 1$. We thus have the situation that from a system-theoretic point of view, the expression

$$\frac{z + a}{z + a}$$

can be considered equal to one if $|a| < 1$ but not otherwise. If equation (2.32) is solved using shift-operator calculus we obtain formally

$$(q + a)y(k) = (q + a)u(k)$$

Notice that we cannot divide by $q + a$ because q is an operator. ∎

The conclusion that we can draw from the simple example is that the algebras of z-transforms and shift operators are different. In z-transforms calculus we can divide with an arbitrary expression, but this is not allowed in shift-operator calculus. The system-theoretic interpretation is that we may throw away some modes in the system with z-transform calculus by cancellation factors. This may make sense if the canceled factors correspond to stable modes, but it may be strongly misleading if the canceled factors are unstable. Another manifestation of this effect will be given in the discussions of the notions of observability and controllability in Chapter 3.

Modified z-transform

The behavior between sampling points can be investigated using the modified z-transform. This is the ordinary z-transform, but a time delay mh, which is a fraction of the sampling period is introduced. The modified z-transform is defined as follows.

DEFINITION 2.2   THE MODIFIED z-TRANSFORM   The modified z-transform of a *continuous-time function* is given by

$$\tilde{F}(z, m) = \sum_{k=0}^{\infty} z^{-k} f(kh - h + mh) \qquad 0 \leq m \leq 1 \tag{2.34}$$

The inverse transform is given by

$$f(nh - h + mh) = \frac{1}{2\pi i} \int_{\Gamma} \tilde{F}(z, m) z^{n-1} dz$$

where the contour Γ encloses all singularities of the integrand. ∎

The modified z-transform is useful for many purposes—for example, the intersample behavior can easily be investigated using these transforms. There are extensive tables of modified z-transforms and many theorems about their properties (see the References).

2.8 Poles and Zeros

For single-input–single-output finite-dimensional systems, poles and zeros can be conveniently obtained from the denominator and numerator of the pulse-transfer function. Poles and zeros have good system-theoretic interpretation. A pole $z = a$ corresponds to a free mode of the system associated with the time function $z(k) = a^k$. Poles are also the eigenvalues of the system matrix Φ. The zeros are related to how the inputs and outputs are coupled to the states.

Zeros can also be characterized by their signal blocking properties. A zero $z = a$ means that the transmission of the input signal $u(k) = a^k$ is blocked by the system. This interpretation can be used to define zeros in terms of the state-space equation. It follows from (2.16) that the input $u(k) = u_0 a^k$ gives the state $x(k) = x_0 a^k$ and zero output if $z = a$ such that

$$\det \begin{pmatrix} zI - \Phi & -\Gamma \\ C & D \end{pmatrix} = 0$$

Poles

Consider a continuous-time system described by the nth-order state-space model

$$\frac{dx}{dt} = Ax + Bu \tag{2.35}$$
$$y = Cx$$

The poles of the system are the eigenvalues of A, which we denote by $\lambda_i(A), i = 1, \dots, n$. The zero-order-hold sampling of (2.35) gives the discrete-time system

$$x(kh + h) = \Phi x(kh) + \Gamma u(kh)$$
$$y(kh) = Cx(kh)$$

Its poles are the eigenvalues of Φ, $\lambda_i(\Phi)$, $i = 1, \dots, n$. Because $\Phi = \exp(Ah)$ it follows from the properties of matrix functions (see Appendix B) that

$$\lambda_i(\Phi) = e^{\lambda_i(A)h} \tag{2.36}$$

Equation (2.36) gives the mapping from the continuous-time poles to the discrete-time poles. Figure 2.5 illustrates a mapping of the complex s-plane into the z-plane, when $z = \exp(sh)$. For instance, the left half of the s-plane is mapped into the unit disc of the z-plane. The map is not bijective—several points in the s-plane are mapped into the same point in the z-plane (see Fig. 2.6). This is an illustration of the aliasing effect discussed in Example 1.4. For poles inside the fundamental strip S_0 in Fig. 2.6, there is a simple relationship between continuous- and discrete-time poles. (Also compare with Example 2.5.)

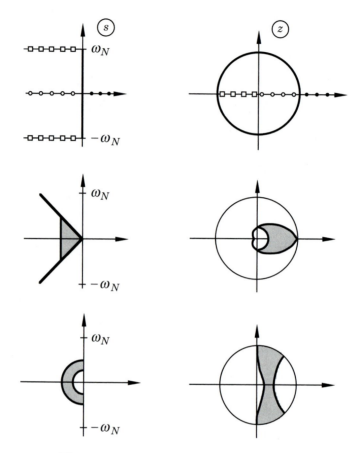

Figure 2.5   The conformal map $z = \exp(sh)$.

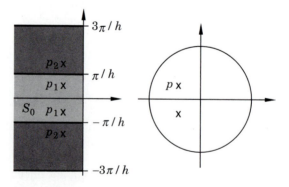

Figure 2.6   Each strip in the left half of the s-plane is mapped into the unit disc. This means that the pair of poles, p_1 and p_2, are both mapped into the pair p.

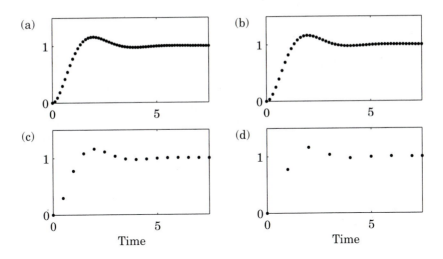

Figure 2.7   Step responses of the discrete-time system in Example 2.16 for different values of h when $\zeta = 0.5$ and $\omega_0 = 1.83$, which gives the rise time $T_r = 1$: (a) $h = 0.125$, (b) $h = 0.25$, (c) $h = 0.5$, and (d) $h = 1.0$.

Example 2.16   Complex poles

Consider the continuous-time system

$$\frac{\omega_0^2}{s^2 + 2\zeta\omega_0 s + \omega_0^2} \tag{2.37}$$

The poles of the corresponding discrete-time system are given by the characteristic equation

$$z^2 + a_1 z + a_2 = 0$$

where

$$a_1 = -2e^{-\zeta\omega_0 h} \cos\left(\sqrt{1 - \zeta^2}\,\omega_0 h\right)$$

$$a_2 = e^{-2\zeta\omega_0 h}$$

(Compare with Table 2.1.) Figure 2.7 shows the step responses of the discrete-time system for different values of the sampling interval when $\omega_0 = 1.83$ and $\zeta = 0.5$. Figure 2.8 gives a more detailed picture of how the continuous-time poles of (2.37) are mapped into the unit circle for different values of ζ and $\omega_0 h$ when the system is sampled.   ■

Zeros

It is not possible to give a simple formula for the mapping of zeros. If a continuous-time transfer function is viewed as a rational function, it has zeros at the zeros of the numerator polynomial and d zeros at infinity, where d is the pole

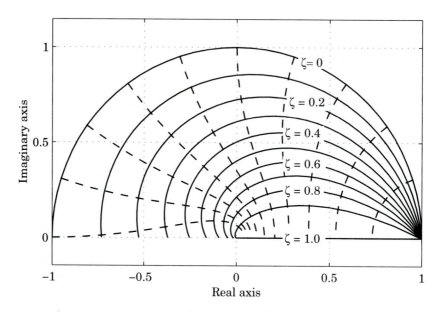

Figure 2.8 Loci of constant ζ (solid) and $\omega_0 h$ (dashed) when (2.37) is sampled.

excess for the continuous-time transfer function—that is, the difference between the number of poles and the number of zeros. The discrete-time system has, in general, $n - 1$ zeros; compare Examples 2.12 and 2.13. The sampling procedure thus gives extra zeros.

For short sampling periods, a discrete-time system will have zeros in

$$z_i \approx e^{s_i h}$$

where the s_i's are the zeros of the continuous-time system. The $r = d - 1$ zeros introduced by the sampling will go to the zeros of the polynomials Z_d in Table 2.4 as the sampling interval goes to zero, because for large s, the transfer function of the continuous-time system is approximately given by $G(s) \approx s^{-d}$.

Example 2.17   Second-order system

Consider the continuous-time transfer function

$$\frac{2}{(s + 1)(s + 2)}$$

Using Table 2.1 gives the zero of the pulse-transfer function

$$z = -\frac{(1 - e^{-2h})e^{-h} - 2(1 - e^{-h})e^{-2h}}{2(1 - e^{-h}) - (1 - e^{-2h})}$$

When h is small

$$z \approx -1 + 3h$$

Table 2.4 Numerator polynomials, Z_d, when sampling s^{-d}.

d	Z_d
1	1
2	$z + 1$
3	$z^2 + 4z + 1$
4	$z^3 + 11z^2 + 11z + 1$
5	$z^4 + 26z^3 + 66z^2 + 26z + 1$

and when h approaches zero, the zero moves to -1. The zero moves toward the origin when h is increased. The zero for small values of h also can be obtained from Table 2.4. The pole excess of the continuous-time system is $d = 2$. The discrete-time system will have a zero at $z = -1$ when h goes to zero.   ∎

Systems with Unstable Inverses

A continuous-time system with a rational transfer function is nonminimum-phase if it has right half-plane zeros or time delays. Analogously, a discrete-time system is often defined to be nonminimum-phase if it has zeros outside the unit disc. That definition implies that a time delay does not make the system nonminimum-phase. On the other hand, time delays do not pose the same severe problems as they do for continuous-time systems. For discrete-systems it is therefore more relevant to talk about systems with or without stable inverses, which are defined as follows.

DEFINITION 2.3   UNSTABLE INVERSE   A discrete-time system has an unstable inverse if it has zeros outside the unit disc.   ∎

A continuous-time system with a stable inverse may become a discrete-time system with an unstable inverse when it is sampled. It follows from Table 2.4 that the inverse system is always unstable if the pole excess of the continuous-time system is larger than 2, and if the sampling period is sufficiently short. Further, a continuous-time nonminimum-phase system will not always become a discrete-time system with an unstable inverse, as shown in the following example.

Example 2.18   Stability of inverse system changes with sampling

The transfer function

$$G(s) = \frac{6(1 - s)}{(s + 2)(s + 3)}$$

has an unstable zero $s = 1$. Sampling the system gives a discrete-time pulse-transfer function with a zero:

$$z_1 = -\frac{8e^{-2h} - 9e^{-3h} + e^{-5h}}{1 - 9e^{-2h} + 8e^{-3h}}$$

For $h \approx 1.25$, $z_1 = -1$; for larger h, the zero is always inside the unit circle and the sampled system has a stable inverse.   ∎

2.9 Selection of Sampling Rate

Proper selection of the sampling rate is a very important issue in computer-controlled systems. Too long a sampling period will make it impossible to reconstruct the continuous-time signal. Too short a sampling period will increase the load on the computer. The problem of sample-rate selection was touched in Sec. 1.3. The choice of the sampling period strongly depends on the purpose of the system. We will return to this question many times in the book. This section only relates the sampling-rate selection to the poles of the open-loop continuous-time system.

It is useful to characterize the sampling period with a variable that is dimension-free and that has a good physical interpretation. For oscillatory systems, it is natural to normalize with respect to the period of oscillation; for nonoscillatory systems, the rise time is a natural normalization factor.

We now introduce N_r as the number of sampling periods per rise time,

$$N_r = \frac{T_r}{h}$$

where T_r is the rise time. For first-order systems, the rise time is equal to the time constant. It is then reasonable to choose N_r between 4 and 10. For a second-order system with damping ζ and natural frequency ω_0, rise time is given by

$$T_r = \omega_0^{-1} e^{\varphi/\tan\varphi}$$

where $\zeta = \cos\varphi$. For a damping around $\zeta = 0.7$, this gives

$$\omega_0 h \approx 0.2 - 0.6$$

where ω_0 is in radians per second.

Figures 2.7 and 2.9 illustrate the choice of the sampling interval for different signals. It is thus reasonable to choose the sampling period so that

$$N_r = \frac{T_r}{h} \approx 4 \text{ to } 10$$

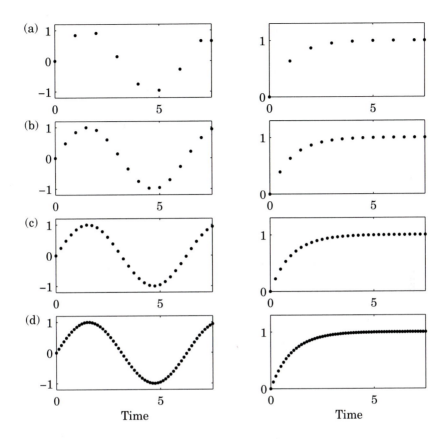

Figure 2.9 Illustration of the sample and hold of a sinusoidal and an exponential signal. The rise times of the signals are $T_r = 1$. The number of samples per rise time is (a) $N_r = 1$, (b) $N_r = 2$, (c) $N_r = 4$, and (d) $N_r = 8$.

Example 2.19 Pole-zero variation with sampling interval

Consider the system

$$G(s) = \frac{1}{(s+1)(s^2 + s + 1)} \tag{2.38}$$

Figure 2.10 shows the step response of the system. Assume that the system is sampled with period h. Figure 2.11 shows how the poles and zeros of the sampled-data system vary with the sampling period. Sampling intervals close to zero give three poles close to 1. Further, the continuous time system has a pole excess of 3. This implies that the zeros for short sampling intervals are close to the roots of

$$z^2 + 4z + 1 = 0$$

See Table 2.4. The poles and zeros approach the origin when the sampling interval is increased. The sampled-data system has a stable inverse if $h > 2.24$. The rules

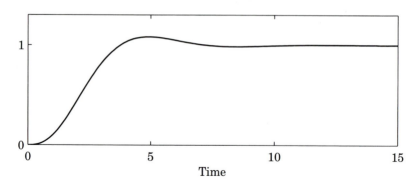

Figure 2.10 Step response of the system (2.38).

of thumb for the choice of the sampling interval give that a reasonable choice is $h = 0.5$. Compare with Figure 2.10. ∎

2.10 Problems

2.1   Consider the system

$$\frac{dx}{dt} = -ax + bu$$

$$y = cx$$

Let the input be constant over periods of length h. Sample the system and discuss how the poles of the discrete-time system vary with the sampling interval h.

2.2   Derive the discrete-time system corresponding to the following continuous-time systems when a zero-order-hold circuit is used:

(a)

$$\frac{dx}{dt} = \begin{pmatrix} 0 & 1 \\ -1 & 0 \end{pmatrix} x + \begin{pmatrix} 0 \\ 1 \end{pmatrix} u$$

$$y = \begin{pmatrix} 1 & 0 \end{pmatrix} x$$

(b)

$$\frac{d^2y}{dt^2} + 3\frac{dy}{dt} + 2y = \frac{du}{dt} + 3u$$

(c)

$$\frac{d^3y}{dt^3} = u$$

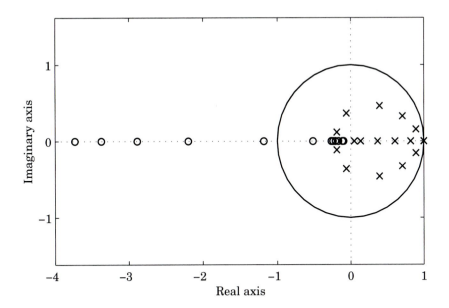

Figure 2.11 Poles (×) and zeros (○) when the system (2.38) is sampled with $h = 0$, 0.2, 0.5, 1, 2, and 3.

2.3 The following difference equations are assumed to describe continuous-time systems sampled using a zero-order-hold circuit and the sampling period h. Determine, if possible, the corresponding continuous-time systems.

(a)

$$y(kh) - 0.5y(kh - h) = 6u(kh - h)$$

(b)

$$x(kh + h) = \begin{pmatrix} -0.5 & 1 \\ 0 & -0.3 \end{pmatrix} x(kh) + \begin{pmatrix} 0.5 \\ 0.7 \end{pmatrix} u(kh)$$

$$y(kh) = \begin{pmatrix} 1 & 1 \end{pmatrix} x(kh)$$

(c)

$$y(kh) + 0.5y(kh - h) = 6u(kh - h)$$

2.4 Consider the harmonic oscillator [see Example A.3 or Problem 2.2(a)]. Compute the step response at 0, h, $2h$, ... when the sampling period is (a) $h = \pi/2$, (b) $h = \pi/4$. Compare with the continuous-time step response.

2.5 Sample the system

$$G(s) = \frac{1}{s^2(s + 2)(s + 3)}$$

using a zero-order-hold circuit and $h = 1$.

2.6 Consider the system in (2.1). Assume that the input is a sum of impulses at the sampling instants, that is,

$$u(t) = \sum \delta(t - kh)u(kh)$$

Determine the discrete-time representation.

2.7 Find the transformation matrix, T, that transforms the state-space representation of the double integrator (2.7) into diagonal or Jordan form.

2.8 Determine the pulse-transfer function of the system

$$x(kh + h) = \begin{pmatrix} 0.5 & -0.2 \\ 0 & 0 \end{pmatrix} x(kh) + \begin{pmatrix} 2 \\ 1 \end{pmatrix} u(kh)$$

$$y(kh) = \begin{pmatrix} 1 & 0 \end{pmatrix} x(kh)$$

2.9 Many physical systems can be described by the form

$$\frac{dx}{dt} = \begin{pmatrix} -a & b \\ c & -d \end{pmatrix} x + \begin{pmatrix} f \\ g \end{pmatrix} u$$

where $a, b, c,$ and d are nonnegative. Derive a formula for the sampled-data system when using a zero-order hold. (*Hint:* Show first that the poles of the system are real.)

2.10 Figure 2.12 shows a system of two tanks, where the input signal is the flow to the first tank and the output is the level in the second tank. Use of the levels as state variables gives the system

$$\frac{dx}{dt} = \begin{pmatrix} -0.0197 & 0 \\ 0.0178 & -0.0129 \end{pmatrix} x + \begin{pmatrix} 0.0263 \\ 0 \end{pmatrix} u$$

$$y = \begin{pmatrix} 0 & 1 \end{pmatrix} x$$

(a) Sample the system with the sampling period $h = 12$.

(b) Verify that the pulse-transfer operator for the system is

$$H_0(q) = \frac{0.030q + 0.026}{q^2 - 1.65q + 0.68}$$

2.11 The normalized motor is described in Example A.2. Show that the sampled system is described by (A.6). Determine the following:

(a) The pulse-transfer function.

(b) The pulse response.

(c) A difference equation relating the input and the output.

(d) The variation of the poles and zeros of the pulse-transfer function with the sampling period.

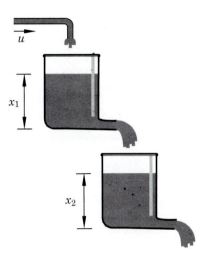

Figure 2.12   The two-tank process.

2.12 A continuous-time system with the transfer function

$$G(s) = \frac{1}{s} e^{-s\tau}$$

is sampled with $h = 1$ when $\tau = 0.5$.

(a) Determine a state-space representation of the sampled system. What is the order of the sampled system?

(b) Determine the pulse-transfer function and the pulse response of the sampled system.

(c) Determine the poles and zeros of the sampled system.

2.13 Solve Problem 2.12 with

$$G(s) = \frac{1}{s+1} e^{-s\tau}$$

and $h = 1$ and $\tau = 1.5$.

2.14 Consider the sampled system

$$y(k+1) = ay(k) + b_3 u(k-3) + b_4 u(k-4)$$

where the sampling interval is 1 s. Show that the system may be obtained by sampling the system

$$\frac{dy(t)}{dt} = -\alpha y(t) + bu(t-\tau)$$

where

$$\tau = 4 - \frac{1}{\ln \alpha} \ln \frac{ab_3 + b_4}{b_3 + b_4}$$

2.15 Consider the system

$$y(k) - 0.5y(k - 1) = u(k - 9) + 0.2u(k - 10)$$

Determine the polynomials $A(q)$, $B(q)$, $A^*(q^{-1})$, and $B^*(q^{-1})$ in the representations

$$A(q)y(k) = B(q)u(k)$$

and

$$A^*(q^{-1})y(k) = B^*(q^{-1})u(k - d)$$

What are d and the order of the system?

2.16 A filter with the pulse-transfer operator

$$H^*(q^{-1}) = b_0 + b_1 q^{-1} + \cdots + b_n q^{-n}$$

is called a finite impulse-response (FIR) filter.

(a) Determine the order of the system.

(b) Determine the poles of the filter and make a state-space representation of the filter.

2.17 Use the z-transform to determine the output sequence of the difference equation

$$y(k + 2) - 1.5y(k + 1) + 0.5y(k) = u(k + 1)$$

when $u(k)$ is a step at $k = 0$ and when $y(0) = 0.5$ and $y(-1) = 1$.

2.18 Verify that

$$\mathcal{Z}\left\{ \frac{1}{2}(kh)^2 \right\} = \frac{h^2 z(z + 1)}{2(z - 1)^3}$$

Compare with Table 2.3 and use that to determine the pulse-transfer function of the double integrator (see Example A.1).

2.19 Use (2.30) to determine the pulse-transfer function of (a) the system in Problem 2.1 and (b) the normalized motor (see Example A.2).

2.20 Show that a curve of constant damping ζ in the s-plane is a logarithmic spiral in the z-plane when using the mapping $z = \exp(sh)$. That is, the distance to the origin can be written as $r = ae^{m\varphi}$, where φ is the angle.

2.21 If $\beta < \alpha$, then

$$\frac{s + \beta}{s + \alpha}$$

is called a *lead network* (i.e., it gives a phase advance). Consider the discrete-time system

$$\frac{z + b}{z + a}$$

(a) Determine when it is a lead network.

(b) Simulate the step response for different pole and zero locations.

2.22 Consider the system

$$\frac{z + b}{(1 + b)(z^2 - 1.1z + 0.4)}$$

The pole location corresponds to a continuous-time system with damping $\zeta = 0.7$. Simulate the system and determine the overshoot for different values of b in the interval $(-1, 1)$.

2.23 Consider the stable continuous-time system

$$G(s) = \frac{s + b}{s + a}$$

where $a \neq b$. Sample the system with the sampling period h. Derive conditions for when the sampled system will have a stable inverse.

2.24 Consider the discrete-time system

$$H(z) = \frac{b_1 z + b_2}{z^{n+1}(z - a)}$$

This system is obtained by sampling a continuous-time system with the transfer function

$$G(s) = \frac{Ke^{-s\tau}}{1 + sT}$$

using the sampling interval h. Show that

$$T = -h/\ln a$$

$$K = \frac{b_1 + b_2}{1 - a}$$

$$\tau = nh - \frac{h}{\ln \alpha} \ln \frac{ab_1 + b_2}{a(b_1 + b_2)}$$

2.25 Use (2.30) to determine the pulse-transfer function associated with

$$G(s) = \frac{1}{s^3}$$

2.26 Use Eq. (2.30) to show that the pulse-transfer function obtained with zero-order-hold samplings of the transfer function

$$G(s) = \frac{1}{s^n}$$

is given by

$$H(z) = \frac{h^n}{n!} \frac{B_n(z)}{(z - 1)^n}$$

where

$$B_n(z) = b_1^n z^{n-1} + b_2^n z^{n-2} + \cdots + b_n^n$$

and

$$b_k^n = \sum_{i=1}^{k} (-1)^{k-i} i^n \binom{n+1}{k-i} \qquad k = 1, 2, \ldots, n$$

Furthermore show that

$$B_1(z) = 1$$
$$B_2(z) = z + 1$$
$$B_3(z) = z^2 + 4z + 1$$
$$B_4(z) = z^3 + 11z^2 + 11z + 1$$
$$B_5(z) = z^4 + 26z^3 + 66z^2 + 26z^2 + 1$$
$$B_6(z) = z^5 + 57z^4 + 302z^3 + 302z^2 + 57z + 1$$

2.27 Derive Eq. (2.31) from (2.30).

2.28 Solve the difference equation

$$y(k) = y(k-1) + y(k-2) \qquad k = 2, 3, \ldots$$

when $y(0) = y(1) = 1$. [The numbers $y(k)$ are called Fibonacci numbers.]

2.29 Determine the poles and zeros (with multiplicity) of the system

$$y(k) - 0.5y(k-1) + y(k-2) = 2u(k-10) + u(k-11)$$

2.30 Which of the following discrete-time systems can be obtained by sampling a causal continuous-time system using a zero-order hold?

$$H_1(q) = \frac{1}{q - 0.8} \qquad H_2(q) = \frac{1}{q + 0.8}$$

$$H_3(q) = \frac{q - 1}{(q + 0.8)^2} \qquad H_4(q) = \frac{2q^2 - 0.7q - 0.8}{q(q - 0.8)}$$

2.31 Determine the pulse-transfer operator obtained by sampling

$$G(s) = \frac{2(s+2)}{(s+1)(s+3)}$$

with $h = 0.02$.

2.32 Sample the continuous-time system

$$\frac{dx(t)}{dt} = \begin{pmatrix} 1 & 0 \\ 1 & 1 \end{pmatrix} x(t) + \begin{pmatrix} 1 \\ 0 \end{pmatrix} u(t - 0.2)$$

using the sampling interval $h = 0.3$. Determine the pulse-transfer operator.

2.33 Consider a linear system with the transfer function

$$G_a(s) = \frac{a}{s + a}$$

Sampling the system gives the pulse-transfer function

$$H_a(z) = \frac{1 - e^{-ah}}{z - e^{-ah}}$$

Letting $a \to \infty$, we get

$$G_\infty(s) = \lim_{a \to \infty} G_a(s) = 1$$

and

$$H_\infty(z) = \lim_{a \to \infty} H_a(z) = \frac{1}{z}$$

Notice that $H_\infty(z)$ is not the pulse-transfer function obtained by sampling the system with the pulse-transfer function $G_\infty(s) = 1$. Determine conditions on the transfer function $G_a(s)$ such sampling commutes with limit operations.

2.11 Notes and References

The early texts on sampled-data systems dealt exclusively with input-output models and transform theory Jury (1958), Ragazzini and Franklin (1958), and Tsypkin (1958). The state-space approach used in this chapter offers significant simplifications. With a zero-order hold, the control signal is constant over the sampling period and the discrete-time model is obtained simply by integrating the state equations over one sampling period. This problem formulation was introduced in Kalman and Bertram (1958). It took some time before this approach found its way into textbooks. Because of its simplicity it is now the predominant approach.

Transformation of state variables and canonical forms is standard material in state-space theory. These results are very similar to the corresponding results for continuous-time systems. A more detailed treatment is given in Kailath (1980). Historically, the input-output approach preceded the state-space approach. A direct treatment from this point of view is given in the classic texts just mentioned. The multivariable case is discussed in Rosenbrock (1970) and Kučera (1979, 1991).

The z-transform is extensively discussed in Jury (1958, 1982) and Doetsch (1971). These references contain large tables of z-transform pairs. A table of zero-order-hold equivalent transfer functions (compare with Table 2.1) is given in Neuman and Baradello (1979).

The relationship between the zeros of continuous and sampled systems is discussed in Åström, Hagander, and Sternby (1984). The theorems for the limiting zeros for large and small sampling periods are given in this paper.

Theorem 2.1 is proved in Wittenmark (1985b). A generalization to the sampling of a system with several time delays is found in Bernhardsson (1993).

Programs for computer algebra such as Maple® and *Mathematica*® are discussed, for instance, in Char (1992) and Wolfram (1988). For MATLAB® and MATRIX$_X$® we refer to the manuals for the programs.

Properties of matrices and transformations are found, for instance, in Gantmacher (1960), Bellman (1970), and Golub and Van Loan (1989).

3

Analysis of

Discrete-Time Systems

3.1 Introduction

Previous chapters have shown how continuous-time systems are transformed when sampled. In this chapter we will develop the key tools for analyzing discrete-time systems. Stability, sensitivity, and robustness are introduced in Secs. 3.2 and 3.3. The concepts of controllability, reachability, and observability, which are useful for understanding discrete-time systems, are discussed in Sec. 3.4. Simple feedback loops and their properties are treated in Sec. 3.5. Simulation is used throughout the text because it is a very important tool for the analysis of sampled-data systems—for instance, in investigating intersample behavior.

3.2 Stability

The concept of stability is very important when analyzing dynamic systems. It is assumed that the notion of stability is known from basic texts in control theory. Only the basic definitions are given here.

Definitions

Stability is first defined with respect to changes in the initial conditions. Consider the discrete-time state-space equation (possibly nonlinear and time-varying)

$$x(k + 1) = f(x(k), k) \tag{3.1}$$

Let $x^0(k)$ and $x(k)$ be solutions of (3.1) when the initial conditions are $x^0(k_0)$ and $x(k_0)$, respectively. Further, let $\| \cdot \|$ denote a vector norm.

DEFINITION 3.1   STABILITY   The solution $x^0(k)$ of (3.1) is *stable* if for a given $\varepsilon > 0$, there exists a $\delta(\varepsilon, k_0) > 0$ such that all solutions with $\|x(k_0) - x^0(k_0)\| < \delta$ are such that $\|x(k) - x_0(k)\| < \varepsilon$ for all $k \geq k_0$.   ∎

DEFINITION 3.2   ASYMPTOTIC STABILITY   The solution $x^0(k)$ (3.1) is *asymptotically stable* if it is stable and if δ can be chosen such that $\|x(k_0) - x^0(k_0)\| < \delta$ implies that $\|x(k) - x^0(k)\| \to 0$ when $k \to \infty$.   ∎

From the definitions, it follows that stability in general is defined for a particular solution and not for the system. The definitions also imply that stability, in general, is a local concept. The interpretation of Definitions 3.1 and 3.2 is that the system is (asymptotically) stable if the trajectories do not change much if the initial condition is changed by a small amount.

Stability of Linear Discrete-Time Systems

Consider the linear system

$$x^0(k + 1) = \Phi x^0(k) \qquad x^0(0) = a^0 \qquad (3.2)$$

To investigate the stability of the solution of (3.2), the initial value is perturbed. Hence

$$x(k + 1) = \Phi x(k) \qquad x(0) = a$$

The difference $\tilde{x} = x - x^0$ satisfies the equation

$$\tilde{x}(k + 1) = \Phi \tilde{x}(k) \qquad \tilde{x}(0) = a - a^0 \qquad (3.3)$$

This implies that if the solution x^0 is stable, then every other solution is also stable. For linear, time-invariant systems, stability is thus a property of the system and not of a special solution.

The system (3.3) has the solution

$$\tilde{x}(k) = \Phi^k \tilde{x}(0)$$

See (2.17). If it is possible to diagonalize Φ, then the solution is a combination of terms λ_i^k, where $\lambda_i, i = 1, \ldots, n$ are the eigenvalues of Φ; see (2.18). In the general case, when Φ cannot be diagonalized, the solution is instead a linear combination of the terms $p_i(k)\lambda_i^k$, where $p_i(k)$ are polynomials in k of order one less than the multiplicity of the corresponding eigenvalue. To get asymptotic stability, all solutions must go to zero as k increases to infinity. The eigenvalues of Φ then have the property

$$|\lambda_i| < 1 \qquad i = 1, \ldots, n$$

which is formulated as the following theorem.

THEOREM 3.1   ASYMPTOTIC STABILITY OF LINEAR SYSTEMS   A discrete-time linear time-invariant system (3.2) is asymptotically stable if and only if all eigenvalues of Φ are strictly inside the unit disk.   ∎

Stability with respect to disturbances in the initial value has already been defined. Other types of stability concepts are also of interest.

Input-Output Stability

DEFINITION 3.3   BOUNDED-INPUT BOUNDED-OUTPUT STABILITY   A linear time-invariant system is defined as *bounded-input–bounded-output* (BIBO) stable if a bounded input gives a bounded output for every initial value.   ∎

From the definition it follows that asymptotic stability is the strongest concept. The following theorem is a result.

THEOREM 3.2   RELATION BETWEEN STABILITY CONCEPTS   Asymptotic stability implies stability and BIBO stability.   ∎

When the word *stable* is used without further qualification in this text, it normally means asymptotic stability.

It is easy to give examples showing that stability does not imply BIBO stability, and vice versa.

Example 3.1   Harmonic oscillator

Consider the sampled harmonic oscillator (see Example A.3)

$$
x(kh + h) = \begin{pmatrix} \cos \omega h & \sin \omega h \\ -\sin \omega h & \cos \omega h \end{pmatrix} x(kh) + \begin{pmatrix} 1 - \cos \omega h \\ \sin \omega h \end{pmatrix} u(kh)
$$
$$
y(kh) = \begin{pmatrix} 1 & 0 \end{pmatrix} x(kh)
$$

The magnitude of the eigenvalues is one. The system is stable because $\|x(kh)\| = \|x(0)\|$ if $u(kh) = 0$. Let the input be a square wave with the frequency ω rad/s. By using the z-transform, it is easily seen that the output contains a sinusoidal function with growing amplitude and the system is not BIBO stable. Figure 3.1 shows the input and output of the system. The input signal is exciting the system at its undamped frequency and the output amplitude is growing.   ∎

Stability Tests

It follows from Theorem 3.1 that a straightforward way to test the stability of a given system is to calculate the eigenvalues of the matrix Φ. There are good numerical algorithms for doing this. Well-established methods are available, for instance, in the package LAPACK, which is easily accessible in most computing centers. The routines are also included in packages like MATLAB®. The eigenvalues of a matrix then can be calculated with a single command.

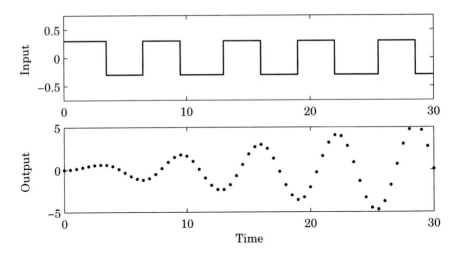

Figure 3.1 Input and output of the system in Example 3.1 when $\omega = 1$, $h = 0.5$, and the initial state is zero.

It is, however, also important to have algebraic or graphical methods for investigating stability. These methods make it possible to understand how parameters in the system or the controller will influence the stability. The following are some of the ways of determining the stability of a discrete-time system:

- Direct numerical or algebraic computation of the eigenvalues of Φ

- Methods based on properties of characteristic polynomials

- The root locus method

- The Nyquist criterion

- Lyapunov's method

Explicit calculation of the eigenvalues of a matrix cannot be done conveniently by hand for systems of order higher than 2. In some cases it is easy to calculate the characteristic equation

$$A(z) = a_0 z^n + a_1 z^{n-1} + \cdots + a_n = 0 \tag{3.4}$$

and investigate its roots. Recall from Sec. 2.6 that the characteristic polynomial is the denominator polynomial of the pulse-transfer function. Stability tests can be obtained by investigating conditions for the zeros of a polynomial to be inside the unit disc.

It is also useful to have algebraic or graphical conditions that tell directly if a polynomial has all its zeros inside the unit disc. Such a criterion, which is the equivalent of the Routh-Hurwitz criterion, was developed by Schur, Cohn, and Jury. This test will be described in detail in the following section. The calculation of the coefficients of the characteristic polynomial from the elements of a matrix is poorly conditioned. If a matrix is given, it is therefore preferable to calculate the eigenvalues directly instead of calculating the characteristic equation.

The well-known root locus method for continuous-time systems can be used for discrete-time systems also. The stability boundary is changed only from the imaginary axis to the unit circle. The rules of thumb for drawing the root locus are otherwise the same. The root locus method and the Nyquist criterion are used to determine the stability of the closed-loop system when the open-loop system is known.

Jury's Stability Criterion

The following test is useful for determining if Eq. (3.4) has all its zeros inside the unit disc. Form the table

$$
\begin{array}{cccc}
a_0 & a_1 & \cdots & a_{n-1} & a_n \\
a_n & a_{n-1} & \cdots & a_1 & a_0 & \alpha_n = \dfrac{a_n}{a_0} \\
\hline
a_0^{n-1} & a_1^{n-1} & \cdots & a_{n-1}^{n-1} \\
a_{n-1}^{n-1} & a_{n-2}^{n-1} & \cdots & a_0^{n-1} & & \alpha_{n-1} = \dfrac{a_{n-1}^{n-1}}{a_0^{n-1}} \\
\hline
\vdots \\
a_0^0
\end{array}
$$

where

$$a_i^{k-1} = a_i^k - \alpha_k a_{k-i}^k$$
$$\alpha_k = a_k^k / a_0^k$$

The first and second rows are the coefficients in (3.4) in forward and reverse order, respectively. The third row is obtained by multiplying the second row by $\alpha_n = a_n/a_0$ and subtracting this from the first row. The last element in the third row is thus zero. The fourth row is the third row in reverse order. The scheme is then repeated until there are $2n + 1$ rows. The last row consists of only one element. The following theorem results.

THEOREM 3.3   JURY'S STABILITY TEST   If $a_0 > 0$, then Eq. (3.4) has all roots inside the unit disc if and only if all a_0^k, $k = 0, 1, \ldots, n - 1$ are positive. If no a_0^k is zero, then the number of negative a_0^k is equal to the number of roots outside the unit disc.   ∎

Remark.   If all a_0^k are positive for $k = 1, 2, \ldots, n - 1$, then the condition $a_0^0 > 0$ can be shown to be equivalent to the conditions

$$A(1) > 0$$
$$(-1)^n A(-1) > 0$$

These conditions constitute necessary conditions for stability and hence can be used before forming the table.

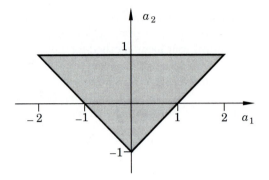

Figure 3.2   The stability area for the second-order equation (3.5) as a function of the coefficients a_1 and a_2.

Example 3.2   Stability of a second-order system

Let the characteristic equation be

$$A(z) = z^2 + a_1 z + a_2 = 0 \qquad (3.5)$$

Jury's scheme is

1	a_1	a_2	
a_2	a_1	1	$\alpha_2 = a_2$

$$
\begin{array}{ll}
1 - a_2^2 & a_1(1 - a_2) \\
a_1(1 - a_2) & 1 - a_2^2
\end{array} \qquad \alpha_1 = \dfrac{a_1}{1 + a_2}
$$

$$1 - a_2^2 - \frac{a_1^2(1 - a_2)}{1 + a_2}$$

All the roots of Eq. (3.5) are inside the unit circle if

$$1 - a_2^2 > 0$$

$$\frac{1 - a_2}{1 + a_2}\left((1 + a_2)^2 - a_1^2\right) > 0$$

This gives the conditions

$$a_2 < 1$$
$$a_2 > -1 + a_1$$
$$a_2 > -1 - a_1$$

The stability area for the second-order equation is shown in Fig. 3.2.   ∎

Nyquist and Bode Diagrams for Discrete-Time Systems

Consider the continuous-time system $G(s)$. The *Nyquist curve* or *frequency response curve* of the system is the map $G(i\omega)$ for $\omega \in [0, \infty)$. This curve is drawn in polar coordinates (Nyquist diagram) or as amplitude and phase curves as a function of the frequency (Bode diagram). In the discrete-time case we have a similar situation. Consider a system with the pulse-transfer function $H(z)$. The Nyquist or frequency curve is given by the map $H(e^{i\omega h})$ for $\omega h \in [0, \pi]$, that is, up to the Nyquist frequency. Notice that it is sufficient to consider the map in the interval $\omega h \in [-\pi, \pi]$ because the function $H(e^{i\omega h})$ is periodic with period $2\pi/h$.

In the continuous-time case, the Nyquist curve $G(i\omega)$ can be interpreted as the stationary amplitude and phase when a sinusoidal signal with frequency ω is applied to the system. In the discrete-time case, higher harmonics are generated; see Example 1.4. This will make the interpretation of $H(e^{i\omega h})$ more complex as is further discussed in Chapter 7.

Example 3.3   Frequency responses

Consider the continuous-time system

$$G(s) = \frac{1}{s^2 + 1.4s + 1} \tag{3.6}$$

Zero-order-hold sampling of the system with $h = 0.4$ gives the discrete-time system

$$H(z) = \frac{0.066z + 0.055}{z^2 - 1.450z + 0.571}$$

The frequency curve is given by $H(e^{i\omega h})$. Figure 3.3 shows the Nyquist diagram and Fig 3.4 shows the Bode diagram for the continuous-time system and for the discrete-time system. The difference between the continuous-time and discrete-time frequency curves will decrease when the sampling period is decreased. The connection between the frequency curves of the discrete-time and continuous-time systems is further discussed in Sec. 7.7.   ■

The Nyquist Criterion

The Nyquist criterion is a well-known stability test for continuous-time systems. It is based on the principle of arguments. The Nyquist criterion is especially useful for determining the stability of the closed-loop system when the open-loop system is given. The test can easily be reformulated to handle discrete-time systems.

Consider the discrete-time system in Fig. 3.5. The closed-loop system has the pulse-transfer function

$$H_{cl}(z) = \frac{Y(z)}{U_c(z)} = \frac{H(z)}{1 + H(z)}$$

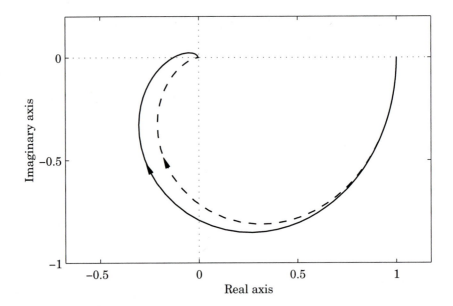

Figure 3.3 The frequency curve of (3.6) (dashed) and for (3.6) sampled with zero-order hold when $h = 0.4$ (solid).

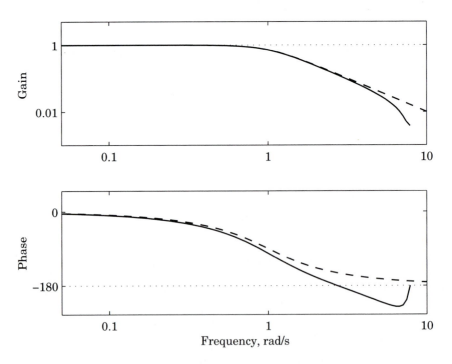

Figure 3.4 The Bode diagram of (3.6) (dashed) and of (3.6) sampled with zero-order hold when $h = 0.4$ (solid).

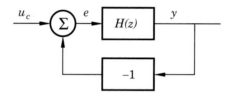

Figure 3.5 A simple unit-feedback system.

The characteristic equation of the closed-loop system is

$$1 + H(z) = 0 \tag{3.7}$$

The stability of the closed-loop system can be investigated from the Nyquist plot of $H(z)$. For discrete-time systems, the stability area in the z-plane is the unit disc instead of the left half-plane. Figure 3.6 shows the path Γ_c encircling the area outside the unit disc. The small indentation at $z = 1$ is to exclude the integrators in the open-loop system. The mapping of the infinitesimal semicircle at $z = 1$ with decreasing arguments from $-\pi/2$ to $\pi/2$ is mapped into the $H(z)$-plane as an infinitely large circle from $-n\pi/2$ to $n\pi/2$, where n is the number of integrators in the open-loop system. If there are poles on the unit circle other than for $z = 1$, those have to be excluded with small semicircles in the same way as for $z = 1$. The map of the unit circle is $H(e^{i\omega h})$ for $\omega h \in (0, 2\pi)$.

   The stability of the closed-loop system now can be determined by investigating how the path Γ_c is mapped by $H(z)$. The principle of arguments states that the number of encirclements N in the positive direction around $(-1, 0)$ by the map of Γ_c is equal to

$$N = Z - P$$

where Z and P are the number of zeros and poles, respectively, of $1 + H(z)$ outside the unit disc. Notice that if the open-loop system is stable, then $P = 0$

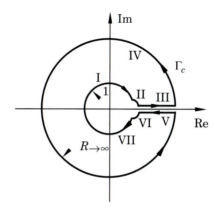

Figure 3.6 The path Γ_c encircling the area outside the unit disc.

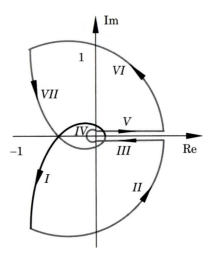

Figure 3.7   The map of Γ_c into the $H(z)$-plane of the system in Example 3.4, when $K = 1$. The solid line is the Nyquist curve.

and thus $N = Z$. The stability of the closed-loop system is then ensured if the map of Γ_c does not encircle the point $(-1, 0)$. If $H_{cl}(z) \to 0$ when $z \to \infty$, the parallel lines III and V do not influence the stability test, and it is sufficient to find the map of the unit circle and the small semicircle at $z = 1$. The Nyquist criterion can be simplified further if the open-loop system and its inverse are stable. Stability of the closed-loop system is then ensured if the point $(-1, 0)$ in the $H(z)$-plane is to the left of the map of $H(e^{i\omega h})$ for $\omega h = 0$ to π—that is, to the left of the Nyquist curve.

Example 3.4   A second-order system

Consider a system with sampling period $h = 1$ and the pulse-transfer function

$$H(z) = \frac{0.25K}{(z - 1)(z - 0.5)}$$

then

$$H(e^{i\omega}) = 0.25K \, \frac{1.5(1 - \cos\omega) - 2\sin^2\omega - i\sin\omega(2\cos\omega - 1.5)}{(2 - 2\cos\omega)(1.25 + \cos\omega)}$$

The map of Γ_c is shown in Fig. 3.7. The solid line is the Nyquist curve, that is, the map of $H(e^{i\omega})$ for $\omega = 0$ to π. Notice that the sampled-data system has a phase shift that is larger than 180° for some frequencies. From the figure it can be found that the Nyquist curve crosses the negative real axis at -0.5. The closed-loop system is thus stable if $K < 2$.   ∎

Relative Stability

Amplitude and phase margins can be defined for discrete-time systems analogously to continuous-time systems.

DEFINITION 3.4   AMPLITUDE MARGIN   Let the open-loop system have the pulse-transfer function $H(z)$ and let ω_o be the smallest frequency such that

$$\arg H(e^{i\omega_o h}) = -\pi$$

and such that $H(e^{i\omega_o h})$ is decreasing for $\omega = \omega_o$. The *amplitude* or *gain margin* is then defined as

$$A_{\mathrm{marg}} = \frac{1}{|H(e^{i\omega_o h})|}$$

∎

DEFINITION 3.5   PHASE MARGIN   Let the open-loop system have the pulse-transfer function $H(z)$ and further let the crossover frequency ω_c be the smallest frequency such that

$$|H(e^{i\omega_c h})| = 1$$

The *phase margin* ϕ_{marg} is then defined as

$$\phi_{\mathrm{marg}} = \pi + \arg H(e^{i\omega_c h})$$

∎

In words, the amplitude margin is how much the gain can be increased before the closed-loop system becomes unstable. The phase margin is how much extra phase lag is allowed before the closed-loop system becomes unstable.

The amplitude and phase margins are easily determined from the Nyquist and Bode diagrams.

Example 3.5   Amplitude margins

Consider the system in Example 3.3. The continuous-time system has an infinite amplitude margin. The closed-loop sampled-data system will, however, be unstable with a proportional controller if the gain is larger than 7.82. The finite-amplitude margin for the discrete-time system is due to the phase lag introduced by the zero-order hold. The difference between the discrete-time system and the continuous-time system will decrease when the sampling interval is decreased.   ∎

The phase margin can be used to select the sampling period. Allowing the phase margin to decrease by 5 to 15° compared with the continuous-time system at the crossover frequency gives one rule of thumb.

Lyapunov's Second Method

Lyapunov's second method is a useful tool for determining the stability of nonlinear dynamic systems. Lyapunov developed the theory for differential equations, but a corresponding theory also can be derived for difference equations. The main idea is to introduce a generalized energy function called the Lyapunov function, which is zero at the equilibrium and positive elsewhere. The equilibrium will be stable if we can show that the Lyapunov function decreases along the trajectories of the system.

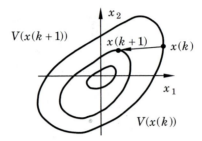

Figure 3.8 Geometric illustration of Lyapunov's theorem.

The first step to show stability is to find the Lyapunov function, which is defined as follows:

DEFINITION 3.6   LYAPUNOV FUNCTION $V(x)$ is a *Lyapunov function* for the system

$$x(k + 1) = f(x(k)) \qquad f(0) = 0 \tag{3.8}$$

if:

1. $V(x)$ is continuous in x and $V(0) = 0$.

2. $V(x)$ is positive definite.

3. $\Delta V(x) = V(f(x)) - V(x)$ is negative definite.

∎

A simple geometric illustration of the definition is given in Fig. 3.8. The level curves of a positive definite continuous function V are closed curves in the neighborhood of the origin. Let each curve be labeled by the value of the function. Condition 3 implies that the dynamics of the system is such that the solution always moves toward curves with lower values. All level curves encircle the origin and do not intersect any other level curve.

From the geometric interpretation it thus seems reasonable that the existence of a Lyapunov function ensures asymptotic stability. The following theorem is a precise statement of this fact.

THEOREM 3.4   STABILITY THEOREM OF LYAPUNOV   The solution $x(k) = 0$ is asymptotically stable if there exists a Lyapunov function to the system (3.8). Further, if

$$0 < \varphi(\|x\|) < V(x)$$

where $\varphi(\|x\|) \to \infty$ as $(\|x\|) \to \infty$, then the solution is asymptotically stable for all initial conditions.

∎

The main obstacle to using the Lyapunov theory is finding a suitable Lyapunov function. This is in general a difficult problem; however, for the linear system of (3.2), it is straightforward to determine quadratic Lyapunov functions. Take $V(x) = x^T P x$ as a candidate for a Lyapunov function. The increment of V is then given by

$$\Delta V(x) = V(\Phi x) - V(x) = x^T \Phi^T P \Phi x - x^T P x$$
$$= x^T (\Phi^T P \Phi - P) x = -x^T Q x$$

For V to be a Lyapunov function, it is thus necessary and sufficient that there exists a positive definite matrix P that satisfies the equation

$$\Phi^T P \Phi - P = -Q \tag{3.9}$$

where Q is positive definite. Equation (3.9) is called the *Lyapunov equation*. It can be shown that there is always a solution to the Lyapunov equation when the linear system is stable. The matrix P is positive definite if Q is positive definite. One way of determining a Lyapunov function for a linear system is to choose a positive definite matrix Q and solve the Lyapunov equation. If the solution P is positive definite then the system is asymptotically stable.

Example 3.6   Lyapunov function

Consider the discrete-time system

$$x(k + 1) = \begin{pmatrix} 0.4 & 0 \\ -0.4 & 0.6 \end{pmatrix} x(k)$$

Using

$$Q = \begin{pmatrix} 1 & 0 \\ 0 & 1 \end{pmatrix}$$

gives the solution of the Lyapunov equation

$$P = \begin{pmatrix} 1.19 & -0.25 \\ -0.25 & 2.05 \end{pmatrix}$$

Figure 3.9 shows the level curves of $V(x) = x^T P x$ and trajectories for some starting values of x. The trajectories are such that for each step, the state is reaching a value of V with a lower value.   ∎

3.3 Sensitivity and Robustness

It is of interest to investigate the sensitivity of a system to perturbations, which may be introduced because of component tolerances. Because the designs of control systems are based on simplified models, it is also interesting to know how accurate the model has to be for the design to be successful. The Nyquist theorem can give good insight into these problems. In this section we will investigate the sensitivity of the closed-loop system to disturbances and perturbations in the components of the system.

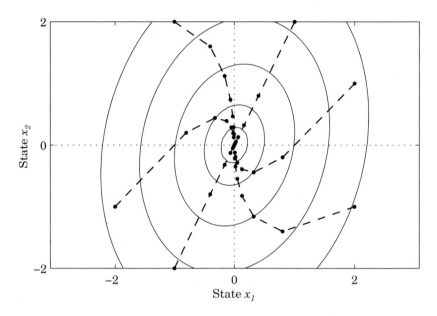

Figure 3.9 Level curves of $V(x)$ and trajectories for different initial values of the system in Example 3.6. The sampling points are indicated by dots.

Sensitivity

We will first determine the sensitivity of a closed-loop system with respect to changes in the open-loop pulse-transfer function. Consider the system in Fig. 3.10. The closed-loop system has a feedforward filter H_{ff} from the reference signal and a feedback controller H_{fb}. There are also an input load disturbance v and measurement noise e. The primary process output is x, and the measured signal is y. The pulse-transfer operator from the inputs to y is given by

$$y = \frac{H_{ff}H}{1+L}\, u_c + \frac{H}{1+L}\, v + \frac{1}{1+L}\, e$$

where the *loop-transfer function* is defined as $L = H_{fb}H$. The closed-loop pulse-

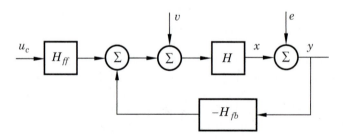

Figure 3.10 Closed-loop system with feedback and feedforward controllers.

transfer function from the reference signal u_c to the output y is

$$H_{cl} = \frac{H_{ff}H}{1+L}$$

The sensitivity of H_{cl} with respect to variations in H is given by

$$\frac{dH_{cl}}{dH} = \frac{H_{ff}}{(1+L)^2}$$

The relative sensitivity of H_{cl} with respect to H thus can be written as

$$\frac{dH_{cl}}{H_{cl}} = \frac{1}{1+L}\frac{dH}{H} = S\frac{dH}{H}$$

The pulse-transfer function S is called the *sensitivity function* and also can be written as

$$S = \frac{1}{1+L} = \frac{d\log H_{cl}}{d\log H} \tag{3.10}$$

The transfer function

$$\mathcal{T} = 1 - S = \frac{L}{1+L} \tag{3.11}$$

is called the *complementary sensitivity function*.

The different transfer functions from the inputs u_c, v, and e to the signals y, x, and u show how the different signals are influenced by the input signals. The sensitivity function can be interpreted as the pulse-transfer function from e to y or as the ratio of the closed-loop and open-loop pulse-transfer functions from v to y. The complementary sensitivity function is the pulse-transfer function with opposite sign from e to x.

Robustness

We will now consider the situation when the design of the controller is based on the nominal model H, but the true open-loop pulse-transfer function is $H^0(z)$. The closeness of H to H^0 needed to make the closed-loop system stable is of concern. Consider the simple closed-loop system in Fig. 3.10 with $H(z)$ replaced by $H^0(z)$. The pulse-transfer function of the closed-loop system is

$$H_{cl}(z) = \frac{H_{ff}H^0(z)}{1+L^0(z)} \tag{3.12}$$

The poles of the closed-loop system are thus the zeros of the function

$$\begin{aligned}
f(z) &= 1 + H_{fb}(z)H^0(z) \\
&= 1 + H_{fb}(z)H(z) + H_{fb}(z)H^0(z) - H_{fb}(z)H(z) \\
&= 1 + H_{fb}(z)H(z) + H_{fb}(z)[H^0(z) - H(z)]
\end{aligned}$$

If

$$\left| H_{fb}\left(H^0(z) - H(z) \right) \right| < |1 + \mathcal{L}(z)| = \left| \frac{H}{H_{cl}} \right| \cdot \left| \frac{H_{ff}}{H_{fb}} \right| \tag{3.13}$$

on the unit circle, then it follows from the principle of variation of the argument that the differences between the number of poles and zeros outside the unit disc for the functions $1 + \mathcal{L}$ and $1 + \mathcal{L}^0$ are the same.

The relative precision needed for stability robustness is obtained by dividing (3.13) by $\mathcal{L}$

$$\left| \frac{H^0(z) - H(z)}{H} \right| \le \left| \frac{1 + \mathcal{L}}{\mathcal{L}} \right| = \left| \frac{1}{\mathcal{T}} \right|$$

where the last equality is obtained from (3.11). The complementary sensitive function thus makes it possible to determine bounds for stability robustness. The following theorem results.

THEOREM 3.5   ROBUSTNESS 1   Consider the closed-loop systems S and S^0 obtained by applying unit negative feedback around systems with pulse-transfer functions H and H^0, respectively. The system S^0 is stable if the following conditions are true:

1. S is stable.

2. H and H^0 have the same number of poles outside the unit disc.

3. The inequality (3.13) is fulfilled for $|z| = 1$.   ∎

The result shows that it is important to know the number of unstable modes in order to design a regulator for the system. The theorem is, however, conservative. The inequality also gives the frequency range in which it is important to have a good description of the process. Notice in particular that the precision requirements are very modest for the frequencies where the loop gain is large. Good precision is needed for frequencies where $H^0(z) \approx 1$.

A closely related result that gives additional insight is obtained as follows. The pulse-transfer function of the closed-loop system given in (3.12) can also be written as

$$H_{cl} = \frac{1}{1 + 1/\mathcal{L}^0}$$

The poles of the closed-loop system are thus given by the zeros of the function

$$f_c(z) = 1 + \frac{1}{H_{fb}(z)H^0(z)}$$

$$= 1 + \frac{1}{H_{fb}(z)H(z)} + \frac{1}{H_{fb}(z)H^0(z)} - \frac{1}{H_{fb}(z)H(z)}$$

It follows from the principle of variation of the argument that the differences between the zeros and poles outside the unit disc of the functions $1 + 1/\mathcal{L}^0$ and $1 + 1/\mathcal{L}$ are the same if

$$\left| \frac{1}{\mathcal{L}^0(z)} - \frac{1}{\mathcal{L}(z)} \right| < \left| 1 + \frac{1}{\mathcal{L}(z)} \right| \tag{3.14}$$

on the unit circle. The following result is thus obtained.

THEOREM 3.6   ROBUSTNESS 2   Consider the closed-loop systems S and S^0 obtained by applying unit negative feedback around systems with the pulse-transfer functions H and H^0, respectively. The system S^0 is stable if the following conditions are true:

1. S is stable.

2. H and H^0 have the same number of zeros outside the unit disc.

3. The inequality (3.14) is fulfilled for $|z| = 1$.

■

The theorem indicates the importance of knowing the number of zeros outside the unit disc. The theorem shows that stability can be maintained in spite of large differences between H and H^0 provided that the loop gain is large.

From the conclusions of Theorems 3.5 and 3.6, the following rules are obtained for design of a feedback system based on approximate or uncertain models.

It is important to know the number of unstable poles and zeros.

It is not important to know the model precisely for those frequencies for which the loop gain can be made large.

It is necessary to make the loop gain small for those frequencies for which the relative error $\Delta H/H$ is large.

It is necessary to have a model that describes the system precisely for those frequencies for which $H^0(z) \approx -1$.

3.4 Controllability, Reachability, Observability, and Detectability

In this section, two fundamental questions for dynamic systems are discussed. The first is whether it is possible to steer a system from a given initial state to any other state. The second is how to determine the state of a dynamic system from observations of inputs and outputs. These questions were posed and answered by Kalman, who also introduced the concepts of controllability and observability. The systems are allowed to be multiple-input–multiple-output systems.

Controllability and Reachability

Consider the system

$$x(k + 1) = \Phi x(k) + \Gamma u(k)$$
$$y(k) = Cx(k) \tag{3.15}$$

Assume that the initial state $x(0)$ is given. The state at time n, where n is the order of the system, is given by

$$x(n) = \Phi^n x(0) + \Phi^{n-1}\Gamma u(0) + \cdots + \Gamma u(n - 1)$$
$$= \Phi^n x(0) + W_c U \tag{3.16}$$

where

$$W_c = \left(\begin{array}{cccc} \Gamma & \Phi\Gamma & \ldots & \Phi^{n-1}\Gamma \end{array} \right)$$
$$U = \left(\begin{array}{ccc} u^T(n - 1) & \ldots & u^T(0) \end{array} \right)^T$$

[Compare with Eq. (2.17).] If W_c has rank n, then it is possible to find n equations from which the control signals can be found such that the initial state is transferred to the desired final state $x(n)$. Notice that the solution is not unique if there is more than one input signal. In the literature, controllability is defined in different ways; the following definition will be used in this text.

DEFINITION 3.7 CONTROLLABILITY   The system (3.15) is controllable if it is possible to find a control sequence such that the origin can be reached from any initial state in finite time.   ∎

A concept related to controllability is reachability, which is defined as follows.

DEFINITION 3.8 REACHABILITY   The system (3.15) is reachable if it is possible to find a control sequence such that an arbitrary state can be reached from any initial state in finite time.   ∎

The two concepts, however, are equivalent if Φ is invertible. Reachability will be discussed here primarily. Controllability does not imply reachability, which is seen from (3.16). If $\Phi^n x(0) = 0$, then the origin will be reached with zero input but the system is not necessarily reachable.

The following theorem follows from the preceding definition and calculations.

THEOREM 3.7 REACHABILITY   The system (3.15) is reachable if and only if the matrix W_c has rank n.   ∎

Remark.   The matrix W_c is usually referred to as the *controllability matrix* because of its analogy with continuous-time systems.

Example 3.7   A controllable system which is not reachable

The system

$$x(k + 1) = \Phi x(k) + \Gamma u(k)$$

where

$$\Phi = \begin{pmatrix} 0 & 0 \\ 1 & 0 \end{pmatrix} \qquad \Gamma = \begin{pmatrix} 1 \\ 0 \end{pmatrix}$$

is reachable because

$$W_c = \begin{pmatrix} 1 & 0 \\ 0 & 1 \end{pmatrix}$$

has full rank. Assume that Γ is changed to $\Gamma^T = \begin{pmatrix} 0 & 1 \end{pmatrix}$; then

$$W_c = \begin{pmatrix} 0 & 0 \\ 1 & 0 \end{pmatrix}$$

and the system is not reachable. The system is, however, controllable because $\Phi^2 = 0$. The origin is reached in two steps for any initial condition by using $u(0) = u(1) = 0$. ∎

By the Cayley-Hamilton theorem it is found from (3.16) that all states that can be reached from the origin are spanned by the columns of the controllability matrix W_c. This implies that the reachable states belong to the linear subspace spanned by the columns of W_c.

Example 3.8   Reachable subspaces

Given the system

$$x(k + 1) = \begin{pmatrix} 1 & 1 \\ -0.25 & 0 \end{pmatrix} x(k) + \begin{pmatrix} 1 \\ -0.5 \end{pmatrix} u(k) \qquad x(0) = \begin{pmatrix} 2 \\ 2 \end{pmatrix}$$

is it possible to find a control sequence such that $x^T(2) = \begin{pmatrix} -0.5 & 1 \end{pmatrix}$? From (3.16),

$$x(2) = \Phi^2 x(0) + \Phi \Gamma u(0) + \Gamma u(1)$$

or

$$\begin{pmatrix} -0.5 \\ 1 \end{pmatrix} = \begin{pmatrix} 3.5 \\ -1 \end{pmatrix} + \begin{pmatrix} 1 \\ -0.5 \end{pmatrix} \left(0.5 u(0) + u(1) \right)$$

which gives the condition

$$0.5 u(0) + u(1) = -4$$

One possible sequence of controls is $u(0) = -2$ and $u(1) = -3$. Assume instead that $x^T(2) = \begin{pmatrix} 0.5 & 1 \end{pmatrix}$. This gives the system of equations

$$\begin{pmatrix} -3 \\ 2 \end{pmatrix} = \begin{pmatrix} 1 \\ -0.5 \end{pmatrix} \left(0.5u(0) + u(1) \right)$$

which does not have a solution. The reason, of course, is that the system is not reachable. The controllability matrix is

$$W_c = \begin{pmatrix} 1 & 0.5 \\ -0.5 & -0.25 \end{pmatrix}$$

By starting at the origin, it is possible to reach only those points of the state space that belong to the subspace spanned by the vector $[1 \quad -0.5]^T$. In the example, it is possible to reach other points due to the effect of the initial value. ∎

Assume that new coordinates are introduced by a nonsingular transformation matrix T (compare with Sec. 2.5). In the new coordinates,

$$\begin{aligned} \tilde{W}_c &= \begin{pmatrix} \tilde{\Gamma} & \tilde{\Phi}\tilde{\Gamma} & \cdots & \tilde{\Phi}^{n-1}\tilde{\Gamma} \end{pmatrix} \\ &= \begin{pmatrix} T\Gamma & T\Phi T^{-1}T\Gamma & \cdots & T\Phi^{n-1}T^{-1}T\Gamma \end{pmatrix} \\ &= TW_c \end{aligned} \qquad (3.17)$$

If W_c has rank n, then $\tilde{W}_c$ also has rank n. This means that the reachability of a system is independent of the coordinates.

Controllable Canonical Form

Assume that Φ has the characteristic equation

$$\det(\lambda I - \Phi) = \lambda^n + a_1\lambda^{n-1} + \cdots + a_n = 0 \qquad (3.18)$$

and that W_c is nonsingular. Then there exists a transformation such that the transformed system is

$$z(k+1) = \begin{pmatrix} -a_1 & -a_2 & \cdots & -a_{n-1} & -a_n \\ 1 & 0 & \cdots & 0 & 0 \\ 0 & 1 & \cdots & 0 & 0 \\ \vdots & \vdots & \ddots & \vdots & \vdots \\ 0 & 0 & \cdots & 1 & 0 \end{pmatrix} z(k) + \begin{pmatrix} 1 \\ 0 \\ 0 \\ \vdots \\ 0 \end{pmatrix} u(k)$$

$$y(k) = \begin{pmatrix} b_1 & \cdots & b_n \end{pmatrix} z(k) \qquad (3.19)$$

which is called the *controllable canonical form*. The advantage of this form is that it is easy to compute the input-output model and to compute a state-feedback-control law. There are simple ways of finding the transformation to controllable canonical form.

For a single-input system it follows from (3.17) that the transformation matrix to the controllable canonical form is $T = W_c^{-1} \tilde{W}_c$, where $\tilde{W}_c$ is the controllability matrix for the representation (3.19). The following example shows that the inverse of the controllability matrix has a simple form.

Example 3.9   The inverse of the controllability matrix

Consider the third-order system

$$
x(k+1) = \begin{pmatrix} -a_1 & -a_2 & -a_3 \\ 1 & 0 & 0 \\ 0 & 1 & 0 \end{pmatrix} x(k) + \begin{pmatrix} 1 \\ 0 \\ 0 \end{pmatrix} u(k)
$$

which is in controllable form. The controllability matrix is

$$
W_c = \begin{pmatrix} \Gamma & \Phi\Gamma & \Phi^2\Gamma \end{pmatrix} = \begin{pmatrix} 1 & -a_1 & a_1^2 - a_2 \\ 0 & 1 & -a_1 \\ 0 & 0 & 1 \end{pmatrix}
$$

The inverse is given by

$$
W_c^{-1} = \begin{pmatrix} 1 & a_1 & a_2 \\ 0 & 1 & a_1 \\ 0 & 0 & 1 \end{pmatrix}
$$

The example can be generalized to the nth-order case, where

$$
W_c^{-1} = \begin{pmatrix}
1 & a_1 & a_2 & \cdots & a_{n-2} & a_{n-1} \\
0 & 1 & a_1 & \cdots & a_{n-3} & a_{n-2} \\
\vdots & \vdots & \vdots & \cdots & \ddots & \vdots \\
0 & 0 & 0 & \cdots & 1 & a_1 \\
0 & 0 & 0 & \cdots & 0 & 1
\end{pmatrix}
$$

■

Trajectory Following

From the preceding definitions and calculations, it is possible to determine a control sequence such that a desired state can be reached after at most n steps of time. Does reachability also imply that it is possible to follow a given trajectory in the state space? Assume that any $x(k)$ is given and that it is necessary to get to $x(k + 1)$. From (3.16) it can be seen that this is possible only if Γ has rank n, that is, it is necessary but not sufficient to have n input signals. For a single-input–single-output system it is, in general, possible to reach desired states only at each nth sample point, provided that the desired points are known n steps ahead.

It is easier to make the output follow a given trajectory. Assume that the trajectory is given by $u_c(k)$. The control signal u then should satisfy

$$y(k) = \frac{B(q)}{A(q)} u(k) = u_c(k)$$

or

$$u(k) = \frac{A(q)}{B(q)} u_c(k) \qquad (3.20)$$

Assume that there are d steps of delay in the system. The generation of $u(k)$ is then causal only if the desired trajectory is known d steps ahead. The control signal then can be generated in real time. The control signal thus is obtained by sending the desired output trajectory through the inverse system A/B. Equation (3.20) has a unique solution if the signal $u_c(k)$ is such that there exists a k_0 such that $u(k) = 0$ for all $k < k_0$ (compare with Sec. 2.6). The signal u is bounded if u_c is bounded and if the system has a stable inverse.

Observability and Detectability

To solve the problem of finding the state of a system from observations of the output, the concept of unobservable states is introduced.

DEFINITION 3.9   UNOBSERVABLE STATES   The state $x^0 \neq 0$ is *unobservable* if there exists a finite $k_1 \geq n - 1$ such that $y(k) = 0$ for $0 \leq k \leq k_1$ when $x(0) = x^0$ and $u(k) = 0$ for $0 \leq k \leq k_1$.   ∎

The system in (3.15) is *observable* if there is a finite k such that knowledge of the inputs $u(0), \dots, u(k-1)$ and the outputs $y(0), \dots, y(k-1)$ is sufficient to determine the initial state of the system. Consider the system in (3.15). The effect of the known input signal always can be determined, and there is no loss of generality to assume that $u(k) = 0$. Assume that $y(0), y(1), \dots, y(n-1)$ are given. This gives the following set of equations:

$$y(0) = Cx(0)$$
$$y(1) = Cx(1) = C\Phi x(0)$$
$$\vdots$$
$$y(n-1) = C\Phi^{n-1}x(0)$$

Using vector notation gives

$$\begin{pmatrix} C \\ C\Phi \\ \vdots \\ C\Phi^{n-1} \end{pmatrix} x(0) = \begin{pmatrix} y(0) \\ y(1) \\ \vdots \\ y(n-1) \end{pmatrix} \qquad (3.21)$$

The state $x(0)$ can be obtained from (3.21) if and only if the *observability matrix*

$$
W_o = \begin{pmatrix} C \\ C\Phi \\ \vdots \\ C\Phi^{n-1} \end{pmatrix}
\tag{3.22}
$$

has rank n. The state $x(0)$ is unobservable if $x(0)$ is in the null space of W_o. If two states are unobservable, then any linear combination is also unobservable; that is, the unobservable states form a linear subspace.

THEOREM 3.8   OBSERVABILITY   The system (3.15) is observable if and only if W_o has rank n.   ∎

DEFINITION 3.10   DETECTABILITY   A system is *detectable* if the only unobservable states are such that they decay to the origin. That is, the corresponding eigenvalues are stable.   ∎

The test of observability given by Theorem 3.8 is equivalent to that of observability for continuous-time systems. It is straightforward to show that the observability matrix is independent of the coordinates in the same way as in the controllability matrix.

Example 3.10   A system with unobservable states

Consider the system

$$
x(k+1) = \begin{pmatrix} 1.1 & -0.3 \\ 1 & 0 \end{pmatrix} x(k)
$$

$$
y(k) = \begin{pmatrix} 1 & -0.5 \end{pmatrix} x(k)
$$

The observability matrix is

$$
W_o = \begin{pmatrix} C \\ C\Phi \end{pmatrix} = \begin{pmatrix} 1 & -0.5 \\ 0.6 & -0.3 \end{pmatrix}
$$

The rank of W_o is 1, and the unobservable states belong to the null space of W_o, that is, $[0.5 \quad 1]$. Figure 3.11 shows the output for four different initial states. All initial states that lie on a line parallel to $[0.5 \quad 1]$ give the same output [see Fig. 3.11(b) and (d)].   ∎

Observable Canonical Form

Assume that the characteristic equation of Φ is (3.18) and that the observability matrix W_o is nonsingular. Then there exists a transformation matrix such that

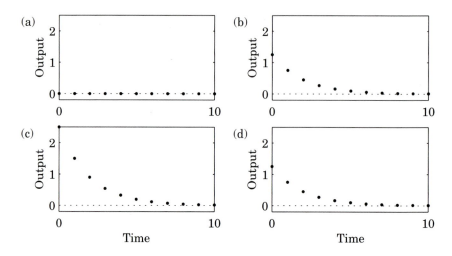

Figure 3.11   The output of the system in Example 3.10 for the initial states (a) $[0.5 \quad 1]^T$, (b) $[1 - 5 \quad 0.5]^T$, (c) $[2.5 \quad 0]^T$, and (d) $[1 \quad -0.5]^T$.

the transformed system is

$$z(k+1) = \begin{pmatrix} -a_1 & 1 & 0 & \cdots & 0 \\ -a_2 & 0 & 1 & \cdots & 0 \\ \vdots & \vdots & \vdots & \ddots & \vdots \\ -a_{n-1} & 0 & 0 & \cdots & 1 \\ -a_n & 0 & 0 & \cdots & 0 \end{pmatrix} z(k) + \begin{pmatrix} b_1 \\ b_2 \\ \vdots \\ b_{n-1} \\ b_n \end{pmatrix} u(k) \qquad (3.23)$$

$$y(k) = \begin{pmatrix} 1 & 0 & \cdots & 0 \end{pmatrix} z(k)$$

which is called the *observable canonical form*. This form has the advantage that it is easy to find the input-output model and to determine a suitable observer. The transformation in this case is given by

$$T = \tilde{W}_o^{-1} W_o$$

where $\tilde{W}_o$ is the observability matrix for the representation (3.23).

 Remark.   The observable and controllable forms are also called *companion forms.*

Example 3.11   A second-order system

 Consider the following system, which is written in observable canonical form:

$$x(k+1) = \begin{pmatrix} -a_1 & 1 \\ -a_2 & 0 \end{pmatrix} x(k) + \begin{pmatrix} b_1 \\ b_2 \end{pmatrix} u(k)$$

$$y(k) = \begin{pmatrix} 1 & 0 \end{pmatrix} x(k)$$

The pulse-transfer operator is

$$
H(q) = \begin{pmatrix} 1 & 0 \end{pmatrix} \begin{pmatrix} q + a_1 & -1 \\ a_2 & q \end{pmatrix}^{-1} \begin{pmatrix} b_1 \\ b_2 \end{pmatrix}
$$

$$
= \frac{1}{q^2 + a_1 q + a_2} \begin{pmatrix} 1 & 0 \end{pmatrix} \begin{pmatrix} q & 1 \\ -a_2 & q + a_1 \end{pmatrix} \begin{pmatrix} b_1 \\ b_2 \end{pmatrix}
$$

$$
= \frac{b_1 q + b_2}{q^2 + a_1 q + a_2} = \frac{b_1 q^{-1} + b_2 q^{-2}}{1 + a_1 q^{-1} + a_2 q^{-2}}
$$

Thus the a_i's and b_i's in the canonical form are defining the polynomials A and B, respectively. This is true for nth-order systems also, in both observable and controllable form.   ∎

Kalman's Decomposition

The reachable and the unobservable parts of a system are two linear subspaces of the state space. Both subspaces are independent of the coordinates in the state space. Kalman showed that it is possible to introduce coordinates such that a system can be partitioned in the following way:

$$
x(k+1) = \begin{pmatrix} \Phi_{11} & \Phi_{12} & 0 & 0 \\ 0 & \Phi_{22} & 0 & 0 \\ \Phi_{31} & \Phi_{32} & \Phi_{33} & \Phi_{34} \\ 0 & \Phi_{42} & 0 & \Phi_{44} \end{pmatrix} x(k) + \begin{pmatrix} \Gamma_1 \\ 0 \\ \Gamma_3 \\ 0 \end{pmatrix} u(k)
$$

$$
y(k) = \begin{pmatrix} C_1 & C_2 & 0 & 0 \end{pmatrix} x(k)
$$

where Φ_{ij}, Γ_i, and C_i are matrices of suitable orders. The state space is partitioned into four parts, which correspond to states that are reachable and observable, not reachable but observable, reachable and not observable, and neither reachable nor observable.

By simple algebraic manipulations, the pulse-transfer operator is given by

$$
H(q) = C_1 (qI - \Phi_{11})^{-1} \Gamma_1
$$

The pulse-transfer operator is thus determined by the reachable and observable part of the system. The following theorem summarizes these results.

THEOREM 3.9   KALMAN'S DECOMPOSITION   A linear system can be partitioned into four subsystems with the following properties:

S_{or} Observable and reachable subsystem

$S_{o\bar{r}}$ Observable but not reachable subsystem

$S_{\bar{o}r}$ Not observable but reachable subsystem

$S_{\bar{o}\bar{r}}$ Neither observable nor reachable subsystem

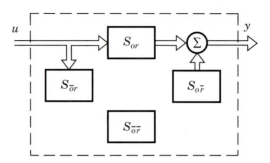

Figure 3.12   Block diagram of the Kalman decomposition when the system is diagonalizable.

Further, the pulse-transfer function of the system is uniquely determined by the subsystem that is both observable and reachable.   ∎

A block diagram for the decomposition is given in Fig. 3.12, which shows how the subsystems are interconnected. The figure also shows that the input-output relationship is given only by the subsystem S_{or}.

Loss of Reachability and Observability Through Sampling

Sampling of a continuous-time system gives a discrete-time system with system matrices that depend on the sampling period. How will that influence the reachability and observability of the sampled system? To get a reachable discrete-time system, it is necessary that the continuous-time system also be reachable, because the allowable control signals for the sampled system—piecewise-constant signals—are a subset of the allowable control signals for the continuous-time system.

   However, it may happen that the reachability is lost for some sampling periods. The conditions for unobservability are more restricted in the continuous-time case because the output has to be zero over a time interval, whereas the sampled-data output has to be zero only at the sampling instants. This means that the continuous output may oscillate between the sampling times and remain zero at the sampling instants. This condition is sometimes called *hidden oscillation*. The sampled-data system thus can be unobservable even if the corresponding continuous-time system is observable.

   The harmonic oscillator can be used to illustrate the preceding discussion.

Example 3.12   Loss of reachability and observability

   The discrete-time model of the harmonic oscillator is given by (see Example A.3)

$$x(kh + h) = \begin{pmatrix} \cos \omega h & \sin \omega h \\ -\sin \omega h & \cos \omega h \end{pmatrix} x(kh) + \begin{pmatrix} 1 - \cos \omega h \\ \sin \omega h \end{pmatrix} u(k)$$

$$y(kh) = \begin{pmatrix} 1 & 0 \end{pmatrix} x(kh)$$

The determinants of the controllability and observability matrices are

$$\det W_c = -2 \sin \omega h (1 - \cos \omega h)$$

and

$$\det W_o = \sin \omega h$$

Both reachability and observability are lost for $\omega h = n\pi$, although the corresponding continuous-time system given by (A.7) is both controllable and observable. ∎

The example shows one obvious way to lose observability and/or reachability. If the sampling period is half the period time (or a multiple thereof) of the natural frequency of the system, then this frequency will not be seen in the output.

The rules of thumb for the choice of the sampling period given in Chapter 2 are such that this situation should not occur. The rules imply about 20 samples per period, not 2.

Observability and/or reachability are lost when the pulse-transfer operator has common poles and zeros. The poles and zeros are functions of the sampling interval. This implies that there will be common factors only for isolated values of the sampling period. A change in sampling period will make the system observable and/or reachable again.

3.5 Analysis of Simple Feedback Loops

In this section the effect of feedback on stability, transient, and steady-state behavior is discussed. Simple feedback systems, as in Fig. 3.10, are primarily considered. Several advantages are obtained by using feedback in continuous-time as well as in discrete-time systems. Feedback, for instance, can do the following:

- Improve the transient behavior of the system

- Decrease the sensitivity to parameter changes in the open-loop system

- Eliminate steady-state errors if there are enough integrators in the open-loop system

- Decrease the influence of load disturbances and measurement errors

The stability of closed-loop systems can be investigated using the tools given in Sec. 3.2. The root locus method is a suitable tool for analyzing simple feedback loops. Because feedback will change the poles of the system, it is important to understand the coupling between the discrete-time poles and the transient behavior of the system. This is treated in Sec. 2.8.

Character of Disturbances

It is customary to distinguish among different types of disturbances, such as load disturbances, measurement errors, and parameter variations.

Load disturbances.   Load disturbances influence the process variables. They may represent disturbance forces in a mechanical system—for example, wind gusts on a stabilized antenna, waves on a ship, load on a motor. In process control, load disturbances may be quality variations in a feed flow or variations in demanded flow. In thermal systems, the load disturbances may be variations in surrounding temperature. Load disturbances typically vary slowly. They may also be periodic—for example, waves in ship-control systems.

Measurement errors.   Measurement errors enter in the sensors. There may be a steady-state error in some sensors due to errors in calibration. However, measurement errors typically have high-frequency components. There may also be dynamic errors because of sensor dynamics. There may also be complicated dynamic interaction between sensors and the process. Typical examples are gyroscopic measurements and measurement of liquid level in nuclear reactors. The character of the measurement errors often depends on the filtering in the instruments. It is often a good idea to look at the instrument and modify the filtering so that it fits the particular problem.

Parameter variations.   Linear theory is used throughout this book. The load disturbance and the measurement noise then appear additively. Real systems are, however, often nonlinear. This means that disturbances can enter in a more complicated way. Because the linear models are obtained by linearizing the nonlinear models, some disturbances then also appear as variations in the parameters of the linear model.

Simple Disturbance Models

There are four different types of disturbances—impulse, step, ramp, and sinusoid—that are commonly used in analyzing control systems. These disturbances are illustrated in Fig. 3.13 and a discussion of their properties follows.

The impulse and the pulse.   The *impulse* and the *pulse* are simple idealizations of sudden disturbances of short duration. They can represent load disturbances as well as measurement errors. For continuous systems, the disturbance is an impulse (a delta function); for sampled systems, the disturbance is modeled as a pulse with unit amplitude and a duration of one sampling period.

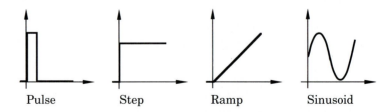

Figure 3.13   Idealized models of simple disturbances.

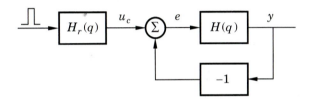

Figure 3.14 Generation of the reference value using a dynamic system with a pulse input.

The pulse and the impulse are also important for theoretical reasons because the response of a linear continuous-time system is completely specified by its impulse response and a linear discrete-time system by its pulse response.

The step.   The step signal is another prototype for a disturbance (see Fig. 3.13). It is typically used to represent a load disturbance or an offset in a measurement.

The ramp.   The ramp is a signal that is zero for negative time and increases linearly for positive time (see Fig. 3.13). It is used to represent drifting measurement errors and disturbances that suddenly start to drift away. In practice, the disturbances are often bounded; however, the ramp is a useful idealization.

The sinusoid.   The sine wave is the prototype for a periodic disturbance. Choice of the frequency makes it possible to represent low-frequency load disturbances, as well as high-frequency measurement noise.

Generation of disturbances.   It is convenient to view disturbances as being generated by dynamic systems (see Fig. 3.14). It is assumed that the input to the dynamic system is a unit pulse δ_k, that is,

$$u_c(k) = H_r(q)\delta_k$$

In order to generate a step, use $H_r(q) = q/(q-1)$; to generate a ramp, use $H_r(q) = q/(q-1)^2$, and a sinusoid from a harmonic oscillator (compare Examples A.1 and A.3 in Appendix A). From an input-output viewpoint, disturbances may be described as impulse responses. Disturbances also may be regarded as the responses of dynamic systems with zero inputs but nonzero initial conditions. In both cases the major characteristics of the disturbances are described by the dynamic systems that generate them. The approach, of course, can be applied to continuous-time, as well as discrete-time, systems.

Steady-State Values

When analyzing control systems, it is important to calculate steady-state values of the output and of the error of the system. Assume a simple feedback system, as shown in Fig. 3.5. To generalize, it can be assumed that -1 in the feedback path is replaced by $-H_{fb}(q)$. The error $e(k)$ is then given by

$$e(k) = \frac{1}{1 + H(q)H_{fb}(q)} u_c(k) = \frac{1}{1 + L(q)} u_c(k) \qquad (3.24)$$

The final-value theorem (Sec. 2.7, Table 2.2) can be used to calculate the steady-state value of $e(k)$. Notice, however, that the stability of the system must be tested before the final-value theorem can be used. If the input signal is a step, the steady-state error can be calculated simply by putting $q = 1$ in (3.24).

The number of integrators in the open-loop system determines the class of reference values that can be followed without steady-state errors. If the open-loop system has p integrators, then the error will be zero in steady state (provided that the closed-loop system is asymptotically stable) for reference signals that are polynomials in k of order less than or equal to $p - 1$.

Example 3.13   Steady-state errors for step and ramp inputs

Consider the system

$$y(k) = H(q)u(k) = \frac{q - 0.5}{(q - 0.8)(q - 1)} u(k)$$

Closing the system, as in Fig. 3.5, gives

$$e(k) = \frac{(q - 0.8)(q - 1)}{(q - 0.8)(q - 1) + q - 0.5} u_c(k)$$

Assume that u_c is a unit step. Because the closed-loop system is stable, the final-value theorem can be used to show that the steady-state error is zero. This can be seen simply by putting $q = 1$. Another way is to observe that the open-loop system contains one integrator, that is, a pole in $+1$. If u_c is a unit ramp, use Table 2.3 in Sec. 2.7 to find the z-transform of the ramp. The steady-state error is then given by

$$\lim_{k \to \infty} e(k) = \lim_{z \to 1} \frac{(z - 0.8)(z - 1)}{(z - 0.8)(z - 1) + z - 0.5} \cdot \frac{z(1 - z^{-1})}{(z - 1)^2} = 0.4$$

∎

Simulation

Simulation is a good way to investigate the behavior of dynamic systems—for example, the intersample behavior of computer-controlled systems. Computer simulation is a very good tool, but it should be remembered that simulation and analysis have to be used together. When making simulations, it is not always possible to investigate all combinations that are unfavorable, for instance, from the point of view of stability, observability, or reachability. These cases can be found through analysis.

It is important that the simulation program be so simple to use that the person primarily interested in the results can be involved in the simulation and in the evaluation of the simulation results.

In the beginning of the 1960s, several digital simulation packages were developed. These packages were basically a digital implementation of analog simulation. The programming was done using block diagrams and fixed-operation modules. Later programs were developed in which the models were given directly as equations.

It is important to have good user-machine communication for simulations; the user should be able to change parameters and modify the model easily. Most simulation programs are interactive, which means that the user interacts with the computer and decides the next step based on the results obtained so far. One way to implement interaction is to let the computer ask questions and the user select from predefined answers. This is called menu-driven interaction. Another possibility is command-driven interaction, which is like a high-level problem-solving language in which the user can choose freely from all commands available in the system. This is also a more flexible way of communicating with the computer, and it is very efficient for the experienced user, whereas a menu-driven program is easier to use for an inexperienced user.

In a simulation package, it is also important to have a flexible way of presenting the results, which are often curves. Finally, to be able to solve the type of problems of interest in this book, it is important to be able to mix continuous- and discrete-time systems.

Examples of simulation packages are MATLAB® with SIMULINK®, MATRIX$_X$®, and Simnon®. Because these packages are readily available we will not describe any of them in detail. However, we urge the reader to use simulation to get a good feel for the behavior of the computer-controlled systems that are described in the text. For the figures in the book we have used MATLAB® with SIMULINK®. Macros for these figures are available through anonymous ftp; see the Preface of the book.

Control of the Double Integrator

The double integrator (Example A.1) will be used as the main example to show how the closed-loop behavior is changed with different controllers. The pulse-transfer operator of the double integrator for the sampling period $h = 1$ is

$$H_0(q) = \frac{0.5(q + 1)}{(q - 1)^2} \tag{3.25}$$

Assume that the purpose of the control is to make the output follow changes in the reference value. Also assume that the process is controlled by a computer using proportional feedback, that is,

$$u(k) = K\left(u_c(k) - y(k)\right) = Ke(k) \qquad K > 0$$

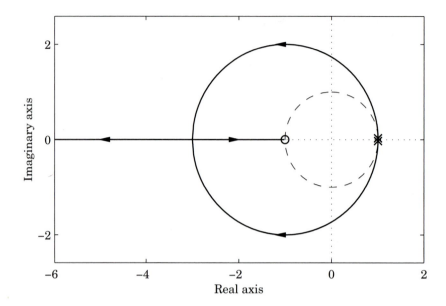

Figure 3.15   The root locus of (3.26) when $K > 0$.

where u_c is the reference value. The characteristic equation of the closed-loop system is

$$(q - 1)^2 + 0.5K(q + 1) = q^2 + (0.5K - 2)q + 1 + 0.5K = 0 \qquad (3.26)$$

Jury's stability test (compare with Example 3.2) gives the following conditions for stability:

$$1 + 0.5K < 1$$
$$1 + 0.5K > -1 + 0.5K - 2$$
$$1 + 0.5K > -1 - 0.5K + 2$$

The closed-loop system is unstable for all values of the gain K. The root locus is shown in Fig. 3.15.

To get a stable system, the controller must be modified. It is known from continuous-time synthesis that derivative action improves stability, so proportional and derivative feedback can be tried also for the discrete-time system. We now assume that it is possible to measure and sample the velocity $\dot{y}$ and use that for feedback; that is,

$$u(k) = K\left(e(k) - T_d\dot{y}(k)\right) \qquad (3.27)$$

(see Fig. 3.16). To find the input-output model of the closed-loop system with

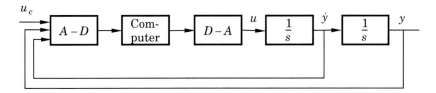

Figure 3.16   Discrete-time controller with feedback from position and velocity of the double integrator.

the controller (3.27), observe that

$$\frac{d\dot{y}}{dt} = u$$

Because u is constant over the sampling intervals,

$$\dot{y}(k+1) - \dot{y}(k) = u(k)$$

or

$$\dot{y}(k) = \frac{1}{q-1} u(k) \tag{3.28}$$

Equations (3.25), (3.27), and (3.28) give the closed-loop system

$$y(k) = \frac{0.5K(q+1)}{(q-1)(q-1+T_dK) + 0.5K(q+1)} u_c(k) \tag{3.29}$$

The system is of second order, and there are two free parameters, K and T_d, that can be used to select the closed-loop poles. The closed-loop system is stable if $K > 0$, $T_d > 0.5$, and $T_dK < 2$. The root locus with respect to K of the characteristic equation of (3.29) is shown in Fig. 3.17 when $T_d = 1.5$.

Let the reference signal be a step. Figure 3.18 shows the continuous-time output for four different values of K. The behavior of the closed-loop system varies from an oscillatory to a well-damped response. When $K = 1$, the poles are in the origin and the output is equal to the reference value after two samples. This is called deadbeat control and is discussed further in Chapters 4 and 5. When $K > 1$, the output and the control signal oscillate because of the discrete-time pole on the negative real axis. The poles are inside the unit circle if $K < 4/3$.

To determine the closed-loop response, it is important to understand the connection between the discrete-time poles and the response of the system. This is discussed in Sec. 2.8. From Fig. 2.8 it can be seen that $K = 0.75$ corresponds to a damping of $\zeta = 0.4$. The distance to the origin is a measure of the speed of the system.

The behavior of the double integrator with some simple controllers has been discussed; the results can be generalized to more complex systems. Also, the importance of analysis and simulation has been illustrated.

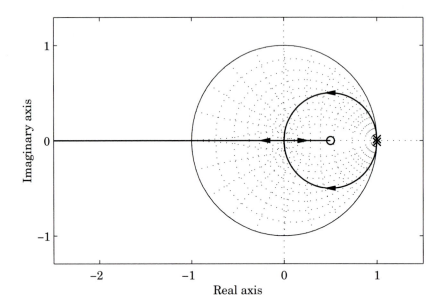

Figure 3.17   The root locus of the characteristic equation of the system in (3.29) with respect to the parameter K when $T_d = 1.5$.

Practical Aspects of the Choice of the Sampling Period

Selection of the sampling period in sampled systems is a fundamental problem that will be discussed several times in this book. The proper choice depends on the properties of the signal, the reconstruction method, and the purpose of the system. In a pure signal-processing problem, the purpose is simply to record a signal digitally and to recover it from its samples. A reasonable criterion for selection may then be the size of the error between the original signal and the reconstructed signal. In signal-processing applications it can be justified to have sampling rates of several hundred samples per period.

A rational choice of the sampling rate in a closed-loop control system should be based on an understanding of its influence on the performance of the control system. It seems reasonable that the highest frequency of interest should be closely related to the bandwidth of the closed-loop system. The selection of sampling rates then can be based on the bandwidth or, equivalently, on the rise time of the closed-loop system. Reasonable sampling rates are 10 to 30 times the bandwidth, or 4 to 10 per rise time, which may seem slow in relation to the typical signal-processing problem. Comparatively low sampling rates can be used in control problems because the dynamics of many controlled systems are of low-pass character and their time constants are typically larger than the closed-loop response times. The contribution to the output from one sampling period then depends on the pulse area; it is comparatively insensitive to the pulse shape.

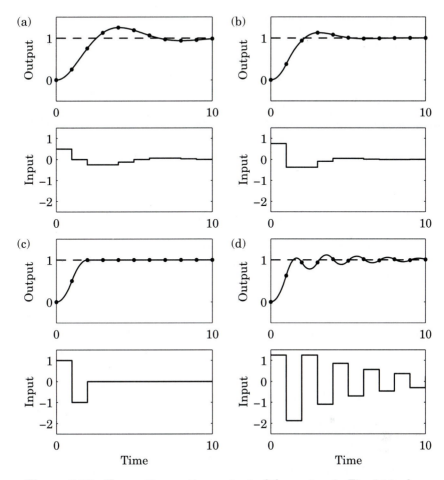

Figure 3.18   The continuous-time output of the system in Fig. 3.16 when $T_d = 1.5$ and (a) $K = 0.5$, (b) $K = 0.75$, (c) $K = 1$, and (d) $K = 1.25$.

Hidden Oscillations

Figure 3.18 shows that the continuous-time output of the process may have oscillations that are not seen at the sampling points. These are called *hidden oscillations*, or *intersample ripple*. Simulation is an effective tool for finding hidden oscillations. The modified z-transform or (2.34) also can be used to calculate the continuous-time output between the sampling instants; however, it is also enlightening to do some analysis.

The intersample ripple is essentially determined by the open-loop dynamics because the system operates in open loop between the sampling points. Two cases can be distinguished.

- Oscillation in the continuous-time output of an open- or closed-loop system when there is no oscillation in the control signal

- Oscillations between the sampling points caused by an oscillation in the control signal

The first case of intersample ripple may occur if observability of the open-loop system is lost due to sampling. The pulse-transfer function then has canceled poles and zeros. The effect of the canceled modes is then not seen at the sampling instants. There may then be hidden oscillations if the continuous-time open-loop system has oscillatory modes and if the sampling period matches the frequency of these modes. This type of hidden oscillation occurs only for certain values of the sampling period. A change in the sampling interval makes the system observable and the oscillation can be seen in the sampled output. The oscillation frequency is often lower in the sampled signal than in the continuous-time signal. To detect this type of intersample ripple, it is necessary to check the observability of the sampled-data system (compare with Example 3.12).

Example 3.14   Hidden oscillation in an open-loop system

Consider a continuous-time system with the transfer function

$$G(s) = \frac{1}{s+1} + \frac{\pi}{(s+0.02)^2 + \pi^2}$$

Sampling this system with $h = 2$ gives the pulse-transfer function

$$H(z) = \frac{1-a}{z-a} + \frac{0.0125}{z-\alpha}$$

where $a = e^{-2}$ and $\alpha = e^{-0.04}$.

The oscillatory part of the continuous-time system has the frequency π and damping of 0.02. The sampling frequency is π, which implies that the oscillation is sampled only once per period.

The discrete-time system is of second order and the continuous-time system is of third order. The cancellation of poles and zeros that are oscillatory is an indication that hidden oscillation may occur. Figure 3.19 shows the step response of the continuous-time system. The sampling points are indicated by dots. The system behaves like a second-order system at the sampling points. Figure 3.19 also shows the sampled output when $h = 1.8$. The oscillation is now clearly seen in the sampled output although it now appears at a lower frequency.   ∎

The second type of hidden oscillation occurs if there are poorly damped zeros in the open-loop system that are canceled by the controller. In this case, the oscillation can be seen in the control signal. This type of hidden oscillation will not be detected if the sampling period is changed, provided that the design is still such that the process zeros are canceled.

Example 3.15   Controller-induced hidden oscillation

The double integrator previously used in this section can be used to show how a controller may introduce hidden oscillations. The model of (3.25) can be written as the difference equation

$$y(k) = 2y(k-1) - y(k-2) + 0.5u(k-1) + 0.5u(k-2)$$

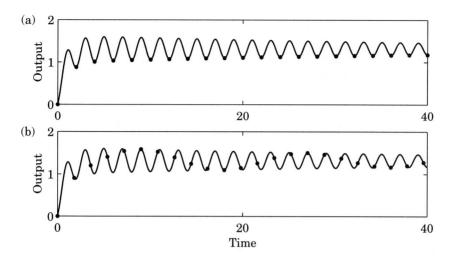

(a)

(b)

Figure 3.19 Step response of the system in Example 3.14. (a) Continuous-time (solid line) and sampled output (dots) when (a) $h = 2$; (b) $h = 1.8$.

Let the purpose of the control be to follow the reference trajectory $u_c(k)$. If the control signal is chosen such that the right-hand side is equal to the reference value at time $k - 1$, the following causal controller is obtained:

$$u(k) = \frac{2q}{q + 1} u_c(k) - \frac{2(2q - 1)}{q + 1} y(k) \qquad (3.30)$$

The closed-loop system is given by

$$y(k) = \frac{q(q - 1)}{(q + 1)(q^2 - 2q + 1 - (-2q + 1))} u_c(k)$$

$$= \frac{q(q + 1)}{q^2(q + 1)} u_c(k) = u_c(k - 1)$$

The output is equal to the reference value after one step. By using the controller in (3.27) with $K = 1$ and $T_d = 1.5$, it took two steps. The step response and the control signal when using the control law (3.30) are shown in Fig. 3.20. At the sampling points, the system has the desired performance, but there is an oscillation in the continuous-time output. This hidden oscillation is caused by the oscillation in the control signal. It is thus important to simulate a system in order to investigate the behavior between the sampling points.

The closed-loop system is of third order, the process has two modes, and the controller has one mode. The zero on the stability boundary is canceled by a pole. This mode is not observable in the closed-loop discrete-time system. This means that observability of the closed-loop system has been lost by an improper choice of the controller. ∎

To summarize, there are no hidden oscillations if the unobservable open-loop modes are not oscillatory and if unstable or poorly damped process zeros are not canceled by the regulator.

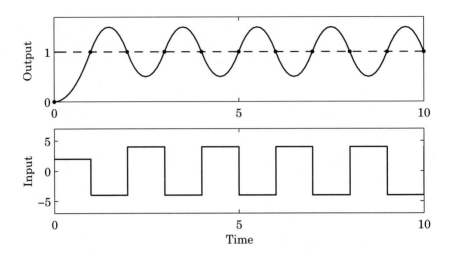

Figure 3.20 The step response and the control signal of the double integrator when the controller of (3.30) is used.

3.6 Problems

3.1 Determine if the following equations have all their roots inside the unit disc:

(a) $z^2 - 1.5z + 0.9 = 0$

(b) $z^3 - 3z^2 + 2z - 0.5 = 0$

(c) $z^3 - 2z^2 + 2z - 0.5 = 0$

(d) $z^3 + 5z^2 - 0.25z - 1.25 = 0$

(e) $z^3 - 1.7z^2 + 1.7z - 0.7 = 0$

3.2 Consider the system in Fig. 3.5 and let

$$H(z) = \frac{K}{z(z - 0.2)(z - 0.4)} \qquad K > 0$$

Determine the values of K for which the closed-loop system is stable.

3.3 Consider the system in Fig. 3.21. Assume that the sampling is done periodically with the period h and that the D-A converter holds the control signal constant over

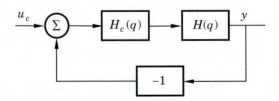

Figure 3.21 Closed-loop system for Problem 3.3.

the sampling interval. The control algorithm is assumed to be

$$u(kh) = K\Big(u_c(kh - \tau) - y(kh - \tau)\Big)$$

where $K > 0$ and T is the computation time. The transfer function of the process is

$$G(s) = \frac{1}{s}$$

(a) How large are the values of the regulator gain, K, for which the closed-loop system is stable if $\tau = 0$ and $\tau = h$?

(b) Compare this system with the corresponding continuous-time systems, that is, when there is a continuous-time proportional controller and a time delay in the process.

3.4   Determine the Nyquist curve for the system

$$H(z) = \frac{1}{z - 0.5}$$

3.5   From the system

$$x(k + 1) = \begin{pmatrix} 1 & 0 \\ 1 & 1 \end{pmatrix} x(k) + \begin{pmatrix} 1 \\ 0 \end{pmatrix} u(k)$$

$$y(k) = \begin{pmatrix} 0 & 1 \end{pmatrix} x(k)$$

the following values are obtained

$$\begin{array}{ll} y(1) = 0 & u(1) = 1 \\ y(2) = 1 & u(2) = -1 \end{array}$$

Determine the value of the state at $k = 3$.

3.6   Is the following system (a) observable, (b) reachable?

$$x(k + 1) = \begin{pmatrix} 0.5 & -0.5 \\ 0 & 0.25 \end{pmatrix} x(k) + \begin{pmatrix} 6 \\ 4 \end{pmatrix} u(k)$$

$$y(k) = \begin{pmatrix} 2 & -4 \end{pmatrix} x(k)$$

3.7   Is the following system reachable?

$$x(k + 1) = \begin{pmatrix} 1 & 0 \\ 0 & 0.5 \end{pmatrix} x(k) + \begin{pmatrix} 1 & 1 \\ 1 & 0 \end{pmatrix} u(k)$$

Assume that a scalar input $u'(k)$ such that

$$u(k) = \begin{pmatrix} 1 \\ -1 \end{pmatrix} u'(k)$$

is introduced. Is the system reachable from $u'(k)$?

3.8 Given the system

$$
x(k+1) = \begin{pmatrix} 0 & 1 & 2 \\ 0 & 0 & 3 \\ 0 & 0 & 0 \end{pmatrix} x(k) + \begin{pmatrix} 0 \\ 1 \\ 0 \end{pmatrix} u(k)
$$

(a) Determine a control sequence such that the system is taken from the initial state $x^T(0) = \begin{pmatrix} 1 & 1 & 1 \end{pmatrix}$ to the origin.

(b) Which is the minimum number of steps that solve the problem in (a)?

(c) Explain why it is not possible to find a sequence of control signals such that the state $\begin{pmatrix} 1 & 1 & 1 \end{pmatrix}^T$ is reached from the origin.

3.9 Verify the formula for W_c^{-1} given in Example 3.9 for an nth-order system.

3.10 The system

$$
x(k+1) = \Phi x(k) + \Gamma u(k)
$$

has been obtained from the system

$$
z(k+1) = Fz(k) + Gu(k)
$$

by a linear transformation

$$
z = Tx
$$

(a) Use the result in Sec. 3.4 to derive a formula for T when $\dim(u) = 1$ and $\dim(u) = r$.

(b) Use the result to solve Problem 2.7.

3.11 Determine the stability and the stationary value of the output for the system described by Fig. 3.21 with

$$
H(q) = \frac{1}{q(q-0.5)}
$$

when u_c is a step function and (a) $H_{fb}(q) = K$ (proportional controller), $K > 0$; and (b) $H_{fb}(q) = Kq/(q-1)$ (integral controller), $K > 0$.

3.12 Consider the system in Problem 3.11. Determine the steady-state error between the command signal, u_c, and the output when u_c is a unit ramp, that is, $u_c(k) = k$. Assume that H_{fb} is (a) a proportional controller and (b) an integral controller.

3.13 Sample the system

$$
G(s) = \frac{s+1}{s^2 + 0.2s + 1}
$$

and determine the sampling intervals for which the response of the system will have hidden oscillations. Verify by simulations.

3.14 Consider the tank system with the pulse-transfer operator given in Problem 2.10(b), that is, when the system is sampled with $h = 12$.

    (a) Introduce a controller as in Fig. 3.21. Let the command input be a step and determine the steady-state error when using a proportional controller K and an integral controller $K/(1 - q^{-1})$.

    (b) Simulate the system using the controllers in (a). Investigate the influence of the controller gain K. Determine K such that the poles of the closed-loop system correspond to a damping of $\zeta = 0.7$.

3.15 Consider the system in Fig. 3.5. Derive a formula for the velocity *error coefficient*. That is an expression for the steady-state error when the reference signal u_c is a unit ramp.

3.16 Determine the values of $K > 0$ for which the system

$$y(k) = K \frac{4q^{-1} + q^{-2}}{1 + q^{-1} + 0.16q^{-2}} u(k)$$

is stable under simple feedback.

3.17 Determine a coordinate transformation $z = Tx$ that transfers the system

$$x(k + 1) = \begin{pmatrix} 1 & 2 \\ 1 & 2 \end{pmatrix} x(k) + \begin{pmatrix} 3 \\ 4 \end{pmatrix} u(k)$$

$$y(k) = \begin{pmatrix} 5 & 6 \end{pmatrix} x(k)$$

to controllable canonical form and to observable canonical form.

3.18 Assume that the continuous-time system (CT)

$$\frac{dx}{dt} = Ax + Bu$$

$$y = Cx$$

is sampled and gives the discrete-time system (DT)

$$x(kh + h) = \Phi x(kh) + \Gamma u(kh)$$

$$y(kh) = Cx(kh)$$

Consider the following statements:

    (a) CT stable $\Rightarrow$ DT stable
    (b) CT unstable $\Rightarrow$ DT unstable
    (c) CT stable inverse $\Rightarrow$ DT stable inverse
    (d) CT unstable inverse $\Rightarrow$ DT unstable inverse
    (e) CT controllable $\Rightarrow$ DT controllable
    (f) CT observable $\Rightarrow$ DT observable
    (g) CT pole excess $r \Rightarrow$ DT pole excess r

Which statements are true for the following cases:

(i)   All sampling intervals $h > 0$.
(ii) All $h > 0$ except for isolated values.
(iii) Neither (i) nor (ii).

3.19 Consider the system

$$x(k + 1) = \begin{pmatrix} 0 & -3 & 2 \\ 3 & -12 & 7 \\ 6 & -21 & 12 \end{pmatrix} x(k) + \begin{pmatrix} 0 \\ 1 \\ 2 \end{pmatrix} u(k)$$

Determine whether

(a) the system is reachable.

(b) the system is controllable.

3.20 Given the system

$$(q^2 + 0.4q)y(k) = u(k)$$

(a) For which values of K in the proportional controller

$$u(k) = K(u_c(k) - y(k))$$

is the closed-loop system stable?

(b) Determine the stationary error $u_c - y$ when u_c is a step and when $K = 0.5$ in the controller in (a).

3.21 Assume that the system

$$y(k) - 1.2y(k - 1) + 0.5y(k - 2) = 0.4u(k - 1) + 0.8u(k - 2)$$

is controlled by

$$u(k) = -Ky(k)$$

(a) Determine for which values of K the closed-loop system is stable.

(b) Assume that there is a computational delay in the controller, that is,

$$u(k) = -Ky(k - 1)$$

For which values of K is the closed-loop system now stable?

3.7 Notes and References

Original papers on tests for checking if a polynomial has all its poles inside the unit circle are Schur (1918) and Cohn (1922). Jury's test is a simplification of the Schur-Cohn test and is found in Jury and Blanchard (1961). A simple proof of Jury's test is found in Åström (1970). The use of the Lyapunov theory for discrete-time control systems is introduced in Kalman and Bertram (1960). Controllability and observability are concepts introduced by Kalman in connection with analysis of optimal control systems. See Kalman (1961) and Kalman, Ho, and Narendra (1963). The hidden oscillations and their cause are discussed in Jury (1957) and Sanchis and Albertos (1995).

General aspects of simulation are discussed in Gordon (1969), Åström (1983a), Kheir (1988), Cellier (1991), and Mattsson, Andersson, and Åström (1993).

4

Pole-Placement Design:
A State-Space Approach

4.1 Introduction

This chapter presents design methods based on internal models of the system. The methods developed in this chapter can be viewed as solutions to specific, idealized control problems. The solutions give insight into the nature of control problems. They also show that many of the concepts introduced earlier are useful. Control design involves compromises between conflicting goals. We capture this by introducing so-called design parameters that have to be chosen by the designer. See Sec. 4.2. In this chapter we will develop a collection of design methods that are called pole placement from the point of view of state feedback. The name pole placement refers to the fact that the design is formulated in terms of obtaining a closed-loop system with specified poles. The methods will be developed gradually. In Sec. 4.3 we will discuss an idealized regulation problem. It is assumed that all state variables are measured and the disturbances are widely spaced impulses. In Sec. 4.4 we will discuss the problem of reconstructing the states from measured outputs. This leads to the introduction of observers. By combining the observers with the state feedback obtained in Sec. 4.3, we obtain a solution to the regulation problem for the case of output feedback in Sec. 4.5. We will also generalize the disturbances by considering disturbances that are obtained as outputs of dynamic systems whose inputs are impulses. In this way we can deal with the classical cases of disturbances that are steps and sinusoids as well as many other cases. This also gives a very natural way to introduce integral action. So far we have only dealt with the regulation problem. In Sec. 4.6 we will discuss the servo problem, that is, how to obtain a system that can also follow command signals. This problem can also be well captured in the state-space formulation. By combining the result with the previous results we obtain a controller that can follow command signals and reject disturbances acting on the system. The controller structure obtained is very interesting because the different tasks of the controller are naturally separated. The design procedure is illustrated in Sec. 4.7 with an application to control of simple robotics system.

4.2 Control-System Design

Many different factors have to be considered in the design of a control system, for example:

- Attenuation of load disturbances

- Reduction of the effect of measurement noise

- Command signal following

- Variations and uncertainties in process behavior

Load disturbances are disturbances that drive the process away from its desired behavior. Measurement noise is a disturbance that corrupts the information about the process obtained from the sensors. Process disturbances can enter the system in many different ways. It is convenient to consider them as if they enter the system in the same way as the control signal; in this way they will excite all modes of the system. For linear systems it also follows from the superposition principle that the assumption is not very critical. The measurement noise will be injected into the process through the control law. The command signal following expresses the property of the system to respond to command signals in a specified way.

Control problems can broadly speaking be classified into *regulation problems* and *servo problems*. The major issue in the regulation problem is to compromise between reduction of load disturbances and the fluctuations created by the measurement noise that is injected in the system due to the feedback. The command signal following is the major issue in servo problems. The major ingredients of a design problem are

- Purpose of the system

- Process model

- Model for disturbances

- Model variations and uncertainties

- Admissible control strategies

- Design parameters

It is difficult to find design methods that consider all the preceding issues mentioned. Most design methods focus on one or two aspects of the problem and the control-system designer then has to check that the other requirements are also

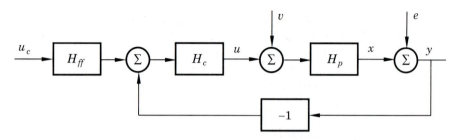

Figure 4.1   Block diagram of a typical control system.

satisfied. To do this it is necessary to consider the signal transmission from command signals, load disturbances, and measurement noise to process variables, measured signals, and control signals. This is illustrated in the block diagram of Fig. 4.1. Compare with Fig. 3.10. In this chapter we will develop a design method based on state models whose purpose is to obtain a specified closed-loop characteristic polynomial of the system. At first sight it may seem unnatural to specify the problem in this way. It will lead, however, to simple design methods that will give considerable insight into the structure of good control systems. The design method is very easy to apply for low-order systems, but it may be difficult to choose the poles properly for systems of high order. The structure of the controller is also the same as the one obtained with more sophisticated design methods, which will be discussed later.

We will start with a simple design problem and gradually make it more and more realistic. The problem is specified as follows.

The process.   It is assumed that the process to be controlled can be described by the model

$$\frac{dx}{dt} = Ax + Bu \tag{4.1}$$

where u represents the control variables, x represents the state vector, and A and B are constant matrices. Further, only the single-input–single-output case will be discussed. Because computer control is considered, the control signals will be constant over sampling periods of constant length. Sampling the system in Eq. (4.1) by the methods described in Sec. 2.3 gives the discrete-time system

$$x(kk + h) = \Phi x(kh) + \Gamma u(kh)$$

where the matrices Φ and Γ are given by

$$\Phi = e^{Ah} \qquad \Gamma = \int_0^h e^{As}\, ds\, B$$

To simplify we will write the system as

$$x(k + 1) = \Phi x(k) + \Gamma u(k) \tag{4.2}$$

The argument of a signal is thus not real time but instead the number of sampling intervals. We call this the *sampling-time convention*. We will use real time whenever there are possibilities for confusion.

Disturbances.   Initially it will be assumed that the disturbances acting on the process are impulses that occur irregularly, and that the impulses are so widely spread that the system settles between the impulses. Because the impulses are far apart and the effect of an impulse is simply to change the process state, the disturbance can be represented by an initial state. Later we will extend the results to much more general disturbances that are generated from dynamic systems whose inputs are impulses. Typical examples are steps, ramps, and sinusoidal signals.

Process uncertainty.   Uncertainties in the elements of the matrices A and B can be dealt with in the state-space formulation, but it is not easy to deal with other forms of unmodeled dynamics. A discussion of process uncertainties therefore will be given later when more appropriate tools have been developed.

The criterion.   When discussing regulation problems it will be assumed that the criterion is to bring the state to zero after perturbations in the initial condition. In the pole-placement formulation, the rate of decay of the state is given indirectly by specifying the poles of the closed-loop system. Servo problems will be dealt with by requiring that the signal transmission from command signal to process variables is close to a behavior specified by a model.

Admissible controls.   Because feedback solutions are desired, it is necessary to specify the information available for generating the control signal. When the properties of the system are specified by its closed-loop poles, it is natural to require that the feedback is linear. Several different versions will be discussed. We will start with the case when all state variables are measured directly without error. The admissible controls are then a linear feedback of the form

$$u(k) = -Lx(k) \qquad (4.3)$$

This assumption will be relaxed later in Sec. 4.5, where it will be assumed that only outputs are available for control. For discrete-time systems it is also of interest to consider the case when there are delays in the measurements.

Design parameters.   In the formal specification of the problem, the design parameters are the sampling period and the desired closed-loop poles. It is rare that a user of a control system can give specifications in terms of these parameters. Therefore, the designer must be able to relate the design parameters to quantities that are more meaningful to the user. For this purpose, it is often useful to consider the time histories of the state variables and the control variables. It is particularly useful to discuss the trade-off between the magnitude of the control signals and the speed at which the system recovers from a disturbance.

4.3 Regulation by State Feedback

A simple regulation problem is discussed in this section. It is assumed that the system is described by Eq. (4.1). Initially we also assume that the sampling period is given so that the process can be described by the discrete-time system (4.2). The disturbances are assumed perturbations in the initial state of the system. The purpose is to find a linear feedback law of the form (4.3) so that the closed-loop system has a specified characteristic equation. This will guarantee that the disturbances decay in a specified way.

     This problem may seem overly simplistic as a representation of a control-system design problem. The solution is, however, very simple and the problem can be generalized successively to make assumptions more and more realistic.

An Example

To introduce the design method and to illustrate the influence of the design parameters, a special case is first discussed.

Example 4.1   Pole placement for the double-integrator plant

By using the sampling-time convention, the sampled double-integrator plant is described by

$$x(k+1) = \begin{pmatrix} 1 & h \\ 0 & 1 \end{pmatrix} x(k) + \begin{pmatrix} h^2/2 \\ h \end{pmatrix} u(k)$$

A general linear feedback can be described by

$$u = -l_1 x_1 - l_2 x_2$$

With this feedback, the closed-loop system becomes

$$x(k+1) = \begin{pmatrix} 1 - l_1 h^2/2 & h - l_2 h^2/2 \\ -l_1 h & 1 - l_2 h \end{pmatrix} x(k)$$

The characteristic equation of the closed-loop system is

$$z^2 + \left(\frac{l_1 h^2}{2} + l_2 h - 2 \right) z + \left(\frac{l_1 h^2}{2} - l_2 h + 1 \right) = 0$$

Assume that the desired characteristic equation is

$$z^2 + p_1 z + p_2 = 0$$

This leads to the following linear equations for l_1 and l_2:

$$\frac{l_1 h^2}{2} + l_2 h - 2 = p_1$$

$$\frac{l_1 h^2}{2} - l_2 h + 1 = p_2$$

These equations have the solution

$$l_1 = \frac{1}{h^2} (1 + p_1 + p_2)$$

$$l_2 = \frac{1}{2h} (3 + p_1 - p_2)$$

(4.4)

In this example it is always possible to find controller parameters that give an arbitrary characteristic equation of the closed-loop system. The linear system of equations for l_1 and l_2 has a solution for all values of p_1 and p_2.   ∎

The General Case

The solution of the pole-placement problem now will be given for systems with one input signal. Let the system be described by (4.2) and let the characteristic polynomial of the matrix Φ be

$$z^n + a_1 z^{n-1} + \cdots + a_n$$

Assume that the system (4.2) is reachable. It then can be transformed to reachable canonical form by changing state variables through the transformation $z = Tx$, and the transformed state equation becomes

$$z(k + 1) = \tilde{\Phi} z(k) + \tilde{\Gamma} u(k)$$

(4.5)

where

$$\tilde{\Phi} = \begin{pmatrix} -a_1 & -a_2 & \cdots & -a_{n-1} & -a_n \\ 1 & 0 & \cdots & 0 & 0 \\ 0 & 1 & \cdots & 0 & 0 \\ \vdots & \ddots & \ddots & \vdots & \vdots \\ 0 & 0 & \cdots & 1 & 0 \end{pmatrix} \qquad \tilde{\Gamma} = \begin{pmatrix} 1 \\ 0 \\ 0 \\ \vdots \\ 0 \end{pmatrix}$$

(4.6)

The coefficients of the characteristic polynomial that determines the closed-loop poles appear explicitly in this representation. It is also easy to see how the characteristic polynomial is modified by state feedback. It follows from (4.6) that the feedback law

$$u = -\tilde{L}z = - \begin{pmatrix} p_1 - a_1 & p_2 - a_2 & \cdots & p_n - a_n \end{pmatrix} z$$

(4.7)

gives a closed-loop system with the characteristic polynomial

$$P(z) = z^n + p_1 z^{n-1} + \cdots + p_n$$

(4.8)

To find the solution to the original problem we simply have to transform back to the original coordinates. This gives

$$u = -\tilde{L}z = -\tilde{L}Tx = -Lx$$

(4.9)

It remains to determine the transformation matrix T. A simple way of determining this matrix is based on a property of the reachability matrices. Let W_c be the reachability matrix of the system (4.2), that is,

$$W_c = \begin{pmatrix} \Gamma & \Phi\Gamma & \cdots & \Phi^{n-1}\Gamma \end{pmatrix} \tag{4.10}$$

and let $\tilde{W}_c$ be the reachability matrix of the system (4.5). The matrices are related through $\tilde{W}_c = TW_c$. The reachability matrix thus transforms in the same way as the coordinates. It thus follows that

$$T = \tilde{W}_c W_c^{-1} \tag{4.11}$$

and a straightforward calculation gives

$$\tilde{W}_c^{-1} = \begin{pmatrix} 1 & a_1 & \cdots & a_{n-1} \\ 0 & 1 & \cdots & a_{n-2} \\ \vdots & \vdots & \ddots & \vdots \\ 0 & 0 & \cdots & 1 \end{pmatrix} \tag{4.12}$$

Compare with Example 3.9. Summarizing, we find that the solution to the design problem is given by a linear state feedback with the gain

$$L = \begin{pmatrix} p_1 - a_1 & p_2 - a_2 & \cdots & p_n - a_n \end{pmatrix} \tilde{W}_c W_c^{-1} \tag{4.13}$$

This equation can be expressed in a slightly different way by the following result.

THEOREM 4.1   POLE-PLACEMENT USING STATE FEEDBACK   Consider the system of (4.2). Assume that there is only one input signal. If the system is reachable there exists a linear feedback that gives a closed-loop system with the characteristic $P(z)$. The feedback is given by

$$u(k) = -Lx(k)$$

with

$$\begin{aligned} L &= \begin{pmatrix} p_1 - a_1 & p_2 - a_2 & \cdots & p_n - a_n \end{pmatrix} \tilde{W}_c W_c^{-1} \\ &= \begin{pmatrix} 0 & \cdots & 0 & 1 \end{pmatrix} W_c^{-1} P(\Phi) \end{aligned} \tag{4.14}$$

where W_c and $\tilde{W}_c$ are the reachability matrices of the systems (4.2) and (4.5), respectively.

Proof.   To prove the result we first observe that

$$P(\tilde{\Phi}) = \tilde{\Phi}^n + p_1\tilde{\Phi}^{n-1} + \cdots + p_n I = (p_1 - a_1)\tilde{\Phi}^{n-1} + \cdots + (p_n - a_n)I$$

where $\tilde{\Phi}$ is the system matrix of the transformed system (4.5). The second equality is obtained by using the Cayley-Hamilton theorem. Introduce e^i as the row vector that has all elements equal to zero except the ith element, which is 1. We have

$$e^i\tilde{\Phi} = e^{i-1}$$

It then follows from Eq. (4.7) that $\tilde{L} = e^n P(\tilde{\Phi})$ and we get

$$L = \tilde{L}T = e^n P(T\Phi T^{-1})T = e^n T P(\Phi) = e^n \tilde{W}_c W_c^{-1}P(\Phi)$$

It follows from (4.12) that $e^n \tilde{W}_c^{-1} = e^n$ and Eq. (4.14) is obtained.

Remark 1.   Equation (4.14) is called *Ackermann's formula*.

Remark 2.   Notice that the pole-placement problem can be formulated as the following abstract problem. Given matrices Φ and Γ, find a matrix L such that the matrix $\Phi - \Gamma L$ has prescribed eigenvalues.

Remark 3.   Notice that it follows from (4.11) and (4.12) that

$$T^{-1} = \begin{pmatrix} \Gamma & \Phi\Gamma + a_1\Gamma & \cdots & \Phi^{n-1}\Gamma + a_1\Phi^{n-2} + \cdots + a_{n-1}\Gamma \end{pmatrix} \qquad (4.15)$$

■

The theorem is illustrated by an example.

Example 4.2   Double integrator

Consider the double-integrator plant in Example 4.1. Assume that the desired characteristic polynomial is given by $P(z) = z^2 + p_1 z + p_2$. We have

$$W_c = \begin{pmatrix} \Gamma & \Phi\Gamma \end{pmatrix} = \begin{pmatrix} h^2/2 & 3h^2/2 \\ h & h \end{pmatrix}$$

and

$$W_c^{-1} = \begin{pmatrix} -1/h^2 & 1.5/h \\ 1/h^2 & -0.5/h \end{pmatrix}$$

The characteristic polynomial of the matrix Φ is $z^2 - 2z + 1$. Hence

$$P(\Phi) = \Phi^2 + p_1\Phi + p_2 I = \begin{pmatrix} 1 + p_1 + p_2 & 2h + p_1 h \\ 0 & 1 + p_1 + p_2 \end{pmatrix}$$

Ackermann's formula (4.14) now gives

$$L = \begin{pmatrix} 0 & 1 \end{pmatrix} W_c^{-1}P(\Phi) = \begin{pmatrix} 1/h^2 & -0.5/h \end{pmatrix} P(\Phi)$$

$$= \begin{pmatrix} \dfrac{1 + p_1 + p_2}{h^2} & \dfrac{3 + p_1 - p_2}{2h} \end{pmatrix}$$

which is the same result obtained by the direct calculation in Example 4.1.   ■

To solve the pole-placement problem it was assumed that the system is reachable. The following example illustrates what happens when this is not the case.

Example 4.3 An unreachable system

The system

$$x(k + 1) = \begin{pmatrix} 0.5 & 1 \\ 0 & 0.3 \end{pmatrix} x(k) + \begin{pmatrix} 1 \\ 0 \end{pmatrix} u(k)$$

is not reachable because

$$\det W_c = \det \begin{pmatrix} 1 & 0.5 \\ 0 & 0 \end{pmatrix} = 0$$

The control law $u = -l_1 x_1 - l_2 x_2$ gives a closed-loop system with the characteristic equation

$$(z - 0.5 + l_1)(z - 0.3) = 0$$

The open-loop pole in 0.5 can be changed to an arbitrary value by changing the parameter l_1. The second pole 0.3, which corresponds to the nonreachable state, cannot be changed. ∎

Practical Aspects

It is easy to solve the pole-placement design problem explicitly. Notice that reachability is a necessary and sufficient condition for solving the problem. To apply the pole-placement design method in practice, it is necessary to understand how the properties of the closed-loop system are influenced by the design parameters—that is, the closed-loop poles and the sampling period. This is illustrated by an example.

Example 4.4 Choice of design parameters

Consider the double-integrator plant. Instead of using the parameters p_1 and p_2 of the characteristic equation we will introduce two other parameters, which have a more direct physical interpretation. If the desired discrete-time system is obtained by sampling a second-order system with the characteristic polynomial $s^2 + 2\zeta \omega s + \omega^2$ we find that

$$p_1 = -2e^{-\zeta \omega h} \cos \left(\omega h \sqrt{1 - \zeta^2} \right)$$

$$p_2 = e^{-2\zeta \omega h}$$

where ω is the natural frequency and ζ is the damping (compare with Example 2.16). The parameter ζ influences the relative damping of the response and ω influences the response speed. To discuss the magnitude of the control signal, it is assumed that the system has an initial position x_0 and an initial velocity v_0. The initial value of the control signal is then

$$u(0) = -l_1 x_0 - l_2 v_0$$

If the sampling period is short, then the expressions for p_1 and p_2 can be approximated using series expansion. The following approximation is then obtained:

$$u(0) \approx -\omega^2 x_0 + 2\zeta \omega v_0$$

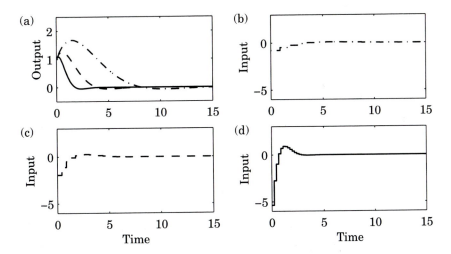

Figure 4.2 Responses of the closed-loop system in Example 4.4. The initial condition is $x^T(0) = [1 \quad 1]$, and the parameter values are $\omega h = 0.44$ and $\zeta = 0.707$. The outputs for sampling periods $\omega = 0.5$ (dashed-dotted), $\omega = 1$ (dashed), and $\omega = 2$ (solid) are shown in (a), and the corresponding control signals are shown in (b), (c), and (d), respectively.

The expression shows that the magnitude of the control signal increases with increasing ω. Thus an increase in the speed of the response of the system will require an increase in the control signals. If the bounds on the control signal and typical disturbances are known, it is possible to determine reasonable values of ω. The consequences of different choices of ω when $x_0 = 1$ and $v_0 = 1$ are illustrated in Fig. 4.2. A larger ω gives a faster system but also larger control signals.

The selection of sampling periods for open-loop systems was discussed in Sec. 2.9. It was suggested that the sampling period can be chosen such that

$$N_r \approx 4 - 10$$

where N_r is the number of samples per rise time. Applying the same rule to closed-loop systems we find that the sampling period should be related to the desired behavior of the closed-loop system. It is convenient to introduce the parameter N defined by

$$N = \frac{2\pi}{\omega h \sqrt{1 - \zeta^2}} \tag{4.16}$$

This parameter gives the number of samples per period of dominating mode of the closed-loop system. Figure 4.3 shows the transient of the system for different values of N. There are small differences between the responses for $N > 10$. The responses obtained for $N > 20$ are indistinguishable in the graph.

Figure 4.3 shows the response to an initial condition when an impulse disturbance has entered the system just before the sampling. In reality the disturbances of course may enter the system at any time. With a long sampling period it will

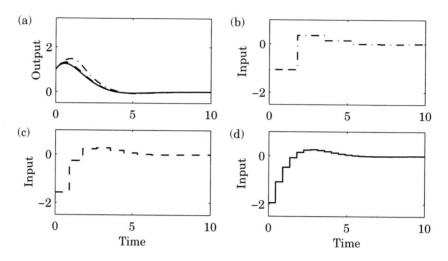

Figure 4.3 Responses of the closed-loop system in Example 4.4. The initial condition is $x^T(0) = [1 \quad 1]$, and the parameter values are $\omega h = 1$ and $\zeta = 0.707$. The outputs obtained for $N = 5$ (dashed-dotted), $N = 10$ (dashed), and $N = 20$ (solid) are shown in (a), and the corresponding control signals are shown in (b), (c), and (d), respectively.

then take a long time before the disturbance is detected. To illustrate this we will repeat the simulation in Fig. 4.3 but it will be assumed that the disturbance comes just after a sampling. This implies that the disturbance acts for a full sampling period before it is detected. Figure 4.4 shows the response of the system when the system is disturbed immediately after the sampling, that is, when $x(0+) = [1 \quad 1]^T$. Notice the significant difference compared with Fig. 4.3. In this case the results for $N = 20$ are much better than the results for $N = 10$. It is reasonable to choose N in the range $N \approx 25$ to 75. This corresponds to $\omega h = 0.12$ to 0.36 for $\zeta = 0.707$. ∎

These examples show that even if we take a discrete-system point of view by only considering what happens at the sampling instants, it is necessary to keep the time-varying nature of sampled systems in mind to make the correct assessment of the results. Particular care should be given to simulations used to assess the performance of the systems. To investigate the effect of the sampling period it is useful to consider cases in which disturbances are introduced both immediately before and immediately after the sampling instants. The differences can be quite noticeable, as is indicated by a comparison of Figs. 4.3 and 4.4. Based on the simulations performed we suggest that the sampling period be chosen as

$$\omega h = 0.1 \text{ to } 0.6 \qquad (4.17)$$

where ω is the desired natural frequency of the closed-loop system. Longer sampling periods can be used in those rare cases in which the sampling can be synchronized to the disturbances.

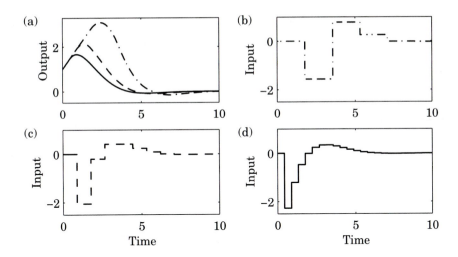

Figure 4.4 Responses of the closed-loop system in Example 4.4. The initial condition is $x^T(0+) = [1 \quad 1]$, and the parameter values are $\omega h = 1$ and $\zeta = 0.707$. The outputs obtained for $N = 5$ (dashed-dotted), $N = 10$ (dashed), and $N = 20$ (solid) are shown in (a), and the control signals are shown in (b), (c), and (d), respectively. The disturbance is immediately after the first sampling. Notice the significant difference compared to Fig. 4.3.

Deadbeat Control

If the desired poles are all chosen to be at the origin, the characteristic polynomial of the closed-loop system becomes

$$P(z) = z^n$$

The Cayley-Hamilton theorem then implies that the system matrix $\Phi_c = \Phi - \Gamma L$ of the closed-loop system satisfies

$$\Phi_c^n = 0$$

This strategy has the property that it will drive all the states to zero in at most n steps after an impulse disturbance in the process state. The control strategy is called *deadbeat control*. Compare with Example 1.3 in Chapter 1.

It follows from Ackermann's formula, Eq. (4.14), that the deadbeat strategy is given by

$$L = \begin{pmatrix} 0 & \cdots & 0 & 1 \end{pmatrix} W_c^{-1} \Phi^n \qquad (4.18)$$

If the matrix Φ is invertible we get

$$L = \begin{pmatrix} 0 & \cdots & 0 & 1 \end{pmatrix} \begin{pmatrix} \Phi^{-n}\Gamma & \Phi^{-n+1}\Gamma & \cdots & \Phi^{-1}\Gamma \end{pmatrix}^{-1} \qquad (4.19)$$

Table 4.1 Control signals for deadbeat control of a double integrator with $x(0) = \text{col}\,[1, 1]$ and different sampling periods.

h	100	10	1	0.1	0.01
$u(0)$	−0.0151	−0.16	−2.5	−115	−10,150
$u(h)$	0.0051	0.06	1.5	105	10,050

In deadbeat control there is only one design parameter—the sampling period. Because the error goes to zero in at most n sampling periods, the settling time is at most nh. The settling time is thus proportional to the sampling period h. The sampling period also influences the magnitude of the control signal, which increases drastically with decreasing sampling period. This fact has given the deadbeat control an undeservedly bad reputation. It is thus important to choose the sampling period carefully when using deadbeat control. The deadbeat strategy is unique to sampled-data systems. There is no corresponding feature for continuous-time systems. The following example demonstrates some properties of deadbeat control.

Example 4.5   Deadbeat control of a double integrator

Consider a double-integrator plant. It follows from Eq. (4.19) that the deadbeat control strategy is given by $u = l_1 x_1 + l_2 x_2$ with

$$l_1 = \frac{1}{h^2} \qquad l_2 = \frac{3}{2h}$$

If the process has the initial state $x(0) = \text{col}\,[x_0, v_0]$, it follows that

$$u(0) = -\frac{x_0}{h^2} - \frac{3v_0}{2h} \qquad u(h) = \frac{x_0}{h^2} + \frac{v_0}{2h}$$

Notice that the magnitude of the control signal increases rapidly with decreasing sampling period. Also notice that for small h, the control signals $u(0)$ and $u(h)$ have opposite signs and approximately equal magnitude. The desired effect is thus obtained as a result of subtracting two large numbers. This is further illustrated in Table 4.1, which gives the control signals for $x_0 = 1$ and $v_0 = 1$. It therefore can be expected that the deadbeat strategy is quite sensitive for small sampling periods. The output and the control signals are shown in Fig. 4.5. In this case the first sampling is at $t = 0+$. The disturbance thus occurs immediately before the sampling. ∎

More General Disturbances

It is highly desirable to handle other disturbances than impulses or equivalently perturbed initial states. One way to do this is to consider disturbances that are generated by sending impulses to dynamic systems. In this way it is possible

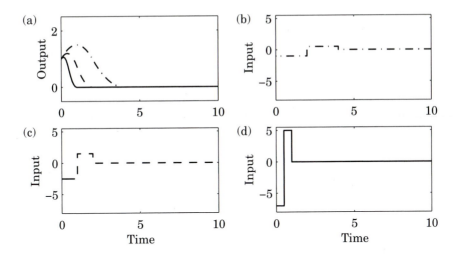

Figure 4.5 Responses of the closed-loop system in Example 4.5 with a dead-beat controller. The initial condition is $x^T(0) = \begin{bmatrix} 1 & 1 \end{bmatrix}$, and the parameter values are $\omega = 1$ and $\zeta = 0.707$. The outputs obtained for sampling periods $h = 2$ (dashed-dotted), $h = 1$ (dashed), and $h = 0.5$ (solid) are shown in (a), and the control signals obtained in the different cases are shown in (b), (c), and (d), respectively.

to capture many different types of disturbances such as steps and sinusoids. To be specific, assume that the system is described by

$$\frac{dx}{dt} = Ax + Bu + v$$

where v is a disturbance described by

$$\frac{dw}{dt} = A_w w$$
$$v = C_w w$$

with given initial conditions. The matrix A_w typically has zeros on the imaginary axis or in the right half plane. A common case is that the disturbance v is a constant. This is captured by $A_w = 0$; another case is sinusoidal disturbances, which correspond to

$$A_w = \begin{pmatrix} 0 & \omega_0 \\ -\omega_0 & 0 \end{pmatrix}$$

It is assumed that w can be measured. This assumption will be relaxed later. We introduce the augmented state vector

$$z = \begin{pmatrix} x \\ w \end{pmatrix}$$

and we find that the system can be described by

$$\frac{d}{dt}\begin{pmatrix} x \\ w \end{pmatrix} = \begin{pmatrix} A & C_w \\ 0 & A_w \end{pmatrix} \begin{pmatrix} x \\ w \end{pmatrix} + \begin{pmatrix} B \\ 0 \end{pmatrix} u \tag{4.20}$$

Thus we have a problem of the same form as the basic pole-placement problem. There is, however, one important difference: The system of (4.20) is not completely reachable. The poles associated with the description of the disturbance—that is, the eigenvalues of A_w—cannot be influenced by the feedback. This is very natural because the disturbances are exogenous variables that are not influenced by the control. Compare with Example 4.3. Sampling the system gives the following discrete-time system:

$$\begin{pmatrix} x(k+1) \\ w(k+1) \end{pmatrix} = \begin{pmatrix} \Phi & \Phi_{xw} \\ 0 & \Phi_w \end{pmatrix} \begin{pmatrix} x(k) \\ w(k) \end{pmatrix} + \begin{pmatrix} \Gamma \\ 0 \end{pmatrix} u(k)$$

The general linear-state feedback is given by

$$u(k) = -Lx(k) - L_w w(k) \tag{4.21}$$

This control law gives the following closed-loop system:

$$\begin{aligned} x(k+1) &= (\Phi - \Gamma L)x(k) + (\Phi_{xw} - \Gamma L_w)w(k) \\ w(k+1) &= \Phi_w w(k) \end{aligned} \tag{4.22}$$

which tells how the closed-loop system is influenced by the control. Notice that the control law in (4.21) can be interpreted as a combination of a feedback term Lx and a feedforward term $L_w w$ from the measured disturbances. If the pair (Φ, Γ) is reachable, the matrix L can be chosen so that the matrix $\Phi - \Gamma L$ has prescribed eigenvalues. This would ensure that the term of the solution that is caused by the initial values decays properly. The matrix Φ_w cannot be influenced by feedback. The effect of the disturbance on the vector x can, however, be reduced by a proper choice of the vector L_w, which should be chosen so that the matrix $\Phi_{xw} - \Gamma L_w$ is small. In some cases it is possible to make this matrix zero. We illustrate this by an example.

Example 4.6   Constant input disturbance

Consider the situation with a constant disturbance that acts on the process input. The matrix Φ_w then becomes the identity and we have $\Phi_{xw} = \Gamma$. The system is described by

$$\begin{aligned} x(k+h) &= (\Phi - \Gamma L)x(k) + \Gamma(1 - L_w)w(k) \\ w(k+h) &= w(k) \end{aligned}$$

The effect of the disturbance v on x is thus eliminated by choosing $L_w = 1$.   ∎

Computational Aspects

The state feedback can be determined by the method used in Example 4.1 for low-order systems. The procedure is simply to introduce a general state feedback with unknown coefficients, determine the characteristic polynomial, and equate it with the desired characteristic polynomial. A set of linear equations for the feedback coefficients is then obtained. The equations can be always solved if the system is reachable. It is also possible to use Ackermann's formula, Eq. (4.14), to calculate the state feedback. This formula is, however, not well suited for very precise numerical calculations. As a rule, any method using computation of powers of matrices should be avoided. There are other ways to compute the feedback matrix L that are better for numerical calculations. These methods also work for multivariable systems. In MATLAB® there is a command `polp` for solving the problem that is based on sound numerical methods.

4.4 Observers

It is unrealistic to assume that all states of a system can be measured, particularly if disturbances are part of the state as in Eq. (4.20). It is therefore of interest to determine the states of a system from available measurements and a model. It is assumed that the system is described by the sampled model

$$
\begin{aligned}
x(k+1) &= \Phi x(k) + \Gamma u(k) \\
y(k) &= C x(k)
\end{aligned}
\tag{4.23}
$$

The problem is thus to calculate or reconstruct the state $x(k)$ from input and output sequences $y(k)$, $y(k-1)$, $\ldots$, $u(k)$, $u(k-l)$, $\ldots$ is considered next. In Sec. 3.4 it was shown that this is possible if the system is observable.

Direct Calculation of the State Variables

The problem was solved in Sec. 3.4 for the special case when there are no inputs. We will now extend this solution slightly and it will be shown that the state can be computed directly from past inputs and outputs. For simplicity it is assumed that there is only one output. The output $y(k) = C x(k)$ obtained at sampling instant k gives one linear equation for determining the state variable. Using information from n different sampling instants k, $k-1$, $\ldots$, $k-n+1$ gives the following linear equations.

$$
\begin{aligned}
y(k-n+1) &= C x(k-n+1) \\
y(k-n+2) &= C \Phi x(k-n+1) + C \Gamma u(k-n+1) \\
&\ \ \vdots \\
y(k) &= C \Phi^{n-1} x(k-n+1) + C \Phi^{n-2} \Gamma u(k-n+1) + \cdots + C \Gamma u(k-1)
\end{aligned}
\tag{4.24}
$$

By introducing the vectors U_{k-1} and Y_k

$$Y_k = \begin{pmatrix} y(k-n+1) \\ y(k-n+2) \\ \vdots \\ y(k) \end{pmatrix} \qquad U_{k-1} = \begin{pmatrix} u(k-n+1) \\ u(k-n+2) \\ \vdots \\ u(k-1) \end{pmatrix}$$

whose components are past inputs and outputs, Eq. (4.24) can be written as

$$Y_k = W_o x(k-n+1) + W_u U_{k-1}$$

where the matrices W_o and W_u are given by

$$W_o = \begin{pmatrix} C \\ C\Phi \\ C\Phi^2 \\ \vdots \\ C\Phi^{n-1} \end{pmatrix} \qquad W_u = \begin{pmatrix} 0 & 0 & \cdots & 0 \\ C\Gamma & 0 & \cdots & 0 \\ C\Phi\Gamma & C\Gamma & \cdots & 0 \\ \vdots & \vdots & \ddots & \vdots \\ C\Phi^{n-2}\Gamma & C\Phi^{n-3}\Gamma & \cdots & C\Gamma \end{pmatrix}$$

The matrix W_o is invertible if the system is observable; we can then solve for $x(k-n+1)$ and obtain

$$x(k-n+1) = W_o^{-1} Y_k - W_o^{-1} W_u U_{k-1}$$

The state has thus been obtained in terms of future outputs and measurements. Repeated use of Eq. (4.23) gives

$$x(k) = \Phi^{n-1} x(k-n+1) + \Phi^{n-2}\Gamma u(k-n+1) + \cdots + \Gamma u(k-1)$$

and we find that

$$x(k) = A_y Y_y + B_u U_k \tag{4.25}$$

where

$$A_y = \Phi^{n-1} W_o^{-1} \qquad B_u = \left(\Phi^{n-1}\Gamma \quad \Phi^{n-2}\Gamma \ldots \Gamma \right) - \Phi^{n-1} W_o^{-1} W_u \tag{4.26}$$

The state vector $x(k)$ is thus a linear combination of $y(k), y(k-1), \ldots, y(k-n+1)$ and $u(k-1), u(k-2), \ldots, u(k-n+1)$. We illustrate by an example.

Example 4.7   Double integrator

For the double integrator we have

$$\Phi = \begin{pmatrix} 1 & h \\ 0 & 1 \end{pmatrix} \qquad \Gamma = \begin{pmatrix} h^2/2 \\ h \end{pmatrix} \qquad C = \begin{pmatrix} 1 & 0 \end{pmatrix}$$

Hence

$$y(k) = x_1(k)$$

$$y(k) = x_1(k-1) + hx_2(k-1) + \frac{h^2}{2}u(k-1)$$

$$= y(k-1) + h\Big(x_2(k) - hu(k)\Big) + \frac{h^2}{2}u(k-1)$$

Solving these equations with respect to x_1 and x_2 we get

$$x_1(k) = y(k)$$

$$x_2(k) = \frac{y(k) - y(k-1)}{h} + \frac{h}{2}u(k-1)$$

The first component x_1 is equal to the measured value and the second component x_2 is obtained by taking differences of the output and adding a fraction of the control signal.   ∎

Reconstruction Using a Dynamic System

Let n be the order of the system. The direct calculation we have just performed gives the state after at most n measurements of input-output pairs. The disadvantage of the method is that it may be sensitive to disturbances; the operations done on the data are typically to form differences, as illustrated in Example 4.7. It is therefore useful to have other alternatives that are less sensitive to noise.

Consider the system (4.23). Assume that the state x is to be approximated by the state $\hat{x}$ of the model

$$\hat{x}(k+1) = \Phi\hat{x}(k) + \Gamma u(k) \tag{4.27}$$

which has the same input as the system of (4.23). If the model is perfect in the sense that the elements of the matrices Φ and Γ are identical to those of the system (4.23) and if the initial conditions are the same, then the state $\hat{x}$ of the model of (4.27) will be identical to the state x of the true system in (4.23). If the initial conditions are different, then $\hat{x}$ will converge to x only if the system (4.27) is asymptotically stable.

Reconstruction by Eq. (4.27) gives the state as a function of past inputs. The reconstruction can be improved by also using the measured outputs. This can be done by introducing a feedback from the difference between the measured and estimated outputs, $y - C\hat{x}$. Hence

$$\hat{x}(k+1 \mid k) = \Phi\hat{x}(k \mid k-1) + \Gamma u(k) + K\Big(y(k) - C\hat{x}(k \mid k-1)\Big) \tag{4.28}$$

where K is a gain matrix. The notation $\hat{x}(k + 1 \mid k)$ is used to indicate that it is an estimate of $x(k + 1)$ based on measurements available at time k, that is, a one-step prediction. Notice that the feedback term $K[y(k) - C\hat{x}(k \mid k - 1)]$ gives no contribution if the output predicted by the model agrees exactly with the measurements. To determine the matrix K we introduce the reconstruction error

$$\tilde{x} = x - \hat{x} \tag{4.29}$$

Subtraction of (4.28) from (4.23) gives

$$\tilde{x}(k + 1 \mid k) = \Phi\tilde{x}(k \mid k - 1) - K\Big(y(k) - C\hat{x}(k \mid k - 1)\Big) = (\Phi - KC)\tilde{x}(k \mid k - 1) \tag{4.30}$$

Hence if K is chosen so that the system (4.30) is asymptotically stable, the error $\tilde{x}$ will always converge to zero. By introducing a feedback from the measurements in the reconstruction, it is thus possible to make the error go to zero even if the system of (4.23) is unstable. The system in (4.28) is called an *observer* for the system of (4.23) because it produces the state of the system from measurements of inputs and outputs. It now remains to find a suitable way to choose the matrix K so that the system (4.28) is stable. Given the matrices Φ and C, the problem is to find a matrix K such that the matrix $\Phi - KC$ has prescribed eigenvalues. Because a matrix and its transpose have the same eigenvalues, the problem is the same as finding a matrix K^T such that $\Phi^T - C^T K^T$ has prescribed eigenvalues. However, this problem was solved in Sec. 4.3 in connection with the pole-placement problem; see Theorem 4.1. If those results are translated, we find then that the problem can be solved if the matrix

$$W_o^T = \Big(\begin{array}{cccc} C^T & \Phi^T C^T & \cdots & (\Phi^{n-1})^T C^T \end{array} \Big)$$

has full rank. Notice that W_o is the observability matrix for the system of (4.23). The result can be expressed by the following.

THEOREM 4.2   OBSERVER DYNAMICS   Consider the discrete system given by Eq. (4.23). Let $P(z)$ be a polynomial of degree n, where n is the order of the system. Assuming that the system is completely observable, then there exists a matrix K such that the matrix $\Phi - KC$ of the observer (4.28) has the characteristic polynomial $P(z)$.   ∎

Computing the Observer Gain

The determination of the matrix K in the observer (4.28) is the same mathematical problem as the problem of determining the feedback matrix L in the pole-placement problem. The practical aspects are also closely related. The selection of the observer poles is a compromise between sensitivity to measurement errors and rapid recovery of initial errors. A fast observer will converge quickly, but it will also be sensitive to measurement errors.

Determining the matrix K is the dual of finding the gain matrix L for pole placement by state feedback. This problem is solved by Ackermann's formula, Theorem 4.1. By using the relations

$$L \to K^T \qquad W_c \to W_o^T \qquad \Phi \to \Phi^T$$

it follows from Equation (4.13) that K is given by

$$K^T = \begin{pmatrix} 0 & \cdots & 0 & 1 \end{pmatrix} \left(W_o^T \right)^{-1} P(\Phi^T)$$

or

$$K = P(\Phi) W_o^{-1} \begin{pmatrix} 0 & \cdots & 0 & 1 \end{pmatrix}^T \tag{4.31}$$

The characteristic polynomial of $\Phi - KC$ is then $P(z)$. The duality with pole placement also implies that K is especially simple to determine if the system is in observable form.

Notice, however, that Ackermann's formula is poorly conditioned numerically. The MATLAB® procedure place is based on better numerical methods. This procedure will also give the observer gain for systems with many measurements.

A Deadbeat Observer

If the observer gain K is chosen so that the matrix $\Phi - KC$ has all eigenvalues zero, the observer is called a *deadbeat observer*. This observer has the property that the observer error goes to zero in finite time, actually in at most n steps, where n is the order of the system. The deadbeat observer is equivalent to the observer given by Eq. (4.25), which was obtained by a direct calculation of the state variables.

We illustrate design of an observer by determining an observer for the double integrator.

Example 4.8   Full-order observer for the double integrator

Consider a double-integrator plant. The matrix $\Phi - KC$ is given by

$$\Phi_o = \Phi - KC = \begin{pmatrix} 1 & h \\ 0 & 1 \end{pmatrix} - \begin{pmatrix} k_1 \\ k_2 \end{pmatrix} \begin{pmatrix} 1 & 0 \end{pmatrix} = \begin{pmatrix} 1 - k_1 & h \\ -k_2 & 1 \end{pmatrix}$$

Thus the characteristic equation is given by

$$z^2 - (2 - k_1)z + 1 - k_1 + k_2 h = 0$$

Let the desired characteristic equation be

$$z^2 + p_1 z + p_2 = 0$$

The following equations are obtained:

$$2 - k_1 = -p_1$$
$$1 - k_1 + k_2 h = p_2$$

These linear equations give

$$k_1 = 2 + p_1$$
$$k_2 = (1 + p_1 + p_2)/h$$

The deadbeat observer is obtained by setting $p_1 = p_2 = 0$. This gives

$$k_1 = 2$$
$$k_2 = 1/h$$

and the observer becomes

$$\hat{x}_1(k + 1) = \hat{x}_1(k) + h\hat{x}_2(k) + 2\Big(y(k) - \hat{x}_1(k)\Big)$$
$$\hat{x}_2(k + 1) = h\hat{x}_2(k) + \frac{1}{h}\Big(y(k) - \hat{x}_1(k)\Big)$$

Straightforward calculations give

$$\hat{x}_1(k + 1) = 2y(k) - y(k - 1)$$
$$\hat{x}_2(k + 1) = \frac{y(k) - y(k - 1)}{h}$$

$\blacksquare$

An Alternative Observer

There are many variations of the observer given by Eq. (4.28). The observer has a delay, because $\hat{x}(k \mid k - 1)$ depends only on measurements up to time $k - 1$. The following observer can be used to avoid the delay:

$$\hat{x}(k \mid k) = \Phi\hat{x}(k - 1 \mid k - 1) + \Gamma u(k - 1)$$
$$+ K\Big[y(k) - C\Big(\Phi\hat{x}(k - 1 \mid k - 1) + \Gamma u(k - 1)\Big)\Big] \qquad (4.32)$$
$$= (I - KC)\Big(\Phi\hat{x}(k - 1 \mid k - 1) + \Gamma u(k - 1)\Big) + Ky(k)$$

The reconstruction error when using this observer is given by

$$\tilde{x}(k \mid k) = x(k) - \hat{x}(k \mid k) = (\Phi - KC\Phi)\tilde{x}(k - 1 \mid k - 1)$$

This equation is similar to (4.30), and from the definition of the observability matrix W_o, it is found that the pair $(\Phi, C\Phi)$ is detectable if the pair (Φ, C) is detectable. This implies that $\Phi - KC\Phi$ can be given arbitrary eigenvalues by selecting K. Further,

$$y(k) - C\hat{x}(k \mid k) = C\tilde{x}(k \mid k) = (I - CK)C\Phi\tilde{x}(k - 1 \mid k - 1)$$

If the system has p outputs, then $I - CK$ is a $p \times p$ matrix; K may be chosen such that $CK = I$ if rank $(C) = p$. This implies that $C\hat{x}(k \mid k) = y(k)$, which means that the output of the system is estimated without error. This will make it possible to eliminate p equations from (4.32), and the order of the observer will be reduced. Reduced-order observers of this type are called *Luenberger observers*.

Example 4.9   Reduced-order observer for the double integrator

The observer (4.32) applied to the double integrator gives the equations

$$\hat{x}(k \mid k) = \begin{pmatrix} 1 - k_1 & h(1 - k_1) \\ -k_2 & 1 - hk_2 \end{pmatrix} \hat{x}(k - 1 \mid k - 1)$$
$$+ \begin{pmatrix} (1 - k_1)h^2/2 \\ h(1 - hk_2/2) \end{pmatrix} u(k - 1) + \begin{pmatrix} k_1 \\ k_2 \end{pmatrix} y(k)$$

If $I - CK = 0$—that is, if $k_1 = 1$—then the first equation is reduced to

$$\hat{x}_1(k \mid k) = y(k)$$

The reduced-order observer is now given by the second equation, which can be simplified to

$$\hat{x}_2(k \mid k) = (1 - hk_2)\hat{x}_2(k - 1 \mid k - 1)$$
$$+ k_2 \Big(y(k) - y(k - 1) \Big) + h(1 - hk_2/2)u(k - 1)$$

By choosing k_2, the reduced-order observer can be given an arbitrary eigenvalue. For instance, if $k_2 = 1/h$, the deadbeat response, then the same result is obtained as when making the direct calculation in Example 4.7.   ∎

4.5 Output Feedback

In Sec. 4.3 the pole-placement problem was solved in the special case when all state variables are measured directly. In Sec. 4.4 the problem of finding the states from the system output was solved. It is now natural to combine the results of these sections to obtain a solution to the pole-placement problem for the case of output feedback. Let the system be described by

$$x(k + 1) = \Phi x(k) + \Gamma u(k)$$
$$y(k) = Cx(k) \tag{4.33}$$

A linear feedback law relating u to y such that the closed-loop system has given poles is desired. The disturbances are first assumed to be impulses or equivalently unknown initial states.

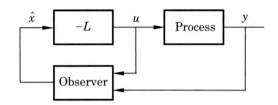

Figure 4.6 Block diagram of a controller obtained by combining state feedback with an observer.

The admissible control law is such that $u(k)$ is a function of $y(k - 1)$, $y(k - 2)$, ..., $u(k - 1)$, $u(k - 2)$,.... If all state variables are measured, it is shown in Sec. 4.3 that the feedback

$$u(k) = -Lx(k)$$

gives the desired poles. When the state cannot be measured, it seems intuitively reasonable to use the control law

$$u(k) = -L\hat{x}(k \mid k - 1) \tag{4.34}$$

where $\hat{x}$ is obtained from the observer

$$\hat{x}(k + 1 \mid k) = \Phi\hat{x}(k \mid k - 1) + \Gamma u(k) + K\Big(y(k) - C\hat{x}(k \mid k - 1)\Big) \tag{4.35}$$

Thus the feedback is a dynamic system of order n. Notice that the dynamics are due to the dynamics of the observer. A block diagram of the feedback is shown in Fig. 4.6.

Analysis of the Closed-Loop System

The closed-loop system has desirable properties. To show this, introduce

$$\tilde{x} = x - \hat{x}$$

It follows from Eqs. (4.33) and (4.34) that the closed-loop system can be described by the equations

$$\begin{aligned} x(k + 1) &= (\Phi - \Gamma L)x(k) + \Gamma L\tilde{x}(k \mid k - 1) \\ \tilde{x}(k + 1 \mid k) &= (\Phi - KC)\tilde{x}(k \mid k - 1) \end{aligned} \tag{4.36}$$

The closed-loop system has order $2n$. The eigenvalues of the closed-loop system are the eigenvalues of the matrices $\Phi - \Gamma L$ and $\Phi - KC$. Notice that the eigenvalues of $\Phi - \Gamma L$ are the desired closed-loop poles obtained by solving the pole-placement problem in Sec. 4.3 and the eigenvalues of $\Phi - KC$ are the poles of the observer given in Sec. 4.4.

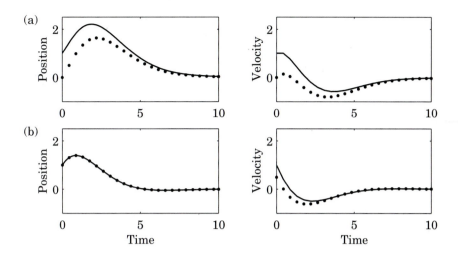

Figure 4.7 Control of the double-integrator plant using estimated states. The states and the estimated states are shown for a (a) second-order observer; (b) reduced-order observer.

This solution to the pole-placement problem has many nice symmetries. The solution to the state feedback and the observer are dual problems. The same numerical algorithm can be used to find the feedback gain L and the observer gain K. It is also attractive that the solution of the full problem can be split into two smaller problems. The separation of the problem is very useful. It also justifies that the closed-loop poles are separated into two groups, one is associated with the state feedback and the other with the observer.

Notice that the observer contains a model of the process internally. This is a special case of the *internal-model principle*, which says that a good controller contains a model of the controlled system.

The controller can also be viewed as a black box that generates the control signal from the process output. The controller described by (4.34) and (4.35) can be represented by the nth-order pulse-transfer function from measured output y to control signal u:

$$H_c(z) = -L(zI - \Phi + \Gamma L + KC)^{-1}K \tag{4.37}$$

Example 4.10   Output feedback of the double integrator

Consider the double-integrator plant. Assume that the feedback vector L is determined as in Examples 4.2 and 4.4 with the closed-loop natural frequency $\omega = 1$, the damping $\zeta = 0.7$, and $h = 0.44$. This gives $L = [0.73, \quad 1.21]$. First assume that the observer is designed as in Example 4.8 with the poles of the observer in $z = 0.75$. Figure 4.7(a) shows the true and the estimated states when the estimated states are used in the control law. Figure 4.7(b) shows the states in full lines and the estimate of the second state in dots when the reduced-order observer in Example 4.9 is used. The observer pole is in $z = 0.75$. ∎

Extensions

The problem discussed can be extended in many different directions. The controller given by Eq. (4.34) has a time delay of one sampling period. The reason for this is that the feedback is based on an observer that gives $\hat{x}(k \mid k-1)$. It is possible to obtain a controller without extra delays by using instead the control law

$$u(k) = -L\hat{x}(k \mid k) \tag{4.38}$$

where $\hat{x}(k \mid k)$ is obtained from the observer given by Eq. (4.32). The properties of the system obtained is analogous to the case that has just been investigated, so the details are left as exercises. See the problem section at the end of this chapter. Notice, however, that the feedback matrix L does not have to be changed when we change the observer. This is a very nice consequence of the separation of the problem into a state feedback and an observer.

More Realistic Disturbance Models

The controller based on a state feedback and an observer is interesting but it is still not very useful in practice. The reason is that the assumption about the disturbances made in Sec. 4.3 has been too simplistic. To generalize the problem it is therefore assumed that the system is described by

$$\frac{dx}{dt} = Ax + Bu + v$$
$$y = Cx$$

where v is a disturbance acting on the process. The disturbance v, which typically has much energy at low frequencies, is modeled as

$$\frac{dw}{dt} = A_w w$$
$$v = C_w w$$

The matrix A_w typically has eigenvalues at the origin or on the imaginary axis. We now introduce the augmented state vector

$$z = \begin{pmatrix} x \\ w \end{pmatrix}$$

The augmented system can be described by

$$\frac{d}{dt} \begin{pmatrix} x \\ w \end{pmatrix} = \begin{pmatrix} A & C_w \\ 0 & A_w \end{pmatrix} \begin{pmatrix} x \\ w \end{pmatrix} + \begin{pmatrix} B \\ 0 \end{pmatrix} u$$
$$y = \begin{pmatrix} C & 0 \end{pmatrix} \begin{pmatrix} x \\ w \end{pmatrix} \tag{4.39}$$

Compare with (4.20). Sampling this system gives the following discrete-time system:

$$\begin{pmatrix} x(k+1) \\ w(k+1) \end{pmatrix} = \begin{pmatrix} \Phi & \Phi_{xw} \\ 0 & \Phi_w \end{pmatrix} \begin{pmatrix} x(k) \\ w(k) \end{pmatrix} + \begin{pmatrix} \Gamma \\ 0 \end{pmatrix} u(k)$$

$$y = \begin{pmatrix} C & 0 \end{pmatrix} \begin{pmatrix} x(k) \\ w(k) \end{pmatrix}$$

The disturbance states w are not reachable from the control signal but the complete state is observable from the output if the system (4.33) is observable. The control law is a linear feedback from all state variables, that is,

$$u(k) = -L\hat{x}(k) - L_w\hat{w}(k) \tag{4.40}$$

where $\hat{x}$ and $\hat{w}$ are obtained from the observer

$$\begin{pmatrix} \hat{x}(k+1) \\ \hat{w}(k+1) \end{pmatrix} = \begin{pmatrix} \Phi & \Phi_{xw} \\ 0 & \Phi_w \end{pmatrix} \begin{pmatrix} \hat{x}(k) \\ \hat{w}(k) \end{pmatrix} + \begin{pmatrix} \Gamma \\ 0 \end{pmatrix} u(k) + \begin{pmatrix} K \\ K_w \end{pmatrix} \varepsilon(k) \tag{4.41}$$

and

$$\varepsilon(k) = y(k) - C\hat{x}(k) \tag{4.42}$$

Notice that the state of the observer is composed of estimates of the states of the process and the disturbances and that the control signal contains a feedback from the estimated disturbance state $\hat{w}$.

The closed-loop system is described by

$$\begin{aligned} x(k+1) &= (\Phi - \Gamma L)x(k) + (\Phi_{xw} - \Gamma L_w)w - \Gamma L\tilde{x}(k) - \Gamma L_w\tilde{w} \\ w(k+1) &= \Phi_w w(k) \\ \tilde{x}(k+1) &= (\Phi - KC)\tilde{x}(k) + \Phi_{xw}\tilde{w}(k) \\ \tilde{w}(k+1) &= \Phi_w\tilde{w}(k) - K_wC\tilde{x}(k) \end{aligned} \tag{4.43}$$

Notice that the disturbance state w is observable but not reachable. The equations for the closed-loop system give useful insight into the behavior of the system. The matrix L ensures that the state x goes to zero at the desired rate after a disturbance. A proper choice of the gain L_w reduces the effect of the disturbance v on the system by feedforward from the estimated disturbances $\hat{w}$. This feedforward control action is particularly effective if the matrix $\Phi_{xw} - \Gamma L_w$ can be made equal to zero. The observer gains K and K_w influence the rate at which the estimation errors go to zero. A block diagram of the system is shown in Fig. 4.8.

Integral Action

The special case of a constant but unknown disturbance acting on the process input is very common. It leads to a solution where the controller has integral

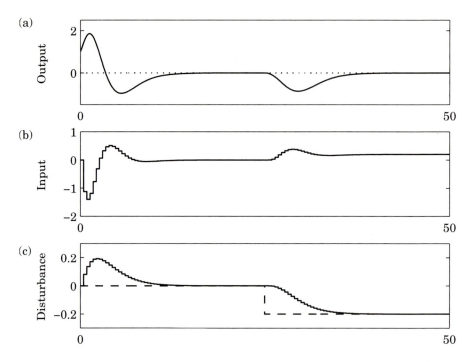

Figure 4.10 Simulation of the system in Example 4.11. (a) Output y, (b) input u, and (c) disturbance v (dashed) and estimated disturbance $\hat{v}$ (solid).

A Naive Approach

A simple way to obtain the desired response to command signals is to replace the regular state feedback $u(k) = -L\hat{x}(k)$ by

$$u(k) = -L\hat{x}(k) + L_c u_c(k) \tag{4.47}$$

where u_c is the command signal. To investigate the response of such a controller we consider the closed-loop system that is described by

$$
\begin{aligned}
x(k+1) &= \Phi x(k) + \Gamma u(k) \\
y(k) &= C x(k) \\
\hat{x}(k+1) &= \Phi \hat{x}(k) + \Gamma u(k) + K\Big(y(k) - C\hat{x}(k)\Big) \\
u(k) &= -L\hat{x}(k) + L_c u_c(k)
\end{aligned}
\tag{4.48}
$$

A block diagram of the system is shown in Fig. 4.11. Eliminating u and introducing the estimation error $\tilde{x} = x - \hat{x}$ we find that the closed-loop system can

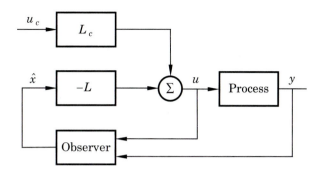

Figure 4.11   Block diagram that shows a simple way of introducing command signals in a controller with state feedback and an observer.

be described by

$$x(k + 1) = (\Phi - \Gamma L)x(k) + \Gamma L\tilde{x}(k) + \Gamma L_c u_c(k)$$
$$\tilde{x}(k + 1) = (\Phi - KC)\tilde{x}(k) \qquad\qquad\qquad (4.49)$$
$$y(k) = Cx(k)$$

Notice that the observer error is not reachable from u_c. This makes sense because it would be highly undesirable to introduce command signals in such a way that they will cause observer errors.

It follows from Eq. (4.49) that the pulse transfer from the command signal to the process output is given by

$$H_{cl}(z) = C(zI - \Phi - \Gamma L)^{-1}\Gamma L_c = L_c \frac{B(z)}{A_m(z)} \qquad\qquad (4.50)$$

This can be compared with the pulse-transfer function of the process

$$H(z) = C(zI - \Phi)^{-1}\Gamma = \frac{B(z)}{A(z)} \qquad\qquad\qquad (4.51)$$

The fact that the polynomial $B(z)$ appears in the numerator of both transfer functions can be seen by transforming both systems to reachable canonical form. Compare with the derivation of Ackermann's formula given by Eq. (4.14).

The closed-loop system obtained with the control law given by Eq. (4.47) has the same zeros as the plant and its poles are the eigenvalues of the matrix $\Phi - \Gamma L$. From the previous discussion we have found that the rejection of disturbances are also influenced by L. Sometimes it is desirable to have a controller where disturbance rejection and command signal response are totally independent. To obtain this we will use a more general controller structure that is discussed later. Before doing this we will show how to introduce integral action in the controller (4.47).

Integral Action

To obtain a controller with integral action we use the same idea as in Sec. 4.5 and introduce a constant disturbance v at the process input. The controller then becomes

$$u(k) = -L\hat{x}(k) - \hat{v}(k) + L_c u_c(k)$$
$$\hat{x}(k+1) = \Phi\hat{x}(k) + \Gamma\Big(\hat{v}(k) + u(k)\Big) + K\Big(y(k) - C\hat{x}(k)\Big) \qquad (4.52)$$
$$\hat{v}(k+1) = \hat{v}(k) + K_w\Big(y(k) - C\hat{x}(k)\Big)$$

These equations can also be written as

$$u(k) = -L\hat{x}(k) - \hat{v}(k) + L_c u_c(k)$$
$$\hat{x}(k+1) = (\Phi - \Gamma L)\hat{x}(k) + \Gamma L_c u_c(k) + K\Big(y(k) - C\hat{x}(k)\Big) \qquad (4.53)$$
$$\hat{v}(k+1) = \hat{v}(k) + K_w\Big(y(k) - C\hat{x}(k)\Big)$$

A comparison with Eq. (4.44) shows that command signal following is obtained by a very simple modification of the systems discussed previously.

A Two-Degree-of-Freedom Controller

Practical control systems often have specifications that involve both servo and regulation properties. This is traditionally solved using a two-degree-of-freedom structure, as shown in Fig. 4.12. Compare with Fig. 3.10. This configuration has the advantage that the servo and regulation problems are separated. The feedback controller H_{fb} is designed to obtain a closed-loop system that is insensitive to process disturbances, measurement noise, and process uncertainties. The feedforward compensator H_{ff} is then designed to obtain the desired servo properties. We will now show how to solve the servo problem in the context of state feedback.

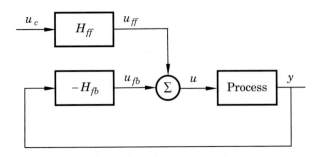

Figure 4.12 Block diagram of a feedback system with a two-degree-of-freedom structure.

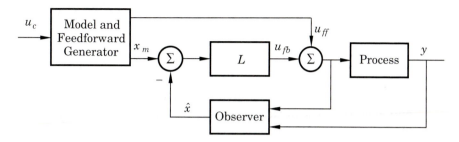

Figure 4.13 A two-degree-of-freedom controller based on state feedback and an observer.

A Controller Structure

In a state-space design it is natural to assume that servo performance is specified in terms of a model that gives the desired response of the output or the state variables to changes in the command signal. This can be specified with the model

$$x_m(k+1) = \Phi_m x_m(k) + \Gamma_m u_c(k)$$
$$y_m(k) = C_m x_m(k)$$

$$(4.54)$$

It is then natural to use the control law

$$u(k) = L\Big(x_m(k) - \hat{x}(k)\Big) + u_{ff}(k)$$

$$(4.55)$$

where x_m is the desired state, and u_{ff} is a control signal that gives the desired output when applied to the open-loop system. The coordinates must be chosen so that the states of the system and the model are compatible. In actual applications it is often useful to choose them so that the components of the state have good physical interpretations.

The term $u_{fb} = L(x_m - \hat{x})$ represents the feedback and u_{ff} represents the feedforward signal. Equation (4.55) has a good physical interpretation. The feedforward signal u_{ff} will ideally produce the desired time variation in the process state. If the estimated process state $\hat{x}$ equals the desired state x_m, the feedback signal $L(x_m - \hat{x})$ is zero. If there is a difference between $\hat{x}$ and x_m, the feedback will generate corrective actions. The feedback term can be viewed as a generalization of error feedback in ordinary control systems, because the error represents deviations of all state variables and not just the output errors. A block diagram of the system is shown in Fig. 4.13.

Generation of the Feedforward Signal

Given the model (4.54) it is straightforward to generate the desired states. It remains to discuss generation of the signal u_{ff}. Let the pulse-transfer functions

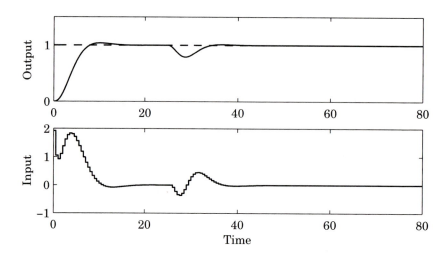

Figure 4.21   The same as Fig. 4.20, but using state feedback from observed states when $\alpha_0 = 2$.

effect of model uncertainty. This will be discussed in the next chapter. Notice that there is no integrator in the controller. The steady-state error will be zero even in the presence of a disturbance because the process dynamics has an integrator.

4.8 Conclusions

The chapter shows how the regulator and servo design problems can be solved using pole placement and observers. The solution has three major components: the feedback matrix L, the observer, and the response model. The feedback matrix L is chosen in such a way that load disturbances decay properly using the techniques discussed in Sec. 4.3. The observer is designed by considering the load disturbances and the measurement noise, as discussed in Sec. 4.5. The major trade-off is between quick convergence and sensitivity to measurement errors. The regulation properties are taken care of by the matrix L and the observer. The response model and the inverse process model are then chosen to obtain the desired servo performance in response to command signals.

The pole-placement design is done here for the single-input–single-output case. With n parameters in the state-feedback vector, it is possible to place n poles arbitrarily, if the system is reachable. In the multivariable case, there are more degrees of freedom. This makes it possible to determine not only the poles, but also some eigenvectors of the closed-loop system. Further details can be found in the references.

4.9 Problems

4.1 A general second-order discrete-time system can be written as

$$x(k+1) = \begin{pmatrix} a_{11} & a_{12} \\ a_{21} & a_{22} \end{pmatrix} x(k) + \begin{pmatrix} b_1 \\ b_2 \end{pmatrix} u(k)$$

$$y(k) = \begin{pmatrix} c_1 & c_2 \end{pmatrix} x(k)$$

Determine a state-feedback controller of the form

$$u(k) = -Lx(k)$$

such that the characteristic equation of the closed-loop system is

$$z^2 + p_1 z + p_2 = 0$$

Use the result to verify the deadbeat controller for the double integrator given in Example 4.5.

4.2 Given the system

$$x(k+1) = \begin{pmatrix} 1.0 & 0.1 \\ 0.5 & 0.1 \end{pmatrix} x(k) + \begin{pmatrix} 1 \\ 0 \end{pmatrix} u(k)$$

$$y(k) = \begin{pmatrix} 1 & 1 \end{pmatrix} x(k)$$

Determine a linear state-feedback controller

$$u(k) = -Lx(k)$$

such that the closed-loop poles are in 0.1 and 0.25.

4.3 Determine the deadbeat controller for the normalized motor in Example A.2. Assume that $x(0) = \begin{bmatrix} 1 & 1 \end{bmatrix}^T$. Determine the sample interval such that the control signal is less than one in magnitude. It can be assumed that the maximum value of $u(k)$ is at $k = 0$.

4.4 Consider the continuous system.

$$\frac{dx}{dt} = \begin{pmatrix} -3 & 1 \\ 0 & -2 \end{pmatrix} x + \begin{pmatrix} 0 \\ 1 \end{pmatrix} u$$

$$y = \begin{pmatrix} 1 & 0 \end{pmatrix} x$$

Sampling the system with $h = 0.2$ gives

$$x(k+1) = \begin{pmatrix} 0.55 & 0.12 \\ 0 & 0.67 \end{pmatrix} x(k) + \begin{pmatrix} 0.01 \\ 0.16 \end{pmatrix} u(k)$$

(a) Determine a state-feedback control law such that the closed-loop characteristic polynomial is

$$z^2 - 0.63z + 0.21$$

(b) Determine the corresponding continuous-time characteristic polynomial and discuss the choice of the sampling period.

(c) Simulate the closed-loop system when $x(0) = [1\ 0]^T$.

4.5 The system

$$x(k + 1) = \begin{pmatrix} 0.78 & 0 \\ 0.22 & 0 \end{pmatrix} x(k) + \begin{pmatrix} 0.22 \\ 0.03 \end{pmatrix} u(k)$$

$$y(k) = \begin{pmatrix} 0 & 1 \end{pmatrix} x(k)$$

represents the normalized motor for the sampling interval $h = 0.25$. Determine observers for the state based on the output by using each of the following.

(a) Direct calculation using (4.25).

(b) A dynamic system that gives $\hat{x}(k + 1 \mid k)$ using (4.28).

(c) The reduced-order observer.

Let the observer be of the deadbeat type; that is, the poles of the observer should be in the origin.

4.6 Determine the full-state observer based on (4.28) for the tank system in Problem 2.10. Choose the observer gain such that the observer is twice as fast as the open-loop system.

4.7 Consider the observer of (4.32) and let the control law be given by

$$u(k) = -L\hat{x}(k \mid k)$$

Show that the resulting controller can be written as

$$w(k + 1) = \Phi_o w(k) + \Gamma_o y(k)$$
$$u(k) = C_o w(k) + D_o y(k)$$

where

$$\Phi_o = (I - KC)(\Phi - \Gamma L) \qquad \Gamma_o = (I - KC)(\Phi - \Gamma L)K$$
$$C_o = -L \qquad\qquad\qquad D_o = -LK$$

4.8 Given the discrete-time system

$$x(k + 1) = \begin{pmatrix} 0.5 & 1 \\ 0.5 & 0.7 \end{pmatrix} x(k) + \begin{pmatrix} 0.2 \\ 0.1 \end{pmatrix} u(k) + \begin{pmatrix} 1 \\ 0 \end{pmatrix} v(k)$$

$$y(k) = \begin{pmatrix} 1 & 0 \end{pmatrix} x(k)$$

where v is a constant disturbance. Determine controllers such that the influence of v can be eliminated in steady state in each case.

(a) The state and v can be measured.

(b) The state can be measured.

(c) Only the output can be measured.

4.9 Consider the two-tank system in Problem 2.10 for $h = 12$ s.

(a) Determine a state-feedback controller such that the closed-loop poles are given by the characteristic equation

$$z^2 - 1.55z + 0.64 = 0$$

This corresponds to $\zeta = 0.7$ and $\omega = 0.027$ rad/s.

(b) Introduce a command signal and determine a controller such that the steady-state error between the command signal and the output is zero in steady state; that is, introduce an integrator in the system.

(c) Simulate the system using the regulators in (a) and (b).

4.10 Consider the double integrator with a load disturbance acting on the process input. The disturbance can be described as a sinusoid with frequency ω_0, but with unknown amplitude and phase. Design a state-feedback controller and an observer such that there is no steady-state error due to the sinusoidal perturbation.

4.11 Consider the discrete-time process

$$x(k + 1) = \begin{pmatrix} 0.9 & 0 \\ 1 & 0.7 \end{pmatrix} x(k) + \begin{pmatrix} 1 \\ 0 \end{pmatrix} u(k)$$

$$y(k) = \begin{pmatrix} 0 & 1 \end{pmatrix} x(k)$$

(a) Determine a state deadbeat controller that gives unit static gain, that is, determine L_c and L in the controller

$$u(k) = L_c u_c(k) - Lx(k)$$

(b) Determine the stability range for the parameters in L, that is, use the controller from (a) and determine how much the other parameters may change before the closed-loop system becomes unstable.

4.12 Consider the system

$$x(k + 1) = \begin{pmatrix} 0.25 & 0.5 \\ 1 & 2 \end{pmatrix} x(k) + \begin{pmatrix} 1 \\ 4 \end{pmatrix} u(k)$$

$$y = \begin{pmatrix} 1 & 0 \end{pmatrix} x(k)$$

(a) Determine the state-feedback controller $u(k) = L_c u_c(k) - Lx(k)$ such that the states are brought to the origin in two sampling intervals.

(b) Is it possible to determine a state-feedback controller that can take the system from the origin to $x(k) = [2 \quad 8]^T$?

(c) Determine an observer that estimates the state such that the estimation error decreases as $p(k) \cdot 0.2^k$.

4.10 Notes and References

Pole placement was one of the first applications of the state-space approach. One of the first to solve the problem was J. Bertram in 1959. The first published solution is given in Rissanen (1960). Treatment of the multivariable case of pole placement can be found, for instance, in Rosenbrock (1970), Wolowich (1974), and Kailath (1980). Observers are also described in the preceding books. The reduced-order observer was first described in a Ph.D. thesis by Luenberger. Easier available references are Luenberger (1964, 1971).

The servo problem and introduction of reference values are discussed in Wittenmark (1985a). Numerical aspects of computing the state feedback and the observer gain are discussed in Miminis and Paige (1982), Petkov, Christov, and Konstantinov (1984), and Miminis and Paige (1988).

5

Pole-Placement Design:
A Polynomial Approach

5.1 Introduction

In this chapter we will discuss the same design problems as in Chapter 4 but we will use polynomial calculations instead of matrix calculations. This gives new insights and new computational methods. In addition we will be able to investigate consequences of errors in the model used to design the controller. The idea of pole placement is to find a controller that gives a closed-loop system with a specified characteristic polynomial. It is natural to explore if this can be done directly by polynomial calculations.

We start by describing a process model and a controller as input-output systems characterized by rational transfer functions. The design problem is then solved in a simple setting in Sec. 5.2. The design problem is identical to the one posed in Secs. 4.2 and 4.5. A polynomial equation is a crucial part of the solution. This equation is investigated in Sec. 5.3, where we give conditions for solvability and algorithms. In Sec. 5.4 we solve more realistic design problems. We consider cancellation of poles and zeros, separation of command signal responses and disturbance responses, and improved responses to disturbances. In Sec. 5.5 we consider the problem of modeling errors, which is much more convenient to deal with in the input-output formulation than in the state-space formulation. In Sec. 5.6 we summarize results and obtain a general design procedure. Some practical aspects are also discussed in that section.

The chapter ends with several design examples that illustrate the procedure. Control of a double integrator is discussed in Sec. 5.7, an harmonic oscillator in Sec. 5.8, and a flexible robot arm in Sec. 5.9. Many other design procedures can be expressed in terms of pole placement. This gives insight and gives a unified view, as is illustrated in Sec. 5.10.

This implies that command signals are introduced in such a way that they do not generate observer errors. Hence

$$T(z) = t_0 A_o(z) \tag{5.7}$$

The response to command signals is then given by

$$Y(z) = \frac{t_0 B(z)}{A_c(z)} U_c(z) \tag{5.8}$$

where the parameter t_0 is chosen to obtain the desired static gain of the system. For example, to have unit gain we have $t_0 = A_c(1)/B(1)$.

Summary

We have thus obtained the following design procedure.

ALGORITHM 5.1   SIMPLE POLE-PLACEMENT DESIGN

Data: A process model is specified by the pulse-transfer function $B(z)/A(z)$, where $A(z)$ and $B(z)$ do not have any common factors. Specifications are given in terms of a desired closed-loop characteristic polynomial $A_{cl}(z)$.

Step 1. Find polynomials $R(z)$ and $S(z)$, such that $\deg S(z) \le \deg R(z)$, which satisfy the equation

$$A(z)R(z) + B(z)S(z) = A_{cl}(z)$$

Step 2. Factor the closed-loop characteristic polynomial as $A_{cl}(z) = A_c(z)A_o(z)$, where $\deg A_o(z) \le \deg R(z)$, and choose

$$T(z) = t_0 A_o(z)$$

where $t_0 = A_c(1)/B(1)$. The control law is

$$R(q)u(k) = T(q)u_c(k) - S(q)y(k)$$

and the response to command signals is given by

$$A_c(q)y(k) = t_0 B(q)u_c(k)$$

                                                                ∎

There are several details that have to be investigated. The most important is the solution of the Diophantine equation (5.4). Before doing this we will, however, consider an example.

Example 5.1   Control of a double integrator

For the double integrator we have

$$A(z) = (z - 1)^2$$

$$B(z) = \frac{h^2}{2} (z + 1)$$

and the Diophantine equation (5.4) becomes

$$(z^2 - 2z + 1)R(z) + \frac{h^2}{2} (z + 1)S(z) = A_{cl}(z)$$

The closed-loop characteristic polynomial A_{cl} is a design parameter. Both its degree and its parameters will be selected to achieve the design goals. It is natural to look for as simple controllers as possible. This means that we will search for polynomials $R(z)$ and $S(z)$ of the lowest order that satisfies the Diophantine equation. The simplest case is $R(z) = 1$ and $S(z) = s_0$, that is, a proportional controller. This gives the equation

$$z^2 - 2z + 1 + \frac{s_0 h^2}{2} (z + 1) = A_{cl}(z)$$

which cannot be solved for an arbitrary $A_{cl}(z)$ of second order. With a first-order controller we have $R(z) = z + r_1$ and $S(z) = s_0 z + s_1$, which gives

$$(z^2 - 2z + 1)(z + r_1) + \frac{h^2}{2} (z + 1)(s_0 z + s_1) = A_{cl}(z)$$

Hence

$$z^3 + \left(r_1 + \frac{h^2}{2} s_0 - 2\right)z^2 + \left(1 - 2r_1 + \frac{h^2}{2} (s_0 + s_1)\right)z + r_1 + s_1 \frac{h^2}{2} = A_{cl}(z)$$

and we find that it is possible to select the controller coefficients r_1, s_0, and s_1 to obtain an arbitrary polynomial $A_{cl}(z)$ of third degree. Choosing

$$A_{cl}(z) = z^3 + p_1 z^2 + p_2 z + p_3$$

and identifying coefficients of powers of equal degree we find that

$$r_1 + \frac{h^2}{2} s_0 = p_1 + 2$$

$$-2r_1 + \frac{h^2}{2} (s_0 + s_1) = p_2 - 1$$

$$r_1 + s_1 \frac{h^2}{2} = p_3$$

This equation has the solution

$$r_1 = \frac{3 + p_1 + p_2 - p_3}{4}$$

$$s_0 = \frac{5 + 3p_1 + p_2 - p_3}{2h^2}$$

$$s_1 = -\frac{3 + p_1 - p_2 - 3p_3}{2h^2}$$

roots inside the unit disc. Because a process pole that is canceled must be a controller zero and vice versa, it follows that the polynomials R, S, and T have the following structure:

$$R = B^+ \bar{R}$$
$$S = A^+ \bar{S} \tag{5.21}$$
$$T = A^+ \bar{T}$$

It follows from Eq. (5.4) that the characteristic polynomial of the closed-loop system is

$$A_{cl} = AR + BS = A^+ B^+ (A^- \bar{R} + B^- \bar{S}) = A^+ B^+ \bar{A}_{cl} \tag{5.22}$$

The polynomials A^+ and B^+, which are canceled, are thus factors of the closed-loop characteristic polynomial A_{cl}. It is natural to factor the characteristic polynomial as $A_{cl} = A_c A_o$, where

$$A_c = B^+ \bar{A}_c$$
$$A_o = A^+ \bar{A}_o \tag{5.23}$$

Cancelling the common factors in Eq. (5.22) we find that polynomials $\bar{R}$ and $\bar{S}$ satisfy

$$A^- \bar{R} + B^- \bar{S} = \bar{A}_{cl} = \bar{A}_c \bar{A}_o \tag{5.24}$$

The minimum-degree causal controller is obtained by choosing the unique solution with $\deg \bar{S} < \deg A^-$. The control (5.2) law can be written as

$$B^+ \bar{R} u = A^+ \bar{T} u_c - A^+ \bar{S} y$$

Hence

$$u = \frac{A^+}{B^+} \left(\frac{\bar{T}}{\bar{R}} u_c - \frac{\bar{S}}{\bar{R}} y \right)$$

This means that we simply cancel the poles and zeros of the process and design a controller for the reduced system as if the canceled poles were not present. Because $T = t_0 A_o$, the pulse-transfer function from the command signal to process output is

$$\frac{BT}{A_{cl}} = \frac{t_0 B^+ B^- A_o}{A_c A_o} = \frac{t_0 B^-}{\bar{A}_c}$$

The canceled factors must correspond to stable modes. If this is not the case the system will have unstable modes that are unreachable or unobservable. In

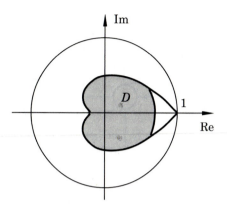

Figure 5.1 A region D such that points in the region have a minimum relative damping and a minimum absolute damping.

practice it is useful to have more stringent requirements on allowable cancellations. Sometimes cancellation may not be desirable at all. In other cases it may be reasonable to cancel zeros that are sufficiently well damped. One way to express this formally is to introduce a region D in the complex plane that corresponds to modes with sufficient relative and absolute damping. Only zeros inside D may be canceled. An example of a region is shown in Fig. 5.1. From Sec. 2.8 we saw that lines with constant relative damping are logarithmic spirals in the z-plane and that lines with constant absolute damping are circles.

Separation of Disturbance and Command Signal Response

In Sec. 4.6 we designed a controller where the response to command signals was completely separated from the response to disturbances. This is a nice property because it gives the designer much freedom. It is straightforward to obtain a similar controller using the polynomial approach. Let the factored model be described by $A = A^+ A^-$ and $B = B^+ B^-$, where A^+ and B^+ are the dynamics that will be canceled. Furthermore let the desired response to command signals be given by

$$y_m = H_m u_c = \frac{B_m}{A_m} u_c \tag{5.25}$$

To obtain perfect model following the polynomial B^- must be a factor of B_m, because B^- cannot be canceled. Hence $B_m = \bar{B}_m B^-$. By introducing

$$R = A_m B^+ \bar{R}$$
$$S = A_m A^+ \bar{S} \tag{5.26}$$
$$T = \bar{B}_m \bar{A}_o \bar{A}_c A^+$$

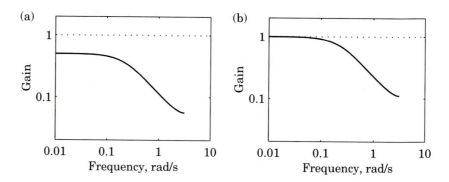

Figure 5.8 Bode diagrams (gain curves) for the transmission of (a) load disturbances and (b) measurement noise for the system in Example 5.9.

Example 5.9   Influence of the observer polynomial 1

Consider a system with the pulse-transfer function

$$H(z) = \frac{0.1}{z - 1}$$

Assume that the desired pulse-transfer function from command to output is given by

$$H_m(z) = \frac{0.2}{z - 0.8}$$

We have thus $A_c(z) = z - 0.8$. The observer polynomial can be chosen as $A_o = 1$, which gives the controller

$$u(k) = 2\big(u_c(k) - y(k)\big)$$

and the desired closed-loop transfer function. The process output is then given by (compare with Fig. 5.3)

$$X(z) = \frac{0.2}{z - 0.8}\, U_c(z) + \frac{0.1}{z - 0.8}\, V(z) - \frac{0.2}{z - 0.8}\, E(z)$$

The Bode diagrams for the transmission of load disturbances and measurement errors are shown in Fig. 5.8. The gain from low-frequency load disturbances is 0.5 and from low-frequency measurement disturbances is 1. High-frequency load and measurement disturbances are well attenuated.   ∎

Figure 5.8 shows that the proportional feedback gives a closed-loop system that is sensitive to load disturbances. A less-sensitive system can be obtained by introducing an observer polynomial of higher degree and constraints on the polynomial R, as shown in the next example.

loop-transfer

and to evalua
also useful to

5.7 Design of a C

In pole-place
whose zeros a
choices it is i
signals, load
rors. This is
integrator pl

Process Mo

Consider a p

where $k = 1$
a compact d

Specificatic

The propert
that the pol

Hence

$A_c(z)$

An observe
action. The
continuous

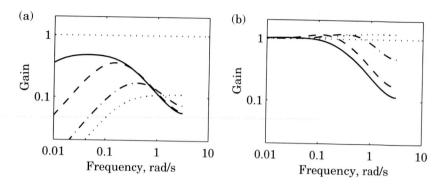

(a)   (b)

Gain   Gain

Frequency, rad/s   Frequency, rad/s

Figure 5.9   Bode diagram for the signal transmission from (a) load disturbances and (b) measurement error to process output for the controller
in Example 5.10. The observer polynomial has $a = 0.99$ (solid), $a = 0.9$
(dashed), $a = 0.5$ (dashed-dotted), and $a = 0$ (dotted).

Example 5.10   Influence of the observer polynomial 2

Consider the same system and the same desired closed-loop response as in Example 5.9. Let the observer polynomial be

$$A_o(z) = z - a$$

The Diophantine equation (5.4) becomes

$$(z - 1)(z + r_1) + 0.1(s_0 z + s_1) = (z - a)(z - 0.8)$$

Hence the following conditions are obtained.

$$-1 + r_1 + 0.1 s_0 = -a - 0.8$$
$$-r_1 + 0.1 s_1 = 0.8a$$

Because there are two linear equations and three unknowns, one extra condition
may be introduced. Choose $r_1 = -1$ to obtain integral action. This gives

$$s_0 = 12 - 10a$$
$$s_1 = 8a - 10$$

The following expression is obtained for the process output.

$$X(z) = \frac{0.2}{z - 0.8} U_c(z) + \frac{0.1(z - 1)}{(z - a)(z - 0.8)} V(z) - \frac{(1.2 - a)z - 1 + 0.8a}{(z - a)(z - 0.8)} E(z)$$

The Bode diagrams for the signal transmission from load disturbances and measurement errors to x are shown in Fig. 5.9; compare this with Fig. 5.8. By changing the observer dynamics the closed-loop system will be less sensitive for low-
frequency load disturbances. The figure shows that the effects of load disturbances
and measurement noise are strongly influenced by the observer polynomial A_o. A
fast observer ($a = 0$) gives a very good rejection of load disturbances, but much
measurement noise is also injected into the system. With a slow observer ($a = 0.9$),
much less measurement noise is injected into the system but the attenuation of
load disturbances is not so good.   ∎

Selection

The choice
design bas
given in th
relation to
poles must
sections. T
closed-loop
order syst
be very di

Validation

After a co
tem obtai
investigat
tivity to r
   In t
Eq. (5.33
Algorithn

(a) Use straightforward calculations to determine the influence of modeling errors. Assume that the design is made for $a = -0.9$ and determine the stability of the closed-loop system if the true process has a pole in a^0.

(b) Use Theorem 3.5 to determine the influence of modeling errors. What happens when α is decreased?

5.8 Consider the system in Problem 5.2. Use (5.52) to determine the maximum value of the control signal as a function of a and α when the command signal is a step.

5.9 A polynomial design for the normalized motor is given in Example 5.5 Simulate the system and investigate the sensitivity of the design method with respect to the choice of the sampling interval. Assume that the closed-loop specifications correspond to a second-order continuous-time system with damping $\zeta = 0.7$ and natural frequency $\omega = 1$ rad/s.

5.10 Consider the system described by

$$A_1(z)x(k) = B_1(z)u(k)$$
$$A_2(z)y(k) = B_2(z)x(k)$$

Assume that the variable to be controlled is $x(k)$, but that the measured variable is $y(k)$. Further assume that A_2 has its roots inside the unit disc. Derive a controller of the form (5.2) such that the closed-loop system is

$$A_c(z)x(z) = B_m(z)u_c(k)$$

What are the restrictions that have to be imposed? How will uncertainties in A_2 and B_2 influence the pulse-transfer function of the closed-loop system?

5.11 Consider the two-tank system in Problem 2.10 for $h = 12$ s.

Notice th
functions
responses
the respo
   Con
disturba

(a) Use polynomial methods to design a controller with an integrator. Assume that the desired closed-loop characteristic equation is

$$z^2 - 1.55z + 0.64 = 0$$

This corresponds to $\zeta = 0.7$ and $\omega = 0.027$ rad/s.

(b) Redesign the controller for different values of ω and study how the magnitude of the control signal varies with ω.

5.12 Consider the control of the normalized motor in Example A.2. Show that velocity feedback can be designed using pole-placement design. (*Hint:* First, design a feedback law with position feedback only. Show then that the control law can be rewritten as a combination of position and velocity feedback.)

5.13 Generalize the results in Problem 5.12 to a general process with several outputs.

5.14 Assume that the desired closed-loop system is given as the continuous-time model

This equ
that the
with Fig
given cor
sensitive
   To

$$G_m(s) = \frac{0.01}{s^2 + 0.14s + 0.01}$$

(a) Choose an appropriate sampling interval.

(b) Determine the corresponding discrete-time transfer operator. Sketch the singularity diagram for the continuous- and the discrete-time systems, respectively.

5.15 Assume that the process has the pulse-transfer operator

$$H(z) = \frac{0.4q + 0.3}{q^2 - 1.6q + 0.65}$$

Use pole placement to design a controller satisfying the following specifications:

- Static gain = 1
- Minimal degree of the observer polynomial
- Cancellation of process zero
- No integrator
- Desired characteristic polynomial

$$A_c = q^2 - 0.7q + 0.25$$

5.16 Consider the process and specifications in the previous problem. Redo the design under the assumption that the controller has an integrator.

5.17 Consider the system

$$H(z) = \frac{z}{(z - 1)(z - 2)}$$

Determine an error-feedback controller that places both poles in the origin, that is, use the controller

$$Ru(k) = -Sy(k) + Tu_c(k)$$

with $S = T$. Show by using the Diophantine equation that there is more than one causal controller that solves the problem. Assume that the observer poles are placed at the origin. Determine two controllers that fulfill the specifications, and determine the closed-loop zeros.

5.13 Notes and References

The polynomial approach for pole placement is treated in Wolowich (1974), Kučera (1979, 1991), and Pernebo (1981). The method discussed in this chapter has been used in connection with adaptive pole-placement algorithms, as in Åström and Wittenmark (1995). The Dahlin-Higham algorithm was derived independently in Dahlin (1968) and Higham (1968). The internal model principle is described in Morari and Zafiriou (1989), and Morari and Lee (1991). The Smith-predictor is introduced in Smith (1957) and the model algorithmic controller is discussed in Richalet et al. (1978).

Solution of the Diophantine equation is discussed in Blankenship (1963), Kučera (1979, 1991), and Ježek (1982). More about the Sylvester matrix can be found in Barnett (1971, 1983).

6

Design: An Overview

6.1 Introduction

This chapter views the control problem in a wider perspective. In practice, more time is often spent formulating control problems than on solving them. It is therefore useful to be aware of these more general problems, although they are seldom discussed in textbooks.

Most control problems arise from design of engineering systems. Such problems are typically large-scale and poorly defined. Typical tasks are design of power plants, chemical processes, rolling mills, industrial robots, aircraft, space vehicles, and biomedical systems. Control theory on the other hand deals with small-scale, well-defined problems. A typical problem is to design a feedback law for a given system, which is described by linear differential equations with constant coefficients, so that the closed-loop system has given poles.

A major difficulty in control-system design is to reconcile the large-scale, poorly defined, real problems with the simple, well-defined problems that control theory can handle. It is, however, in this intermediate area that a control engineer can use creativity and ingenuity effectively. This situation is not peculiar to control engineering. Similar situations are encountered in almost all fields of engineering design. Control is, however, one field of engineering in which a comparatively sophisticated theory is needed to understand the problems.

It is useful to have some perspective on the design process and a feel for the role of theory in the design process. First, a good engineering design must satisfy a large number of specifications, and there often are many equally good solutions to a design problem. A good design is often a compromise based on reasonable trade-offs between cost and performance. Sadly enough, it is often true that the best is the worst enemy of the good. Consequently, when words like *optimal* are used in this context, they should be taken with a grain of salt.

Another aspect is that design is often arrived at by interaction between customer and vendor. Many subjective factors—such as pride, tradition, and ambition—enter into this interaction. This situation with regard to customer preference is particularly confused when technology is changing. Typical examples are discussions concerning pneumatic or electronic controllers or analog versus digital control, which have been abundant in the trade journals.

What theory *can* contribute to the design process is to give insight and understanding. In particular, theory can often pinpoint fundamental limitations on control performance. There are also some idealized design problems, which can be solved theoretically. Such solutions can often give good insight into suitable structures and algorithms.

It is also useful to remember that control problems can be widely different in nature. They can range from design of a simple loop in a given system to design of an integrated control system for a complete process. The approach to design can also be widely different for mass-produced systems, and one-of-a-kind systems. For mass-produced systems, a substantial effort can be made to obtain a cheap system that will give good performance. For unique systems, it is often much better to install a flexible standard system and to tune it in situ.

The relation between process design and control design is also important. Control systems have traditionally been introduced into given processes to simplify or improve their operation. It has, however, become clear that much can be gained by considering process design and control design in one context. The availability of a control system always gives the designer an extra degree of freedom, which frequently can be used to improve performance or economy. Similarly, there are many situations where difficult control problems arise because of improper process design. An understanding of control also makes it possible to design a process so that difficult control problems are avoided.

Some operational aspects of control systems are first discussed in Sec. 6.2. This includes interfaces to the process, the operator, and the computer. Various aspects of design, commissioning, and process operation are also given. The problems of structuring are discussed in Sec. 6.3. The basic problem is to decompose a large, complicated problem into a set of smaller, simpler problems. This includes choice of control principles, and selection of control variables and measured variables. The common structuring principles—top-down, bottom-up, middle-out, and outside-in—are also discussed. The top-down approach is treated in Sec. 6.4. This includes choice of control principles and selection and grouping of control signals and measurements. The bottom-up approach is discussed in Sec. 6.5, including a discussion of the elementary control structures, feedback, feedforward, prediction, estimation, optimization, and adaptation. Combinations of these concepts are also discussed. The design of simple loops is discussed in Sec. 6.6. Design methods for simple loops are also reviewed.

6.2 Operational Aspects

It is useful to understand how the control system interacts with its environment. This section discusses the interfaces between process and controller design. Commissioning, operation, and modification of the system are also discussed.

Process and Controller Design

In the early stages of automation, the control system was always designed when the process design was completed. This still happens in many cases. Because process design is largely based on static considerations, it can lead to a process that is difficult to control. For this reason, it is very useful to consider the control design jointly with the process design. The fact that a process will be controlled automatically also gives the process designers an additional degree of freedom, which can be used to make better trade-offs. The process and the controller should therefore be designed together. An example illustrates the idea.

Example 6.1   Elimination of disturbances by mixing

Elimination of inhomogeneities in a product stream is one of the major problems in process control. One possibility for reducing the variations is to introduce large storage tanks and thus increase the material stored in the process. A system with large mixing tanks has slow dynamics. It will take a long time to change product quality in such a system. One consequence is that the product may be off the specifications for a considerable time during a change in quality. Another possibility for eliminating inhomogeneities is to measure the product quality and to reduce the variations by feedback control. In this case, it is possible to use much smaller tanks and to get systems with a faster response. The control system does, however, become more complicated. Because the total system will always have a finite bandwidth, small mixing tanks must be used to eliminate rapid variations.   ∎

Stability Versus Controllability (Maneuverability)

It frequently happens that stability and controllability have contradictory requirements. This has been evident in the design of vehicles, for instance. The Wright brothers succeeded in the design of their aircraft because they decided to make a maneuverable, but unstable, aircraft, whereas their competitors were instead designing stable aircrafts. In ship design, a stable ship is commonly difficult to turn, but a ship that turns easily tends to be unstable. Traditionally, the tendency has been to emphasize stability. It is, however, interesting to see that if a control system is used, the basic system can instead be designed for controllability. The required stability can then be provided by the control system. An example from aircraft design is used to demonstrate that considerable savings can be obtained by this approach.

Example 6.2   Design of a supersonic aircraft

For a high-performance aircraft, which operates over a wide speed range, the center of pressure moves aft with increasing speed. For a modern supersonic fighter, the shift in center of pressure can be about 1 m. If the aircraft is designed so that it is statically stable at subsonic speeds, the center of mass will be a few decimeters in front of the center of pressure at low speed. At supersonic speeds, the distance between the center of mass and the center of pressure will then increase to about 1 m. Thus there will he a very strong stabilizing torque, which tends to keep the airplane on a straight course. The torque will be proportional to the product of the thrust and the distance between the center of mass and the center of pressure.

To maneuver the plane at high speeds, a large rudder is then necessary. A large rudder will, however, give a considerable drag.

There is a considerable advantage to change the design so that the center of mass is in the middle of the range of variation of the center of pressure. A much smaller rudder can then be used, and the drag induced by the rudder is then decreased. The drag reduction can be over 10%. Such an airplane will, however, be statically unstable at low speeds—that is, at takeoff and landing! The proper stability, however, can be obtained by using a control system. Such a control system must, of course, be absolutely reliable.

Current thinking in aircraft design is moving in the direction of designing an aircraft that is statically unstable at low speeds and providing sufficient stability by using a control system. Similar examples are common in the design of other vehicles.   ∎

There are analogous cases also in the control of chemical processes. The following is a typical case.

Example 6.3   Exothermic chemical reactor

To obtain a high yield in an exothermic chemical reactor, it may be advantageous to run the reactor at operating conditions in which the reactor is open-loop unstable. Obviously, the safe operation then depends critically on the control system that stabilizes the reactor.   ∎

Controllability, Observability, and Dynamics

When designing a process, it is very important to make sure that all the important process variables can be changed conveniently. The word controllability is often used in this context, although it is interpreted in a much wider sense than in the formal controllability concepts introduced in Sec. 3.4.

To obtain plants that are controllable in the wide sense, it is first necessary to have a sufficient number of actuators. If there are four important process variables that should be manipulated separately, there must be at least four actuators. Moreover, the system should be such that the static relationship between the process variables and the actuators is one-to-one. To achieve good control, the dynamic relationship between the actuators and the process variables should ideally be such that tight control is possible. This means that time delays and nonminimum phase relations should be avoided. Ideally the dynamic relations should be like an integrator or a first-order lag. It is, however, often difficult to obtain such processes. Nonminimum phase loops are therefore common in the dynamics of industrial processes.

Simple dynamic models are often very helpful in assessing system dynamics at the stage of process design. Actuators should be designed so that the process variables can be changed over a sufficient range with a good resolution. The relationships should also be such that the gain does not change too much over the whole operating range. A common mistake in flow systems is to choose a control valve that is too large. This leads to a very nonlinear relation between valve opening and flow. The flow changes very little when the valve opening is reduced until the valve is almost closed. There is then a drastic change in flow over a very small range of valve position.

The process must also have appropriate sensors, whose signals are closely related to the important process variables. Sensors need to be located properly to give signals that are representative for the important process variables. For example, care must be taken not to position sensors in pockets where the properties of the process fluid may not be typical. Time delays must also be avoided. Time lags can occur due to factors such as transportation or encapsulation of temperature sensors.

Simple dynamic models, combined with observability analysis, are very useful to assess suggested arrangements of sensors and actuators. It is also very useful for this purpose to estimate time constants from simple dynamic models.

Controller Design or On-Line Tuning

Another fact that drastically influences the controller design is the effort that can be spent on the design. For systems that will be produced in large numbers, it may be possible to spend much engineering effort to design a controller. A controller with fixed parameters not requiring any adjustments can then be designed. In many cases, however, it is not economically feasible to spend much effort on controller design. For such applications it is common to use a standard general-purpose controller with adjustable parameters. The controller is installed and appropriate parameters are found by tuning.

The possibilities for designing flexible general-purpose controllers have increased drastically with computer control. When a controller is implemented on a computer, it is also possible to provide the system with computer-aided tools that simplify design and tuning. In process control, the majority of the loops for control of liquid level, temperature, flow, and pressure are designed by rules of thumb and are tuned on line. Systematic design techniques are, however, applied to control of composition and pH, as well as to control of multivariable, nonlinear, and distributed systems like distillation columns.

Interaction Among Process, Controller, and Operator

The controller and the process must, of course, work well together. A controller is normally designed for steady-state operation, which is one operating state. It is, however, necessary to make sure that the system will work well also during startup and shutdown and under emergency conditions, such as drastic process failures. During normal conditions it is natural to design for maximum efficiency. At a failure, it may be much more important to recover and quickly return to a safe operating condition.

In process control, it has been customary to use automatic regulation for steady-state operation. In other operating modes, the controller is switched to manual and an operator takes over. With an increased level of automation, good control over more operating states is, however, required.

6.3 Principles of Structuring

As mentioned earlier, real control problems are large and poorly defined, and control theory deals with small well-defined problems. According to the dictionary, *structuring* can mean to construct a systematic framework for something. In this context, however, structuring is used to describe the process of bridging the gap between the real problems and the problems that control theory can handle.

The problems associated with structuring are very important for control-system design. Unfortunately, these problems cannot yet be put into a complete systematic framework. For this reason they are often avoided both in textbooks and in research. As an analogy, structuring can be said to have the same relation to control-system design as grammar has to composition. It is clearly impossible to write well without knowing grammar. It is also clear that a grammatically flawless essay is not necessarily a good essay. Structuring of control systems must be based on the scientific principles given by control theory. However, structuring also contains elements of creativity, ingenuity, and art. Perhaps the best way to introduce structuring is to teach it as a *craft*.

The problem of structuring occurs in many disciplines. Formal approaches have also been developed. The terminology used here is borrowed from the fields of computer science and problem solving, where structuring of large programs has been the subject of much work. There are two major approaches, called top-down and bottom-up.

The *top-down* approach starts with the problem definition. The problem is then divided into successively smaller pieces, adding more and more details. The procedure stops when all pieces correspond to well-known problems. It is a characteristic of the top-down approach that many details are left out in the beginning. More and more details are added as the problem is subdivided. The buzz word *successive refinement* is therefore often associated with the top-down approach.

The *bottom-up* approach starts instead with the small pieces, which represent known solutions for subproblems. These are then combined into larger and larger pieces, until a solution to the large problem is obtained.

The top-down approach is often considered to be more systematic and more logical. It is, of course, not possible to use such an approach unless the details of the system are known very well. Similarly, it is not easy to use the bottom-up approach unless the characteristics of the complete problem are known. In practice, it is common to use combinations of the approaches. This is sometimes called an *inside-out–outside-in* approach.

Structuring is an iterative procedure. It will be a long time before a fully systematic approach to structuring is obtained. It is difficult to appreciate the structuring problems unless problems of reasonable size and complexity are considered. Therefore, most of the work on structuring is done in industry. It also appears that many industries have engineers who are very good at structuring. Students are therefore advised to learn what the "structuring masters" are doing, in the same way as painters have always learned from the grand masters.

6.4 A Top-Down Approach

This section describes a top-down approach to control-system design. This involves the selection of control principles, choice of control variables and measured variables, and pairing these variables.

Control Principles

A control principle gives a broad indication of how a process should be controlled. The control principle thus tells how a process should respond to disturbances and command signals. The establishment of a control principle is the starting point for a top-down design. Some examples of control principles are given next.

Example 6.4   Flow control

> When controlling a valve, it is possible to control the valve position, the flow, or both. It is simplest and cheapest to control the valve position. Because flow is, in general, a nonlinear function of the valve opening, this leads to a system in which the relationship between the control variable (valve position) and the physical variable (flow) is very nonlinear. The relationship will also change with such variables as changing pressure and wear of the valve. These difficulties are avoided if both valve position and flow are controlled. A system for flow control is, however, more complicated because it requires a flow meter.   ∎

Example 6.5   Composition control

> When controlling important product-quality variables, it is normally desired to keep them close to prescribed values. This can be done by minimizing the variance of product-quality variations. If a flow is fed to a large storage tank with mixing, the quality variations in the mixing tank should be minimized. This is not necessarily the same as minimizing quality variations in the flow into the tank.   ∎

Example 6.6   Control of a drum boiler

> Consider a turbine and a generator, which are driven by a drum boiler. The control system can have different structures, as illustrated in Fig. 6.1, which shows three control modes: boiler follow, turbine follow, and sliding pressure control. The system has two key control variables, the steam valve and the oil flow. In the boiler follow mode, the generator speed, ω, is controlled directly by feedback to the turbine valve, and the oil flow is controlled to maintain the steam pressure, p. In the turbine follow mode, the generator speed is used instead to control the oil flow to the boiler, and the steam valve is used to control the drum pressure. In sliding pressure control, the turbine valve is fully open, and oil flow is controlled from the generator speed.
>
> The boiler follow mode admits a very rapid control of generator speed and power output because it uses the stored energy in the boiler. There may be rapid pressure and temperature variations, however, that impose thermal strains on the turbine and the boiler. In the turbine follow mode, steam pressure is kept constant and thermal stresses are thus much smaller. The response to power demand will, however, be much slower. The sliding pressure control mode may be regarded as a compromise between boiler follow and turbine follow.   ∎

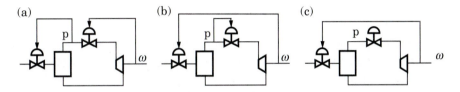

Figure 6.1   Control modes for a boiler-turbine unit: (a) boiler follow, (b) turbine follow, and (c) sliding pressure.

Example 6.7   Ship control

When designing an autopilot for a highly maneuverable ship, there are many alternatives for design. One possibility is to design the autopilot so that the captain can order a turn to a new course with a specified turning rate. Another possibility is to specify the turning radius instead of the turning speed. The advantage of specifying the turning radius is that the path of the ship will be independent of the speed of the ship. Control of the turning radius leads to a more complicated system, because it is necessary to measure both turning rate and ship speed. ■

Example 6.8   Material-balance control

Many processes involve flow and storage of materials. Although the processes are very different, they all include material storage. The reason for introducing these is to smooth out variations in material flow. It is therefore not sensible to control these systems in such a way that the storages have constant mass. Instead the criteria should be to maintain the following:

- Inventories between maximum and minimum limits

- An exact long-term material balance between input and output

- Smooth flow rates ■

Example 6.9   Constraint control

When designing systems, it is frequently necessary to consider several operating conditions. This means that constraints for safety or economical conditions may need to be considered. It may also be necessary to consider constraints during start-up and shutdown. The control during these situations is usually done with logical controllers. Today the logical control and the analog control are often done within the same equipment, programmable logic control (PLC) systems. This means that there are good possibilities to integrate different functions of the control system.
  ■

The choice of a control principle is an important issue. A good control principle can often simplify the control problem. The selection often involves technical and economical trade-offs. The selection of a control principle is often based on investigations of models of the process. The models used for this purpose are typically internal models derived from physical principles. It is therefore difficult to define general rules for finding control principles.

Choice of Control Variables

After the control principle has been chosen, the next logical step is to choose the control variables. The choice of control variables can often be limited for various practical reasons. Because the selection of control principle tells what physical variables should be controlled, it is natural to choose control variables that have a close relation to the variables given by the control principle. Because mathematical models are needed for the selection of control principles, these models also can be used for controllability studies when choosing control variables.

Choice of Measured Variables

When the control principle is chosen, the primary choice of measured variables is also given. If the variables used to express the control principle cannot be measured, it is natural to choose measured variables that are closely related to these control variables. Mathematical models and observability analysis can be very helpful in making this choice. Typical examples are found in chemical-process control, where temperatures—which are easy to measure—are used instead of compositions, which are difficult and costly to measure.

Pairing of Inputs and Outputs

A large system will typically have a large number of inputs and outputs. Even if a control principle, which involves only a few variables, is found initially,

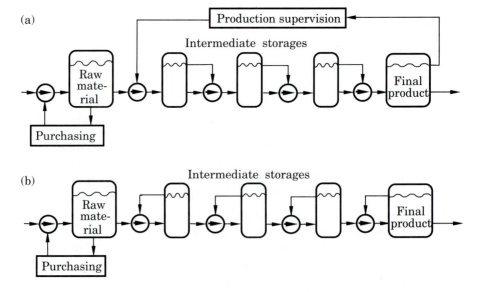

Figure 6.2 Material-balance control (a) in the direction of the flow and (b) in the direction opposite to the flow.

many variables typically must be considered once the variables that can be manipulated and measured are introduced. With a top-down approach, a system should be broken down into small subsystems. It is then desirable to group different inputs and outputs together, so that a collection of smaller systems is obtained. If possible, the grouping should be done so that (1) there are only weak couplings between the subsystems; and (2) each subsystem is dynamically well behaved, that is, time constants are of the same magnitude and time delay, nonminimum phase, and severe variations in process dynamics are avoided.

There are no general rules for the grouping. Neither are there any good ways of deciding if it is possible to find a grouping with the desired properties. Trial and error, combined with analysis of models, is one possibility. The following example illustrates the pairing problem.

Example 6.10   Material-balance control

A system with material flow is shown in Fig. 6.2. The system consists of a series of tanks. The flows between the tanks are controlled by pumps. The figure illustrates two different control structures. In one structure, the flow *out of* each tank is controlled from the tank level. This is called control *in the direction of the flow*. To maintain balance between production and demand, it is necessary to control the flow into the first tank by feedback from the last tank level. In the other approach, the flow into each tank is controlled by the tank level. This is called control *in the direction opposite to the flow*. This control mode is superior, because all control loops are simple first-order systems and there are no stability problems. With control in the direction of the flow, there may be instabilities due to the feedback around all tanks. It can also be shown that control in the direction opposite to the flow can be done by using smaller storage tanks.   ∎

6.5 A Bottom-Up Approach

In the bottom-up approach, a choice of control variables and measurements comes first. Different controllers are then introduced until a closed-loop system, with the desired properties, is obtained. The controllers used to build up the system are the standard types based on the ideas of feedback, feedforward, prediction and estimation, optimization, and adaptation. Because these techniques are familiar from elementary courses, they will be discussed only briefly.

Feedback

The feedback loops used include, for example, simple PID controllers and their cascade combinations. When digital computers are used to implement the controllers, it is also easy to use more sophisticated control, such as Smith-predictors for dead-time compensation, state feedback, and model reference control. Feedback is used in the usual context. Its advantage is that sensitivity to disturbances and parameter variations can be reduced. Feedback is most effective when the process dynamics are such that a high bandwidth can be used. Many

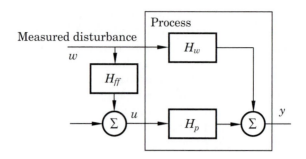

Figure 6.3 Reduction of disturbances by feedforward.

systems that are difficult to implement using analog techniques may be easy to implement using computer-control technology.

Feedforward

Feedforward is another control method. It is used to eliminate disturbances that can be measured. The basic idea is to use the measured disturbance to anticipate the influence of the disturbance on the process variables and to introduce suitable compensating control actions. See Fig. 6.3. The advantage compared to feedback is that corrective actions may be taken before the disturbance has influenced the variables. If the transfer functions relating the output y to the disturbance w and the control u are H_w and H_p, the transfer function H_{ff} of the feedforward compensator should ideally be

$$H_{ff} = -H_p^{-1}H_w$$

If this transfer function is unstable or nonrealizable, a suitable approximation is chosen instead. The design of the feedforward compensator is often based on a simple static model. The transfer function H_{ff} is then simply a static gain.

   Because feedforward is an open-loop compensation, it requires a good process model. With digital control, it is easy to incorporate a process model. Thus it can be anticipated that use of feedforward will increase with digital control. The design of a feedforward compensator is in essence a calculation of the inverse of a dynamic system.

Selector Control

There are many cases in which it is desirable to switch control modes, depending on the operating condition. This can be achieved by a combination of logic and feedback control. The same objective can, however, also be achieved with a combination of feedback controllers. A typical example is control of the air-to-fuel ratio in boiler control. In ship boilers it is essential to avoid smoke puffs when the ship is in the harbor. To do this it is essential that the air flow leads the oil flow when load is increased and that the air flow lags the oil flow when

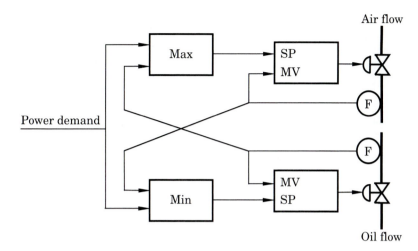

Figure 6.4 System with selectors for control of the air-to-fuel ratio in a boiler.

the load is decreased. This can be achieved with the system shown in Fig. 6.4, which has two selectors. The maximum selector gives an output signal that at each instant of time is the largest of the input signals, and the minimum selector chooses the smallest of the inputs. When the power demand is increased, the maximum selector chooses the demand signal as the input to the air-flow controller, and the minimum selector chooses the air flow as the set point to the fuel-flow controller. The fuel will thus follow the actual air flow.

When the power demand is decreased, the maximum selector will choose the fuel flow as the set point to the air-flow controller, and the minimum selector will choose the power demand as the set point to the fuel-flow controller. The air flow will thus lag the fuel flow.

Control using selectors is very common in industry. Selectors are very convenient for switching between different control modes.

Prediction and Estimation

State variables and parameters often cannot be measured directly. In such a case it is convenient to pretend that the quantities are known when design-ing a feedback. The unknown variables can then be replaced by estimates or predictions. In some cases such a solution is in fact optimal. The notions of pre-dictions and estimation are therefore important. Estimators for state variables in linear systems can easily be generated by analog techniques. They can also easily be implemented using a computer. Parameter estimators are more diffi-cult to implement with analog methods. They can, however, easily be done with a computer. Prediction and estimation are thus easier to use with computer control.

Optimization

Some control problems can be conveniently expressed as optimization problems. With computer-control systems, it is possible to include optimization algorithms as elements of the control system.

Combinations

When using a bottom-up approach, the basic control structures are combined to obtain a solution to the control problem. It is often convenient to make the combinations hierarchically. Many combinations, like cascade control, state feedback, and observers, are known from elementary control courses. Very complicated control systems can be built up by combining the simple structures. An example is shown in Fig. 6.5. This way of designing control using the bottom-up

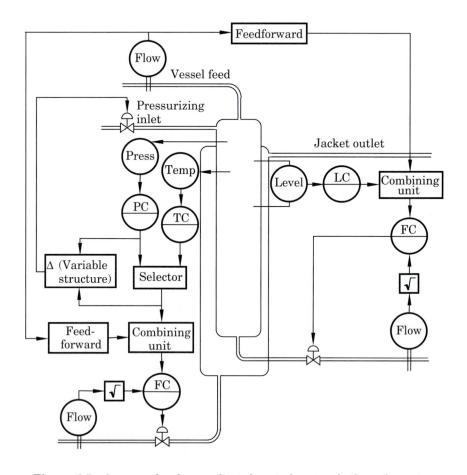

Figure 6.5 An example of a complicated control system built up from simple control structures. (Redrawn from Foxboro Company with permission.)

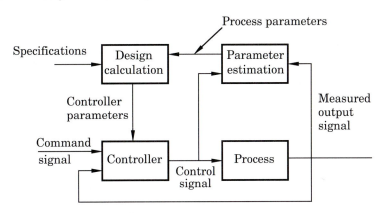

Figure 6.6 Block diagram of an adaptive controller obtained by combining a parameter estimator with a design calculation.

approach is in fact the technique predominantly used in process control. Its success depends largely on the experience and skill of the designer.

An adaptive system, which is obtained by combining a parameter estimator with a design procedure, is shown in Fig. 6.6.

6.6 Design of Simple Loops

If a top-down approach is used, the design procedure will end in the design of simple loops containing one or several controls, or measurements. If a bottom-up approach is used, the design will start with the design of simple loops. Therefore, the design of simple loops is an important step in both approaches. The design of simple loops is also one area in which there is substantial theory available, which will be described in detail in the book. To give some perspective, an overview of design methods for simple loops is given in this section. The prototype problems of controller and servo design will be discussed.

Simple Criteria

A simple way to specify regulation performance is to give allowable errors for typical disturbances. For example, it can be required that a step disturbance give no steady-state error, and that the error due to a ramp disturbance be a fraction of the ramp velocity. These specifications are typically expressed in terms of the steady-state behavior, as discussed in Sec. 3.5. The error coefficients give requirements only on the low-frequency behavior. The bandwidth of the system should therefore be specified, in addition to the error coefficients.

Another more complete way to specify regulation performance is to give conditions on the transfer function from the disturbances to the process output.

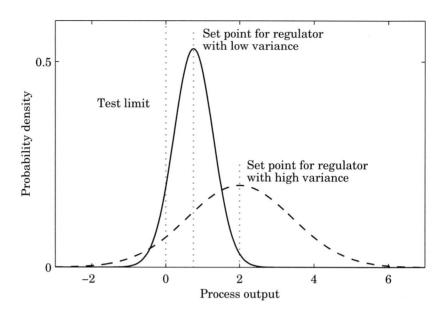

Figure 6.7 Expressing regulation performance in terms of variation in quality variables.

Specifications for the Controller Problem

The purpose of regulation is to keep process variables close to specified values in spite of process disturbances and variations in process dynamics.

Minimum-variance control. For regulation of important quality variables, it is often possible to state objective criteria for regulation performance. A common situation is illustrated in Fig. 6.7, which shows the distribution of the quality variables. It is often specified that a certain percentage of the production should be at a quality level above a given value. By reducing the quality variations, it is then possible to move the set point closer to the target. The improved performance can be expressed in terms of reduced consumption of energy or raw material or increased production. It is thus possible to express reductions in quality variations directly in economic terms.

For processes with a large production, reductions of a fraction of a percent can amount to a large sum of money. For example, a reduction in moisture variation of 1% in paper-machine control can amount to savings of $100,000 per year.

If the variations in quality can be expressed by Gaussian distributions, the criterion would simply be to minimize the variance of the quality variables. In these problems, the required control actions are irrelevant as long as they do not cause excessive wear or excessively large signals. A control strategy that minimizes the variance of the process output is called minimum-variance control.

Optimal control.   Minimum-variance control is a typical example of how a control problem can be specified as an optimization problem. In a more general case, it is not appropriate to minimize the variance of the output. Instead there will be a criterion of the type

$$\mathrm{E} \int_{t_0}^{t_1} g\big(x(s), u(s)\big)\, ds$$

where x is the state variable, u is the control variable, and E denotes the mean value. An example of such a criterion is given next.

Example 6.11   Ship steering

It can be shown that the relative increase in resistance due to deviations from a straight-line course can be approximately expressed as

$$\frac{\Delta R}{R} = \frac{k}{T} \int_0^T \big(\phi^2(t) + \rho\,\delta^2(t)\big)\, dt$$

where ϕ is the heading deviation, δ is the rudder angle, R is the resistance, and ρ is a parameter. Typical parameter values for a tanker are $k = 0.014$ and $\rho = 0.1$.
∎

Techniques for Controller Design

Regulation problems are often solved by feedback, but feedforward techniques can be very useful if disturbances can be measured.

   If the specifications are given in terms of the transfer function, relating the output to the disturbance, it is natural to apply methods that admit control of this transfer function. One method is pole placement, which allows specification of the complete transfer function. This straightforward design technique was discussed in detail in Chapters 4 and 5. It is often too restrictive to specify the complete closed-loop transfer function, which is a drawback.

   Another possibility is to use a frequency-response method, which admits control of the frequency response from the disturbance to the output. Such problems are most conveniently expressed in terms of continuous-time theory. The controllers obtained can then be translated to digital-control algorithms using the techniques described in Chapter 8.

   If the criteria are expressed as optimization criteria, it is natural to use design techniques based on optimization. Techniques based on minimizing the variance of the process output and other types of quadratic criteria are discussed in Chapters 11 and 12.

The Servo Problem

In the servo problem, the task is to make the process variables respond to changes in a command signal in a given way. Servo performance is typically specified in terms of requirements on the step response or the frequency response. Typical specifications for step responses include settling time and overshoot. Specifications can also be given in the frequency domain, for example, in terms of bandwidth. An alternative is to use a model that gives the desired response to command signals.

It is often very advantageous to use a two-degree-of-freedom configuration, because this admits a complete decoupling of the responses to disturbances and command signals. For such systems the feedback is first designed to solve the regulation problem and the feedforward is then designed to solve the servo problem. Examples of this were given in Secs. 4.6 and 5.4.

6.7 Conclusions

This chapter presents an overview of the design problems. There is a large step from the large and poorly defined problems of the real world to the small and well-defined problems that control theory can handle. Problems of structuring are discussed.

The notion of the control principle is introduced in order to apply the top-down approach. It is also shown how a bottom-up approach can be used to build complex systems from simple control structures such as feedback, feedforward, estimation, and optimization. Finally, specifications and approaches to the design of simple loops are discussed.

A chemical process consists of many unit operations, such as performed by reactors, mixers, and distillation columns. In a bottom-up approach to control-system design, control loops are first designed for the individual unit operations. Interconnections are then added to obtain a total system. In a top-down approach, control principles—such as composition control and material-balance control—are first postulated for the complete plant. In the decomposition, these principles are then applied to the individual units and loops.

In process control the majority of the loops for liquid level, flow, and pressure control are most frequently designed empirically and tuned on-line. However, control of composition and pH, as well as control of nonlinear distributed large systems with strong interaction, are often designed with care.

Control systems can be quite complicated because design is a compromise between many different factors. The following issues must typically be considered:

- Command signals
- Load disturbances
- Measurement noise
- Model uncertainty
- Actuator saturation
- State constraints
- Controller complexity

There are few design methods that consider all these factors. The design methods discussed in this book will typically focus on a few of the issues. In a good design it is often necessary to grasp all factors. To do this it is often necessary to investigate many aspects by simulation. The relation between process design and controller design should also be considered.

6.8 Problems

6.1 Consider the material-balance problem shown in Fig. 6.2. Assume that each tank (storage) is an integrator and that each controller is a proportional controller. Discuss the influence on the two systems when there is a pulse disturbance out from the raw material storage.

6.2 Identify and discuss the use of (a) cascade control, (b) feedforward, and (c) nonlinear elements in Fig. 6.5.

6.9 Notes and References

The problem discussed in this chapter touches on several aspects of problem solving. A reader with general interests may enjoy reading Polya (1945), which takes problems from the mathematical domain, and Wirth (1979), which applies to computer programming. There is some work on the structuring problem in the literature on process control; see, for instance, Buckley (1964), Bristol (1980), Balchen and Mummé (1988), and Shinskey (1988). Buckley (1978) contains much useful material of general interest although it deals with a very specific problem. Foss (1973) is more general in scope.

There are only a few areas in which control design and process design have been considered jointly. Design of high-performance aircrafts is a notable example. See Boudreau (1976) and Burns (1976).

Specifications of controller performance for simple loops are discussed in depth in standard texts on servomechanisms; see, for instance, Franklin, Powell, and Emami-Naeini (1994) and Dorf and Bishop (1995).

7

Process-Oriented Models

7.1 Introduction

Mathematical models for a sampled-data system from the point of view of the computer are developed in Chapter 2. These models are quite simple. The variables that represent the measured signal and the control signal are considered at the sampling instants only. These variables change in a given time sequence in synchronization with the clock. The signals are naturally represented in the computer as sequences of numbers. Thus the time-varying nature of sampled-data systems can be ignored, because the signals are considered only at times that are synchronized with the clock in the system. The sampled-data system can then be described as a time-invariant discrete-time system. The model obtained is called th *stroboscopic model*.

The stroboscopic model has the great advantage of being simple. Most of the problems in analysis and design of sampled-data systems can fortunately be handled by this model. The model will also give a complete description of the system as long as it is observed from the computer, but sometimes this is not enough. The main deficiency is that the model does not tell what happens between the sampling instants. Therefore it is useful to have other models that give a more detailed description. Such models are needed when the computer-controlled system is observed from the process, for example, if a frequency response is performed by cutting the loop on the analog side. The models required are necessarily more complicated than those discussed in Chapter 3 because the periodic nature of the system must be dealt with explicitly to describe the intersample behavior.

A detailed description of the major events in a computer-controlled system is given in Sec. 7.2. Section 7.3 give a discussion of sampling and reconstructing continuous-time signals. The alias problem encountered in Chapters 1 and 2 is analyzed in Sec. 7.4. Control of a system using predictive first-order-hold is discussed in Sec. 7.5. The key problem when making process-oriented models is the description of the sampling process. This is described using the modulation model in Sec. 7.6. Section 7.7 deals with the frequency response of sampled-data systems—several unexpected things can happen. The results give more insight into the aliasing problem. An algebraic system theory for sampled-data systems is outlined in Sec. 7.8. Multirate systems are discussed in Sec. 7.9.

7.2 A Computer-Controlled System

A schematic diagram of a computer-controlled system is given in Fig. 7.1. In Chapter 2 the loop is cut inside the computer between the A-D and D-A converters —for example, at C in the figure. In this chapter the loop is instead cut on the analog side —for example, at A in the figure. The discussions of this chapter require a more detailed description of the sequence of operations in a computer-controlled system. The following events take place in the computer:

1. Wait for a clock pulse.

2. Perform analog-to-digital conversion.

3. Compute control variable.

4. Perform digital-to-analog conversion.

5. Update the state of the regulator.

6. Go to step 1.

Because the operations in the computer take some time, there is a time delay between steps 2 and 4. The relationships among the different signals in the system are illustrated in Fig. 7.2. When the control law is implemented in a computer it is important to structure the code so that the calculations required in step 3 are minimized (see Chapter 9).

It is also important to express the synchronization of the signals precisely. For the analysis the sampling instants have been arbitrarily chosen as the time when the D-A conversion is completed. Because the control signal is discontinuous, it is important to be precise about the limit points. The convention of

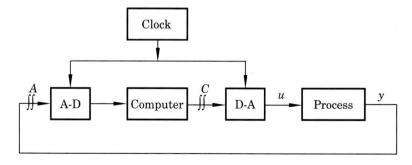

Figure 7.1   Schematic diagram of a computer-controlled system.

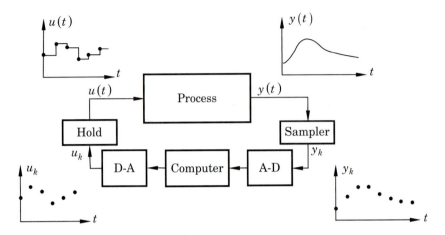

Figure 7.2 Relationships among the measured signal, control signal, and their representations in the computer.

continuity from the right was adopted. Notice that the real input signal to the process is continuous because of the nonzero settling time of the D-A converter and the actuators.

7.3 Sampling and Reconstruction

In this section we will discuss sampling and reconstruction of continuous-time signals. The periodic nature of sampled-data systems are considered.

The Sampling Theorem

Very little is lost by sampling a continuous-time signal if the sampling instants are sufficiently close, but much of the information about a signal can be lost if the sampling points are too far apart. This was illustrated in Examples 1.4 and 3.14.

It is, of course, essential to know precisely when a continuous-time signal is uniquely given by its sampled version. The following theorem gives the conditions for the case of periodic sampling.

THEOREM 7.1 SHANNON'S SAMPLING THEOREM   A continuous-time signal with a Fourier transform that is zero outside the interval $(-\omega_0, \omega_0)$ is given uniquely by its values in equidistant points if the sampling frequency is higher than $2\omega_0$. The continuous-time signal can be computed from the sampled signal by the interpolation formula

$$f(t) = \sum_{k=-\infty}^{\infty} f(kh) \frac{\sin\left(\omega_s(t-kh)/2\right)}{\omega_s(t-kh)/2} = \sum_{k=-\infty}^{\infty} f(kh) \operatorname{sinc} \frac{\omega_s(t-kh)}{2} \qquad (7.1)$$

where ω_s is the sampling angular frequency in radians per second (rad/s).

Proof.   Let the signal be f and let F be its Fourier transform.

$$F(\omega) = \int_{-\infty}^{\infty} e^{-i\omega t} f(t)\, dt$$

$$f(t) = \frac{1}{2\pi} \int_{-\infty}^{\infty} e^{i\omega t} F(\omega)\, d\omega \tag{7.2}$$

Introduce

$$F_s(\omega) = \frac{1}{h} \sum_{k=-\infty}^{\infty} F(\omega + k\omega_s) \tag{7.3}$$

The proof is based on the observation that the samples $f(kh)$ can be regarded as the coefficients of the Fourier series of the periodic function $F_s(\omega)$. This is shown by a direct calculation. The Fourier expansion of F_s is

$$F_s(\omega) = \sum_{k=-\infty}^{\infty} C_k e^{-ikh\omega} \tag{7.4}$$

where the coefficients are given by

$$C_k = \frac{1}{\omega_s} \int_0^{\omega_s} e^{ikh\omega} F_s(\omega)\, d\omega$$

By using the definition of the Fourier coefficients and the relations given in (7.2) and (7.3), it is straightforward to show that

$$C_k = f(kh) \tag{7.5}$$

It thus follows that the sampled signal $\{f(kh), k = \ldots, -1, 0, 1, \ldots\}$ uniquely determines the function $F_s(\omega)$. Under the assumptions of the theorem the function F is zero outside the interval $(-\omega_0, \omega_0)$. If $\omega_s > 2\omega_0$, it follows from (7.3) that

$$F(\omega) = \begin{cases} hF_s(\omega) & |\omega| \le \dfrac{\omega_s}{2} \\[2mm] 0 & |\omega| > \dfrac{\omega_s}{2} \end{cases} \tag{7.6}$$

The Fourier transform of the continuous-time signal is thus uniquely given by F_s, which in turn is given by the sampled function $\{f(kh), k = \ldots, -1, 0, 1, \ldots\}$.

The first part of the theorem is thus proved. To show Eq. (7.1), notice that it follows from (7.2) and (7.6) that

$$
\begin{aligned}
f(t) &= \frac{1}{2\pi} \int_{-\infty}^{\infty} e^{i\omega t} F(\omega)\, d\omega \\
&= \frac{h}{2\pi} \int_{-\omega_s/2}^{\omega_s/2} e^{i\omega t} F_s(\omega)\, d\omega \\
&= \frac{h}{2\pi} \int_{-\omega_s/2}^{\omega_s/2} e^{i\omega t} \sum_{k=-\infty}^{\infty} e^{-ikh\omega} f(kh)\, d\omega
\end{aligned}
$$

where the last equality follows from (7.4) and (7.5). Interchanging the order of integration and summation,

$$
\begin{aligned}
f(t) &= \sum_{k=-\infty}^{\infty} f(kh) \frac{h}{2\pi} \int_{-\omega_s/2}^{\omega_s/2} e^{i\omega t - i\omega kh}\, d\omega \\
&= \sum_{k=-\infty}^{\infty} f(kh) \left. \frac{h}{2\pi(t-kh)} e^{i\omega t - i\omega kh} \right|_{-\omega_s/2}^{\omega_s/2} \\
&= \sum_{k=-\infty}^{\infty} f(kh) \frac{\sin\left(\omega_s(t-kh)/2\right)}{\pi(t-kh)/h}
\end{aligned}
$$

Because $\omega_s h = 2\pi$, Eq. (7.1) now follows.   ∎

Remark 1.   The frequency $\omega_N = \omega_s/2$ plays an important role. This frequency is called the *Nyquist frequency*.

Remark 2.   Notice that Eq. (7.1) defines the reconstruction of signals whose Fourier transforms vanish for frequencies larger than the Nyquist frequency $\omega_N = \omega_s/2$.

Remark 3.   Because of the factor $1/h$ in Eq. (7.3), the sampling operation has a gain of $1/h$.

Reconstruction

The inversion of the sampling operation, that is, the conversion of a sequence of numbers $\{f(t_k) : k \in Z\}$ to a continuous-time function $f(t)$ is called *recon-struction*. In computer-controlled systems, it is necessary to convert the control actions calculated by the computer as a sequence of numbers to a continuous-time signal that can be applied to the process. In digital filtering, it is similarly necessary to convert the representation of the filtered signal as a sequence of numbers into a continuous-time function. Some different reconstructions are discussed in this section.

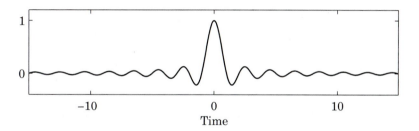

Figure 7.3 The impulse response of the Shannon reconstruction given by (7.7) when $h = 1$.

Shannon Reconstruction

For the case of periodic sampling of band-limited signals, it follows from the sampling theorem that a reconstruction is given by (7.1). This reconstruction is called the *Shannon reconstruction*. Equation (7.1) defines an inverse of the sampling operation, which can be considered as a linear operator. It is, however, not a causal operator because the value of f at time t is expressed in terms of past values $\{f(kh) : k \leq t/h\}$ as well as future values $\{f(kh) : k > t/h\}$. The characteristics of the Shannon reconstruction are given by the function

$$h(t) = \frac{\sin(\omega_s t/2)}{\omega_s t/2} \tag{7.7}$$

See Fig. 7.3. This reconstruction will introduce a delay. The weight is 10% after about three samples and less than 5% after six samples. The delay implies that the Shannon reconstruction is not useful in control applications. It is, however, sometimes used in communication and signal-processing applications, where the delay can be acceptable. Other drawbacks of the Shannon reconstruction are that it is complicated and that it can be applied only to periodic sampling. It is therefore useful to have other reconstructions.

Zero-Order Hold (ZOH)

In previous chapters zero-order-hold sampling has been used. This causal reconstruction is given by

$$f(t) = f(t_k) \qquad t_k \leq t < t_{k+1} \tag{7.8}$$

This means that the reconstructed signal is piecewise constant, continuous from the right, and equal to the sampled signal at the sampling instants. Because of its simplicity, the zero-order hold is very common in computer-controlled systems. The standard D-A converters are often designed in such a way that the old value is held constant until a new conversion is ordered. The zero-order hold also has the advantage that it can be used for nonperiodic sampling. Notice, however, that the reconstruction in (7.8) gives an exact inverse of the sampling

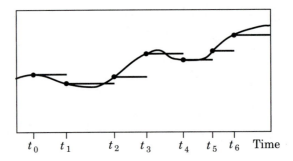

Figure 7.4 Sampling and zero-order-hold reconstruction of a continuous-time signal.

operation only for signals that are right continuous and piecewise constant over the sampling intervals. For all other signals, the reconstruction of (7.8) gives an error (see Fig. 7.4).

Higher-Order Holds

The zero-order hold can be regarded as an extrapolation using a polynomial of degree zero. For smooth functions it is possible to obtain smaller reconstruction errors by extrapolation with higher-order polynomials. A first-order causal polynomial extrapolation gives

$$f(t) = f(t_k) + \frac{t - t_k}{t_k - t_{k-1}} \left(f(t_k) - f(t_{k-1}) \right) \qquad t_k \le t < t_{k+1}$$

The reconstruction is thus obtained by drawing a line between the two most recent samples. The first-order hold is illustrated in Fig. 7.5.

Predictive First-Order Hold

A drawback of the zero- and first-order hold is that the output is discontinuous.

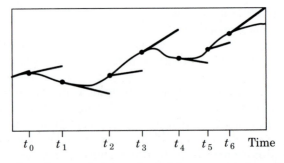

Figure 7.5 Sampling and first-order-hold reconstruction of a continuous-time signal.

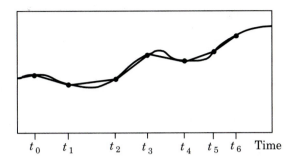

Figure 7.6 Sampling and predictive first-order hold reconstruction of a continuous-time signal.

A way to avoid this problem is to use a predictive first-order-hold. The inter-sample behavior with this hold circuit is a linear interpolation of the sampled values; see Figure 7.6. Mathematically the reconstruction can be described by

$$f(t) = f(t_k) + \frac{t - t_k}{t_{k+1} - t_k}\left(f(t_{k+1}) - f(t_k)\right) \qquad t_k \le t < t_{k+1} \qquad (7.9)$$

Notice that this requires that $f(t_{k+1})$ is available at time t_k. For general applications the predictive first-order hold is not realizable. The value $f(t_{k+1})$ can be replaced by a prediction. This can be done very conveniently in a feedback loop, as will be discussed in Section 7.5.

7.4 Aliasing or Frequency Folding

If a continuous-time signal that has the Fourier transform F is sampled periodically, it follows from (7.4) and (7.5) that the sampled signal $f(kh), k = \ldots, -1, 0, 1, \ldots$ can be interpreted as the Fourier coefficients of the function F_s, defined by (7.3).

The function F_s can thus be interpreted as the Fourier transform of the sampled signal. The function of (7.3) is periodic with a period equal to the sampling frequency ω_s. If the continuous-time signal has no frequency components higher than the Nyquist frequency, the Fourier transform is simply a periodic repetition of the Fourier transform of the continuous-time signal (see Fig. 7.7).

It follows from (7.3) that the value of the Fourier transform of the sampled signal at ω is the sum of the values of the Fourier transform of the continuous-time signal at the frequencies $\omega + n\omega_s$. After sampling, it is thus no longer possible to separate the contributions from these frequencies. The frequency ω can thus be considered to be the *alias* of $\omega + n\omega_s$. It is customary to consider only positive frequencies. The frequency ω is then the alias of $\omega_s - \omega$, $\omega_s + \omega$, $2\omega_s - \omega$, $2\omega_s + \omega, \ldots$, where $0 \le \omega < \omega_N$. After sampling, a frequency thus cannot be distinguished from its aliases. The fundamental alias for a frequency

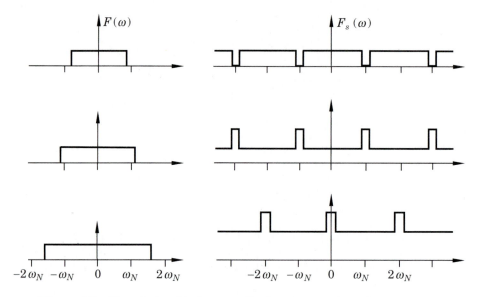

Figure 7.7   The relationship between the Fourier transform for continuous and sampled signals for different sampling frequencies. For simplicity it has been assumed that the Fourier transform is real.

$\omega_1 > \omega_N$ is given by

$$\omega = |(\omega_1 + \omega_N) \bmod (\omega_s) - \omega_N| \tag{7.10}$$

Notice that although sampling is a linear operation, it is not time-invariant. This explains why new frequencies will be created by the sampling. This is discussed further in Sec. 7.7.

An illustration of the aliasing effect is shown in Fig. 7.8. Two signals with the frequencies 0.1 Hz and 0.9 Hz are sampled with a sampling frequency of 1 Hz ($h = 1$ s). The figure shows that the signals have the same values at the sampling instants. Equation (7.10) gives that 0.9 has the alias frequency 0.1. The aliasing problem was also seen in Fig. 1.11.

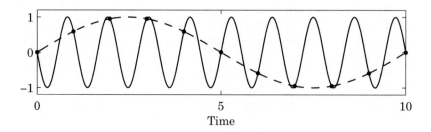

Figure 7.8   Two signals with different frequencies, 0.1 Hz and 0.9 Hz, may have the same value at all sampling instants.

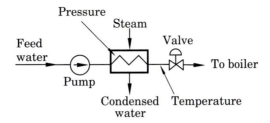

Figure 7.9   Process diagram for a feed-water heating system of a boiler.

Example 7.1   Aliasing

Figure 7.9 is a process diagram of feed-water heating in a boiler of a ship. A valve controls the flow of water. There is a backlash in the valve positioner due to wear. This causes the temperature and the pressure to oscillate. Figure 7.10 shows a sampled recording of the temperature and a continuous recording of the pressure.

From the temperature recording one might believe that there is an oscillation with a period of about 38 min. The pressure recording reveals, however, that the oscillation in pressure has a period of 2.11 min. Physically the two variables are coupled and should oscillate with the same frequency.

The temperature is sampled every other minute. The sampling frequency is $\omega_s = 2\pi/2 = 3.142$ rad/min and the frequency of the pressure oscillation is $\omega_0 = 2\pi/2.11 = 2.978$ rad/min. The lowest aliasing frequency is $\omega_s - \omega_0 = 0.1638$ rad/min. This corresponds to a period of 38 min, which is the period of the recorded oscillation in the temperature.   ∎

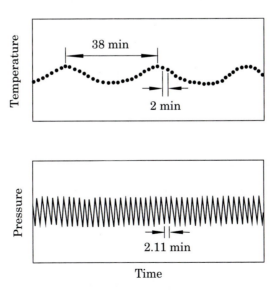

Figure 7.10   Recordings of temperature and pressure.

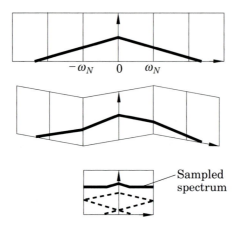

Figure 7.11   Frequency folding.

Frequency Folding

Equation (7.3) can also be given another interpretation. The graph of the spectrum of the continuous-time signal is first drawn on a paper. The paper is then folded at abscissas that are odd multiples of the Nyquist frequency, as indicated in Fig. 7.11. The sampled spectrum is then obtained by adding the contributions, with proper phase, from all sheets.

Prefiltering

A practical difficulty is that real signals do not have Fourier transforms that vanish outside a given frequency band. The high-frequency components may

Table 7.1   Damping ζ and natural frequency ω for Butterworth, ITAE (Integral Time Absolute Error), and Bessel filters. The higher-order filters with arbitrary bandwidth ω_B are obtained by cascading filters of the form (7.12).

Order	Butterworth		ITAE		Bessel	
	ω	ζ	ω	ζ	ω	ζ
2	1	0.71	0.99	0.71	1.27	0.87
4	1	0.38	1.49	0.32	1.60	0.62
		0.92	0.84	0.83	1.43	0.96
6	1	0.26	1.51	0.24	1.90	0.49
		0.71	1.13	0.60	1.69	0.82
		0.97	0.92	0.93	1.61	0.98

Table 7.2 Approximate time delay T_d of Bessel filters of different orders.

Order	T_d
2	$1.3/\omega_B$
4	$2.1/\omega_B$
6	$2.7/\omega_B$

appear to be low-frequency components due to aliasing. The problem is particularly serious if there are periodic high-frequency components. To avoid the alias problem, it is necessary to filter the analog signals before sampling. This may be done in many different ways.

Practically all analog sensors have some kind of filter, but the filter is seldom chosen for a particular control problem. It is therefore often necessary to modify the filter so that the signals obtained do not have frequencies above the Nyquist frequency.

Sometimes the simplest solution is to introduce an analog filter in front of the sampler. A standard analog circuit for a second-order filter is

$$G_f(s) = \frac{\omega^2}{s^2 + 2\zeta\omega s + \omega^2} \tag{7.11}$$

Higher-order filters are obtained by cascading first- and second-order systems. Examples of filters are given in Table 7.1. The table gives filters with bandwidth $\omega_B = 1$. The filters get bandwidth ω_B by changing the factors (7.11) to

$$\frac{\omega^2}{(s/\omega_B)^2 + 2\zeta\omega(s/\omega_B) + \omega^2} \tag{7.12}$$

where ω and ζ are given by Table 7.1. The Bessel filter has a linear phase curve, which means that the shape of the signal is not distorted much. The Bessel filters are therefore common in high-performance systems.

The filter must be taken into account in the design of the regulator if the desired crossover frequency is larger than about $\omega_B/10$, where ω_B is the bandwidth of the filter. The Bessel filter can, however, be approximated with a time delay, because the filter has linear phase for low frequencies. Table 7.2 shows the delay for different orders of the filter. Figure 7.12 shows the Bode plot of a sixth-order Bessel filter and a time delay of $2.7/\omega_B$. This property implies that the sampled-data model including the antialiasing filter can be assumed to contain an additional time delay compared to the process. Assume that the bandwidth of the filter is chosen as

$$|G_{aa}(i\omega_N)| = \beta$$

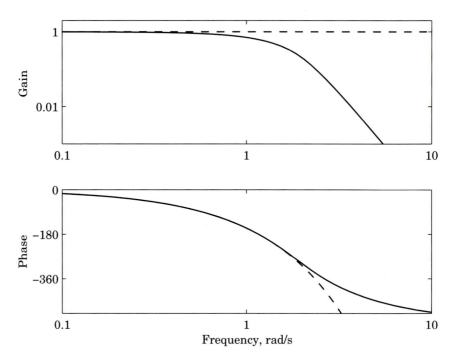

Figure 7.12   Bode plot of a sixth-order Bessel filter (solid) when $\omega_B = 1$ and a time delay $T_d = 2.7$ (dashed).

where ω_N is the Nyquist frequency, and $G_{aa}(s)$ is the transfer function of the antialiasing filter. Table 7.3 gives some values of T_d, as a function of β. First, the attenuation β is chosen. The table then gives the bandwidth of the filter in relation to the Nyquist frequency. The delay measured in the units of the

Table 7.3   The time delay T_d as a function of the desired attenuation at the Nyquist frequency for fourth- and sixth-order Bessel filters. The sampling period is denoted h.

β	Fourth Order		Sixth Order	
	ω_B/ω_N	T_d/h	ω_B/ω_N	T_d/h
0.001	0.1	5.6	0.2	4.8
0.01	0.2	3.2	0.3	3.1
0.05	0.3	2.1	0.4	2.3
0.1	0.4	1.7	0.4	2.0
0.2	0.5	1.4	0.5	1.7
0.5	0.7	0.9	0.7	1.2
0.7	1.0	0.7	1.0	0.9

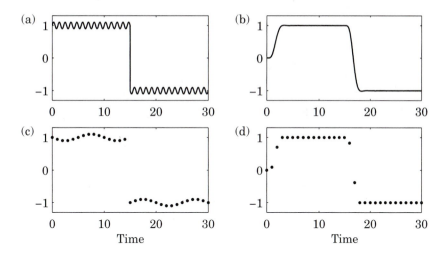

Figure 7.13 Usefulness of a prefilter. (a) Signal plus sinusoidal disturbance. (b) The signal is filtered through a sixth-order Bessel filter. (c) Sample
and hold of the signal in (a). (d) Sample and hold of the signal in (b).

sampling period is also obtained, that is, if a small value of β is desired, then
the bandwidth of the filter must be low and the corresponding delay is long.

Example 7.2   Prefiltering

   The usefulness of a prefilter is illustrated in Fig. 7.13. An analog signal composed
   of a square wave with a superimposed sinusoidal perturbation (0.9 Hz) is shown
   in (a). The result of sampling the analog signal with a period of 1 Hz is shown in
   (c). The Nyquist frequency is 0.5 Hz. The disturbance with the frequency 0.9 Hz
   has the alias 0.1 Hz [see (7.10)]. This signal is clearly noticeable in the sampled
   signal. The output of a prefilter, a sixth-order Bessel filter with a bandwidth of
   0.25 Hz, is shown in (b), and the result obtained by sampling with the prefilter
   is shown in (d). Thus the amplitude of the disturbance is reduced significantly by
   the prefilter.   ∎

Example 7.3   Product-stream sampling

   In process control there is one situation in which prefiltering cannot be used:
   namely when a product stream is sampled and sent to an instrument for analysis. Examples are samples taken for mass spectrographs, gas chromatographs,
   and laboratory analysis. In such cases it is advisable to take many samples and
   to mix them thoroughly before sending them to the analyzer. This is equivalent to
   taking several samples and taking the mean value.   ∎

When Can Dynamics of Antialiasing Filters Be Neglected?

We have mentioned that the dynamics in the antialiasing filter often must be
taken into account. The following analysis gives additional insight. The phase

lag at the frequency ω_0 introduced by a second-order Butterworth filter (7.11) is

$$\alpha \approx \frac{2.6}{\sqrt{\beta}} \frac{\omega_0}{\omega_s}$$

where ω_s is the sampling frequency, and β is the attenuation of the filter at the Nyquist frequency. For a Bessel filter of sixth order the relation is

$$\alpha \approx \frac{8.6}{\beta^{1/6}} \frac{\omega_0}{\omega_s}$$

Our rules for selecting the sampling rates in digital systems require that $\omega_0 h$ is in the range of 0.2 to 0.6. With $\omega_0 h = 0.2$ the preceding equation implies that the phase lag of the second-order antialiasing filter is

$$\alpha = \frac{0.083}{\sqrt{\beta}}$$

With $\beta = 0.1$ we get $\alpha = 0.26$ rad, or $15°$. With the sixth-order Bessel filter as an antialiasing filter and $\beta = 0.1$, we get $\alpha = 0.4$ rad, or $23°$. These calculations show that it is necessary to take the dynamics of the antialiasing filter into account for practically all digital designs. Approximating the filter by a delay is a convenient way of doing that.

Postsampling Filters

The signal from the D-A converter is piecewise constant. This may cause difficulties for systems with weakly damped oscillatory modes because they may be excited by the small steps in the signal. In such a case it is useful to introduce a special postsampling filter that smoothes the signal before applying it to the actuator. In some cases this can be achieved by suitable modification of the actuator dynamics. In extreme cases it may be advisable to design special D-A converters that will give piecewise linear control signals.

7.5 Designing Controllers with Predictive First-Order Hold

Design of computer-controlled systems based on a first-order hold was discussed in Chapters 4 and 5. In this section it will be shown that the methods used in these chapters can easily be generalized to deal with systems where the D-A conversion is based on a predictive first-order hold.

The reason for using other hold devices is that the control signal changes stepwise, which implies that high-frequency signals are injected into the process. This is not a serious drawback for processes that attenuate high-frequencies effectively. It can, however, be a severe drawback for systems with poorly damped oscillatory poles. For hydraulic systems it may also create severe hydraulic transients. The remedy is to replace the first-order hold by a hold circuit that gives smooth control signals. A simple fix is to introduce a smoothing filter after the D-A converter, as was discussed in Sec. 7.4. Another possibility is to use the predictive first-order hold. This device, which was discussed in Sec. 7.3, generates an output that is piecewise linear.

Implementation of a Predictive First-Order Hold

The predictive first-order hold is described by Eq. (7.9). The system can be implemented by switched operational amplifiers. It is, however, often more convenient to implement an approximation with a multirate sampled system. The time interval (t_k, t_{k+1}) is then subdivided into a N equal parts of length $\Delta_t = (t_{k+1} - t_k)$ and the output of the hold circuit is incremented by

$$\Delta_u = \frac{f(t_{k+1}) - f(t_k)}{N(t_{k+1} - t_k)}$$

at each time increment Δ_t. If N is large, the output from the hold circuit is then a staircase function with very small steps, which is very close to the output given by Eq. (7.9). If necessary the output can also be filtered.

Predictive First-Order-Hold Sampling: A State-Space Approach

We will now consider a system in which the sampling period is constant and equal to h. In Chapter 2 we showed that the behavior of the system at the sampling interval at the sampling instants $t = kh$ could be conveniently described by a difference equation. The key idea was that the system equations could be integrated over one sampling interval if the shape of the input signal was known. In Chapter 2 the calculations were based on the assumption that sampling was made by a first-order hold, which implies that the control signal is constant over the sampling intervals. It is straightforward to repeat the calculations in Chapter 2 for the case when the control signal is affine over a sampling interval. The modifications required can also be obtained as follows.

Consider a continuous-time system described by

$$\frac{dx}{dt} = Ax(t) + Bu(t)$$

$$y(t) = Cx(t) + Du(t)$$

(7.13)

Assume that the input signal is linear between the sampling instants. Integration of (7.13) over one sampling period gives

$$x(kh + h) = e^{Ah}x(kh)$$

$$+ \int_{kh}^{kh+h} e^{A(kh+h-s)}B\left[u(kh) + \frac{s - kh}{h}\left(u(kh + h) - u(kh)\right)\right] ds \quad (7.14)$$

Hence

$$x(kh + h) = \Phi x(kh) + \Gamma u(kh) + \frac{1}{h}\Gamma_1\Big(u(kh + h) - u(kh)\Big)$$
$$= \Phi x(kh) + \frac{1}{h}\Gamma_1 u(kh + h) + \Big(\Gamma - \frac{1}{h}\Gamma_1\Big)u(kh)$$

where

$$\Phi = e^{Ah}$$
$$\Gamma = \int_0^h e^{As}\,ds\,B \tag{7.15}$$
$$\Gamma_1 = \int_0^h e^{As}(h - s)\,ds\,B$$

The pulse-transfer function that corresponds to ramp-invariant sampling thus becomes

$$H(z) = D + C(zI - \Phi)^{-1}\Big(\frac{z}{h}\Gamma_1 + \Gamma - \frac{1}{h}\Gamma_1\Big) \tag{7.16}$$

It follows from (7.14) that the matrices Φ, Γ, and Γ_1 satisfy the differential equations

$$\frac{d\Phi(t)}{dt} = \Phi(t)A$$
$$\frac{d\Gamma(t)}{dt} = \Phi(t)B$$
$$\frac{d\Gamma_1(t)}{dt} = \Gamma(t)$$

These equations can also be written as

$$\frac{d}{dt}\begin{pmatrix} \Phi(t) & \Gamma(t) & \Gamma_1(t) \\ 0 & I & It \\ 0 & 0 & I \end{pmatrix} = \begin{pmatrix} \Phi(t) & \Gamma(t) & \Gamma_1(t) \\ 0 & I & It \\ 0 & 0 & I \end{pmatrix}\begin{pmatrix} A & B & 0 \\ 0 & 0 & I \\ 0 & 0 & 0 \end{pmatrix}$$

This implies that the matrices Φ, Γ, and Γ_1 can be obtained as

$$\begin{pmatrix} \Phi & \Gamma & \Gamma_1 \end{pmatrix} = \begin{pmatrix} I & 0 & 0 \end{pmatrix}\exp\left(\begin{pmatrix} A & B & 0 \\ 0 & 0 & I \\ 0 & 0 & 0 \end{pmatrix}h\right) \tag{7.17}$$

The calculation of ramp-invariant systems is illustrated by some examples.

Example 7.4   Ramp-invariant sampling of an integrator

Consider a system with the transfer function $G(s) = 1/s$. In this case we have $A = D = 0$ and $B = C = 1$. Using (7.17) we get

$$\begin{pmatrix} \Phi & \Gamma & \Gamma_1 \end{pmatrix} = \begin{pmatrix} 1 & 0 & 0 \end{pmatrix} \exp\left(\begin{pmatrix} 0 & 1 & 0 \\ 0 & 0 & 1 \\ 0 & 0 & 0 \end{pmatrix} h \right)$$

$$= \begin{pmatrix} 1 & h & \frac{1}{2}h^2 \end{pmatrix}$$

The pulse-transfer function becomes

$$H(z) = \frac{\frac{1}{2}zh + h - \frac{1}{2}h}{z - 1} = \frac{h}{2} \frac{z + 1}{z - 1}$$

This pulse-transfer function corresponds to the trapezoidal formula for computing an integral. Also notice that Tustin's transformation gives the same result in this case. ∎

Example 7.5   Ramp-invariant sampling of a double integrator

Consider a system with the transfer function $G(s) = 1/s^2$. This system has the realization

$$\frac{dx}{dt} = \begin{pmatrix} 0 & 1 \\ 0 & 0 \end{pmatrix} x + \begin{pmatrix} 0 \\ 1 \end{pmatrix} u$$

$$y = \begin{pmatrix} 1 & 0 \end{pmatrix} x$$

for the matrix

$$\tilde{A} = \begin{pmatrix} A & B & 0 \\ 0 & 0 & I \\ 0 & 0 & 0 \end{pmatrix} = \begin{pmatrix} 0 & 1 & 0 & 0 \\ 0 & 0 & 1 & 0 \\ 0 & 0 & 0 & 1 \\ 0 & 0 & 0 & 0 \end{pmatrix}$$

and its matrix exponential

$$e^{\tilde{A}h} = \begin{pmatrix} 1 & h & h^2/2 & h^3/6 \\ 0 & 1 & h & h^2/2 \\ 0 & 0 & 1 & h \\ 0 & 0 & 0 & 1 \end{pmatrix}$$

Hence from (7.17)

$$\Phi = \begin{pmatrix} 1 & h \\ 0 & 1 \end{pmatrix} \qquad \Gamma = \begin{pmatrix} h^2/2 \\ h \end{pmatrix} \qquad \Gamma_1 = \begin{pmatrix} h^3/6 \\ h^2/2 \end{pmatrix}$$

The pulse-transfer function is now obtained from (7.16), that is,

$$H(z) = \begin{pmatrix} 1 & 0 \end{pmatrix} \begin{pmatrix} z - 1 & -h \\ 0 & z - 1 \end{pmatrix}^{-1} \left(\begin{pmatrix} h^2/6 \\ h/2 \end{pmatrix} z + \begin{pmatrix} h^2/2 \\ h \end{pmatrix} - \begin{pmatrix} h^2/6 \\ h/2 \end{pmatrix} \right)$$

$$= \frac{h^2}{6} \frac{z^2 + 4z + 1}{(z - 1)^2}$$

∎

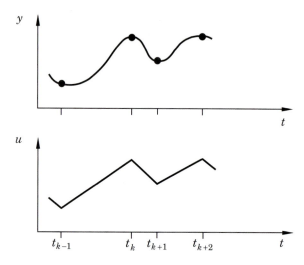

Figure 7.14   Inputs and outputs of a process with predictive first-order-hold sampling.

Predictive First-Order-Hold Sampling: An Input-Output Approach

The pulse-transfer function obtained by predictive first-order-hold sampling of a system with a transfer function $G(s)$ can also be obtained by a direct calculation. Figure 7.14 shows the inputs and the outputs of a system with predictive zero-order-hold sampling. Let u be the input of the system and let y denote the output. The piecewise affine input u can be generated as the output of an integrator whose input is the piecewise constant signal

$$v(t) = \frac{u(kh + h) - u(kh)}{h} \tag{7.18}$$

Because this signal is constant over the sampling intervals, the results of Chapter 2 can be applied and we find that the z-transform of the output is given by

$$Y(z) = \mathcal{S}_{\text{zoh}}\left(\frac{G(s)}{s}\right) V(z) \tag{7.19}$$

where $\mathcal{S}_{\text{zoh}}$ denotes the map of transfer functions to pulse-transfer functions through zero-order-hold sampling. This operator is given by Eq. (2.30). Combining Eqs. (7.18) and (7.19) we get

$$Y(z) = \mathcal{S}_{\text{zoh}}\left(\frac{G(s)}{s}\right) \frac{z-1}{h} U(z)$$

We have thus obtained the input-output relation for sampling with a predictive first-order-hold that can be expressed as follows.

$$\mathcal{S}_{\text{pfoh}}\left(G(s)\right) = \frac{z-1}{h} \mathcal{S}_{\text{zoh}}\left(\frac{G(s)}{s}\right) \tag{7.20}$$

By using Eq. (2.30) it follows that the pulse-transfer function obtained by the predictive first-order-hold sampling of a continuous system with the transfer function $G(s)$ can be expressed by

$$H(z) = \frac{(z-1)^2}{zh} \frac{1}{2\pi i} \int_{\gamma-i\infty}^{\gamma+i\infty} \frac{e^{sh}}{z - e^{sh}} \frac{G(s)}{s^2} \, ds \qquad (7.21)$$

We illustrate the results with an example.

Example 7.6   Predictive first-order-hold sampling of an integrator

An integrator has the transfer function $G(s) = 1/s$. The zero-order-hold sampling of the integrator is

$$\frac{h^2}{2} \frac{z+1}{(z-1)^2}$$

It then follows from Eq. (7.20) that

$$H(z) = \frac{h}{2} \frac{z+1}{z-1}$$

This is the same result obtained in Example 7.4.   ∎

Example 7.7   Predictive first-order-hold sampling of a double integrator

A double integrator has the transfer function $G(s) = 1/s^2$. It follows from Table 2.1 that the zero-order-hold sampling of $1/s^3$ is

$$\frac{h^3}{6} \frac{z^2 + 4z + 1}{(z-1)^3}$$

It then follows from Eq. (7.20) that

$$H(z) = \frac{h^2}{6} \frac{z^2 + 4z + 1}{(z-1)^2}$$

Notice that in this case the orders of the numerator and denominator polynomials are the same. This is due to the predictive nature of the hold.   ∎

Control Design

We have thus found that predictive first-order-hold sampling is similar to zero-order-hold sampling. In both cases the behavior of a system at the sampling instants can be described as a time-invariant discrete-time system. The methods for designing controllers obtained can then be used with minor modifications. We will illustrate this by giving the results for pole-placement control.

Consider a system with the pulse-transfer function $H(z) = B(z)/A(z)$ obtained by predictive first-order-hold sampling. A general linear controller with a two-degree-of-freedom structure can be described by the triple $(R(z), S(z), T(z))$.

With a predictive hold the controller must generate the signal $u(kh + h)$ at time kh. This means that the controller polynomials have the property

$$\deg R(z) \geq \deg S(z) + 1$$
$$\deg R(z) \geq \deg T(z) + 1$$

(7.22)

Specifying a desired closed-loop characteristic polynomial A_{cl} we find that the Diophantine equation associated with the design problem becomes

$$A(z)R(z) + B(z)S(z) = A_{cl}(z)$$

(7.23)

and the control design can then be done in the same way as in Chapter 5. The only difference is that the order condition (7.22) is different. We illustrate the procedure by an example.

Example 7.8 Pole-placement design of a double integrator

In Example 7.7 we derived the pulse-transfer function for a double integrator under predictive first-order-hold sampling. It follows from this example that the system is characterized by

$$A(z) = (z - 1)^2$$
$$B(z) = \frac{h^2}{6} (z^2 + 4z + 1)$$

Assuming that a controller with integral action is desired we find that the Diophantine equation (7.23) becomes

$$(z - 1)^3 \bar{R}(z) + \frac{h^2}{6} (z^2 + 4z + 1)S(z) = A_{cl}(z)$$

where $R(z) = (z - 1)\bar{R}(z)$. The minimum-degree solution of this equation has the property $\deg S(z) = 2$. It then follows from the order condition (7.22) that $\deg R(z) = 3$ and consequently that $\deg \bar{R}(z) = 2$. The minimum-degree solution thus gives a closed-loop system of order five. The previous Diophantine equation becomes

$$(z - 1)^3(z^2 + r_1 z + r_2) + \frac{h^2}{6} (z^2 + 4z + 1)(s_0 z^2 + s_1 z + s_2) = A_{cl}(z)$$

The solution of this equation was discussed in Sec. 5.3. ∎

7.6 The Modulation Model

A characteristic feature of computer-controlled systems with zero-order hold is that the control signal is constant over the sampling period. This fact is used in Chapter 2 to describe how the system changes from one sampling instant to the next by integrating the system equations over one sampling period; this section attempts to describe what happens between the sampling instants. Other

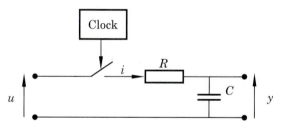

Figure 7.15   Schematic diagram of a sample-and-hold circuit.

mathematical models are then needed, because it is no longer sufficient to model signals as sequences (functions that map Z to R); instead they must be modeled as continuous-time functions (functions that map R to R).

The central theme is to develop the modulation model. This model is more complicated than the stroboscopic model discussed in Chapter 2. The main difficulty is that the periodic nature of sampled-data systems must be taken into account. The system can be described as an amplitude modulator followed by a linear system. The modulation signal is a pulse train. A further idealization is obtained by approximating the pulses by impulses. The model has its origin in early work on sampled-data systems by MacColl (1945), Linvill (1951), and others.

In the special case of computer control with a unit-gain algorithm and negligible time delay, the combined action of the A-D converter, the computer, and the D-A converter can be described as a system that samples the analog signal and produces another analog signal that is constant over the sampling periods. Such a circuit is called a *sample-and-hold circuit*. An A-D converter can also be described as a sample-and-hold circuit. The hold circuit keeps the analog voltage constant during the conversion to a digital representation. A more detailed model for the sample-and-hold circuit will first be developed.

A Model of the Sample-and-Hold Circuit

A schematic diagram of an analog sample-and-hold circuit is shown in Fig. 7.15. It is assumed that the circuit is followed by an amplifier with very high input impedance. The circuit works as follows: When the sampling switch is closed, the capacitor is charged to the input voltage via the resistor R. When the sampling switch is opened, the capacitor holds its voltage until the next closing.

To describe the system, a function m, which describes the closing and opening of the sampling switch, is introduced. This function is defined by

$$m(t) = \begin{cases} 1 & \text{if switch is closed} \\ 0 & \text{if switch is open} \end{cases}$$

The current is then given by

$$i = \frac{u - y}{R} m$$

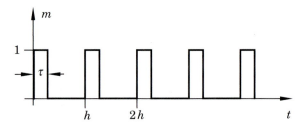

Figure 7.16 Graph of the modulation function m with period h and pulse width τ.

The current is thus *modulated* by the function m, which is called the *modulation function*. If the input impedance of the circuit that follows the sample-and-hold circuit is high, the voltage over the capacitor is given by

$$C\,\frac{dy(t)}{dt} = i(t) = \frac{u(t) - y(t)}{R}\,m(t) \qquad (7.24)$$

The differential equation of (7.24) is a linear *time-varying system*. The time variation is caused by the modulation. If the sampling period h is constant and if the switch is closed for τ seconds at each sampling, the function m has the shape shown in Fig. 7.16. Because m is a periodic function the system becomes a *periodic system*.

Once a mathematical model of the circuit is obtained the response of the circuit to an input signal u can be investigated. It follows directly from Eq. (7.24) that the voltage across the capacitor is constant when the switch is open, that is, when $m(t) = 0$. When the switch is closed, the voltage y approaches the input signal u as a first-order dynamic system with the time constant RC. The time constant of the RC circuit must be considerably shorter than the pulse width; otherwise, there is no time to charge the capacitor to the input voltage when the switch is closed.

A simulation of the sample-and-hold circuit is shown in Fig. 7.17. With the chosen parameters, the pulse width is so long that the input signal changes significantly when the switch is closed.

Figure 7.18 shows what happens when the pulse width is shorter. The results shown in Fig. 7.18 represent a reasonable choice of parameter values. The sample-and-hold circuit quickly reaches the value of the input signal and then remains constant over the sampling period.

Practical Samplers

In practice, a sampler is not implemented, as shown in Fig. 7.15. They are today made using semiconductor technology, but the circuits can still be described by Eq. (7.24). To avoid difficulties with noise and ground loops, it is important to have the computer galvanically isolated from the process signals. This can be achieved using the flying capacitor technique, which combines electrical insulation with sample-and-hold action in an elegant way. A capacitor is charged

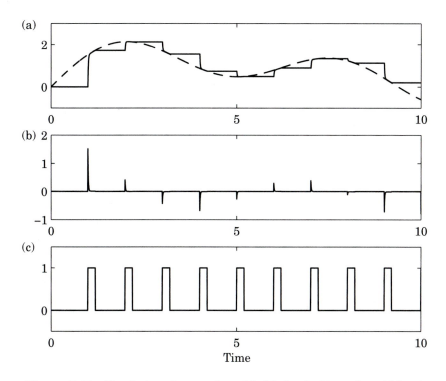

Figure 7.17 Simulation of a sample-and-hold circuit. The pulse width τ is 0.2 s and the time constant is $RC = 0.01$ s. (a) The continuous-time signal (dashed) and the output of the sample-and-hold circuit (solid); (b) the current $i(t)$ in the sample-and-hold circuit; (c) the modulation function $m(t)$.

to the input voltage when it is connected to the input line. When the capacitor is connected to the D-A converter it holds its voltage. Electrical isolation is obtained because the capacitor is connected either to the process or to the D-A converter of the control computer. In practice it is common to charge the capacitor via an operational amplifier. The flying capacitor circuit can also be described by Eq. (7.24).

A Mathematical Idealization

The pulse-modulation scheme is easy to simulate but difficult to analyze. A more easily used mathematical idealization will therefore be introduced. It seems reasonable to design the sample-and-hold circuit so that the pulse width τ is much shorter than the sampling period. It also seems reasonable to choose the time constant RC to be shorter than the pulse width. The current through the capacitor will then consist of short pulses. Both the height and the time integral of a pulse are proportional to the difference $u - y$ between the input voltage u and the capacitor voltage y at the sampling instant.

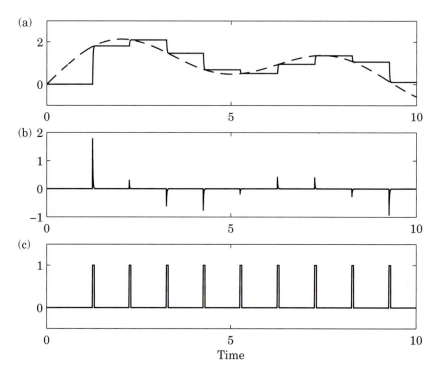

Figure 7.18 Simulation of a sample-and-hold circuit. The pulse width is 0.05 s and the time constant is $RC = 0.01$ s. (a) The continuous-time signal (dashed) and the output of the sample-and-hold circuit (solid); (b) the current $i(t)$ in the sample-and-hold circuit; (c) the modulation function $m(t)$.

In the idealization, the current pulses are replaced by impulses. For simplicity the integral of the impulse is chosen to be proportional to the value of the input signal u at the sampling instant. The capacitor is then replaced by an integrator. Because the pulses were chosen to be proportional to u and not to $u - y$, it is necessary to reset the integral to zero when a new pulse arrives. The current is then represented as

$$u^* = um \tag{7.25}$$

where

$$m(t) = \sum_{k=-\infty}^{\infty} \delta(t - kh) \tag{7.26}$$

and δ is a delta function [compare with (7.24)]. The signal u^* is called the *sampled representation* of the continuous signal u. It is useful to remember that u^* is related to the current through the capacitor of the sample-and-hold circuit in Fig. 7.15.

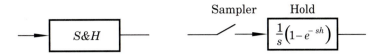

Figure 7.19 Block diagram of a sample-and-hold circuit and its idealized representation.

The signal u^* can be thought of as a modulation of u with a carrier signal in the form of an impulse train. The model is therefore called the *impulse-train modulation model*. The signal u^* carries the same information as the sequence $\{u(kh), k = \dots, -1, 0, 1, \dots\}$. Notice, however, that u^* is a (generalized) time function. The signal u^* is introduced to represent a sampled signal in a form that can be processed by linear filtering.

The Hold Circuit

The hold circuit can be represented as an integrator that is automatically reset to zero after one sampling period. Such a system has the transfer function

$$G_{\text{zoh}}(s) = \frac{1}{s}\left(1 - e^{-sh}\right) \tag{7.27}$$

The impulse response of the transfer function $1/s$ is a unit step and the impulse response of $(1/s)\exp(-sh)$ is a unit step that is delayed h time units. Subtraction of these impulse responses gives the impulse response as a pulse of unit height and duration h.

Notice that the steady-state gain of the hold circuit is $G_{\text{zoh}}(0) = h$. Section 7.3 shows that ideal sampling could be said to have a gain $1/h$. The combination of a sampler with a hold circuit thus has unit steady-state gain. For very fast sampling, the sample-and-hold circuit thus acts as a continuous-time system with unit transfer function.

The idealized model of a sample-and-hold circuit is thus obtained by combining a sampler with impulse modulation given by (7.25) and (7.26) with a hold circuit given by (7.27). A block-diagram representation of the system is shown in Fig. 7.19. Because the impulse modulator is a periodic system it follows that the sample-and-hold circuit is also a periodic system.

Input-Output Relationships

Once a convenient representation of a sample-and-hold circuit is obtained, the response of a sampled-data system to an arbitrary input signal can be computed. Consider the system shown in Fig. 7.20(a), which is composed of a sample-and-hold circuit connected to a time-variant linear dynamic system with the transfer function G. This is a typical representation of a sampler and a D-A converter connected to a process. Use of the impulse-modulation model of the sample-and-hold circuit allows the system to be represented as in Fig. 7.20(b).

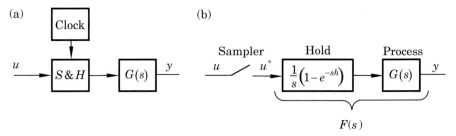

Figure 7.20 (a) Schematic diagram of a sample-and-hold circuit connected to a linear system and (b) its representation using the idealized model of a sample-and-hold circuit.

Let u be the input, y the output, and F the transfer function of the combination of the zero-order-hold circuit and the process, that is,

$$F(s) = \frac{1}{s}\left(1 - e^{-sh}\right)G(s) \tag{7.28}$$

The input-output relationship is easily determined using transform theory. The Laplace transform of u^* is given by

$$U^*(s) = \int_0^\infty e^{-st}u^*(t)\,dt = \sum_{k=0}^\infty e^{-skh}\,u(kh)$$

The Laplace transform of the output signal is then given by

$$Y(s) = F(s)\sum_{k=0}^\infty e^{-skh}\,u(kh) \tag{7.29}$$

It is thus straightforward to calculate the Laplace transform of the output signal. Notice, however, that it is not possible to factor out the Laplace transform of the signal u on the right-hand side of (7.29). This means that the input-output relationship of the system cannot be characterized by an ordinary transfer function. This is because the system is not time-invariant. How to get around this problem is discussed in Sec. 7.8.

7.7 Frequency Response

Many powerful methods for analysis and design of control systems are based on frequency response. The key idea is to use the fact that a linear time-invariant system can be completely characterized by its steady-state response to sinusoidal signals. It would be highly desirable to extend these results to sampled-data systems. We have in fact done this intuitively, for example, when plotting the sensitivity function, $S(e^{i\omega h})$, for discrete-time systems in Chapter 5. There are, however, some difficulties in interpreting the results that we must be aware of. The frequency response of a discrete-time system is a well-defined quantity. When dealing with a sampled systems it is, however, also necessary to consider what happens between the sampling instants. There is an essential difficulty because sampled systems are time-varying. One consequence of this is that a sinusoidal input of frequency generates outputs with many frequencies. This was illustrated, for example, in Example 1.4.

In this section we will explore sampled systems from the point of view of frequency response. We will first investigate how sinusoids propagate through a sampled systems. The results are useful when organizing and interpreting experiments with frequency response. We will then briefly outline how frequency response of a sampled system can be defined rigorously. The section ends with a few practical remarks.

A Special Case

When performing the frequency-response test, it is natural to cut the loop on the analog side, for example, at A in Fig. 7.1. To simplify the analysis consider the special case in which the output of the D-A converter is equal to the input of the A-D converter. The action of the computer on the signals can then be described as a sample-and-hold circuit. It follows from Fig. 7.19 that a sample-and-hold circuit can be represented as a sampler followed by a hold circuit. The problem is thus reduced to calculation of the response of a sampler followed by a linear time-invariant system.

Equation (7.25) gives the sampled representation u^* of the input signal u. A formal Fourier series representation of a sequence of delta functions gives

$$m(t) = \sum_{k=-\infty}^{\infty} \delta(t - kh) = \frac{1}{h}\left(1 + 2\sum_{k=1}^{\infty} \cos k\omega_s t\right)$$

where h is the sampling period, and ω_s is the corresponding sampling frequency in radians per second.

Assume that the input to the system is

$$u(t) = \sin(\omega t + \varphi) = \mathrm{Im}\left(\exp i(\omega t + \varphi)\right)$$

The series expansion of the output $u^* = um^*$ of the sampler then becomes

$$u^*(t) = \frac{1}{h}\left[\sin(\omega t + \varphi) + 2\sum_{k=1}^{\infty} \cos(k\omega_s t)\sin(\omega t + \varphi)\right]$$

$$= \frac{1}{h}\left[\sin(\omega t + \varphi) + \sum_{k=1}^{\infty}\left(\sin(k\omega_s t + \omega t + \varphi) - \sin(k\omega_s t - \omega t - \varphi)\right)\right]$$

The signal u^* has a component with the frequency ω of the input signal. This component is multiplied by $1/h$ because the steady-state gain of a sampler is

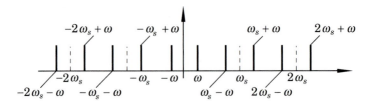

Figure 7.21 Frequency content of the sampled input signal u^* when $u = \sin(\omega t + \varphi)$.

$1/h$. The signal also has components corresponding to the *sidebands* $k\omega_s \pm \omega$. The frequency content of the output u^* of the sampler is shown in Fig. 7.21. The output signal y is obtained by linear filtering of the signal u^* with a system having the transfer function $F(s)$. The output thus has components with the fundamental frequency ω and the sidebands $k\omega_s \pm \omega$.

For $\omega \neq k\omega_N$, where ω_N is the Nyquist frequency, the fundamental component of the output is

$$y(t) = \frac{1}{h} \, \text{Im} \left(F(i\omega)e^{i(\omega t + \varphi)} \right)$$

For $\omega = k\omega_N$, the frequency of one of the sidebands coincides with the fundamental frequency. Two terms thus contribute to the component with frequency ω. This component is

$$y(t) = \frac{1}{h} \, \text{Im} \left(F(i\omega)e^{i(\omega t + \varphi)} - F(i\omega)e^{i(\omega t - \varphi)} \right)$$

$$= \frac{1}{h} \, \text{Im} \left(\left(1 - e^{2i\varphi}\right) F(i\omega)e^{i(\omega t - \varphi)} \right)$$

$$= \frac{1}{h} \, \text{Im} \left(2e^{i(\pi/2 - \varphi)} \sin \varphi \, F(i\omega)e^{i(\omega t + \varphi)} \right)$$

If the input signal is a sine wave with frequency ω, it is found that the output contains the fundamental frequency ω and the sidebands $k\omega_s \pm \omega, k = 1, 2, \ldots$ (compare with the discussion of aliasing in Sec. 7.4). The transmission of the fundamental frequency is characterized by

$$\hat{F}(i\omega) = \begin{cases} \dfrac{1}{h} F(i\omega) & \omega \neq k\omega_N \\[2mm] \dfrac{2}{h} F(i\omega)e^{i(\pi/2 - \varphi)} \sin \varphi & \omega = k\omega_N \end{cases}$$

For $\omega \neq k\omega_N$, the transmission is simply characterized by a combination of the transfer functions of the sample-and-hold circuit and the system G. The factor $1/h$ is due to the steady-state gain of the sampler.

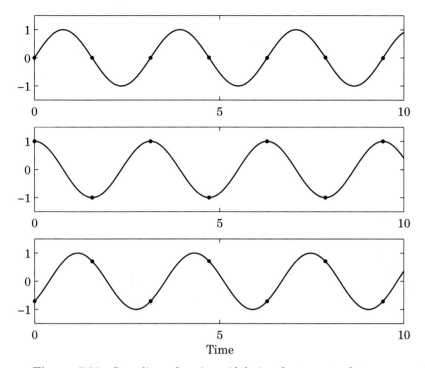

Figure 7.22 Sampling of a sinusoidal signal at a rate that corresponds to the Nyquist frequency. Notice that the amplitude and the phase of the sampled signal depend strongly on how the sine wave is synchronized to the sampling instants.

The fact that the signal transmission at the Nyquist frequency ω_N critically depends on φ—that is, how the sinusoidal input signal is synchronized with respect to the sampling instants—is illustrated in Fig. 7.22.

There may be interference between the sidebands and the fundamental frequency that can cause the output of the system to be very irregular. A typical illustration of this was given in Example 1.4. In this case the fundamental component has the frequency 4.9 Hz and the Nyquist frequency is 5 Hz. The interaction between the fundamental component and the lowest sideband, which has the frequency 5.1 Hz, will produce beats with the frequency 0.1 Hz. This is clearly seen in Fig. 1.12.

If the sideband frequencies are filtered out, the sampled system appears as a linear time-invariant system except at frequencies that are multiples of the Nyquist frequency, $\omega_s/2$. At this frequency the amplitude ratio and the phase lag depend on the phase shift of the input relative to the sampling instants.

If an attempt is made to determine the frequency response of a sampled system using frequency response, it is important to filter out the sidebands efficiently. Even with perfect filtering, there will be problems at the Nyquist frequency. The results depend critically on how the input is synchronized with the clock of the computer.

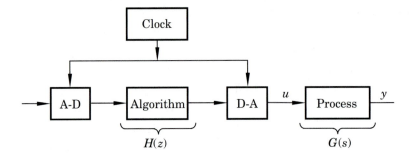

Figure 7.23   Open-loop computer-controlled system.

The General Case

It is easy to extend the analysis to the general case of the system shown in Fig. 7.1. The corresponding open-loop system is shown in Fig. 7.23.

It consists of an A-D converter, the computer, a D-A converter, and the process. It is assumed that the D-A converter holds the signal constant over a sampling interval. It is also assumed that the calculations performed by the computer can be expressed by the pulse-transfer function $H(z)$ and that the process is described by the transfer function $G(s)$.

If a sinusoid

$$v(t) = \sin(\omega t + \varphi) = \text{Im}\left(\exp i(\omega t + \varphi)\right)$$

is applied to the A-D converter, then the computer will generate a sequence of numbers that in steady state can be described by

$$w(kh) = \text{Im}\left(H(e^{i\omega h})e^{i(\omega kh + \varphi)}\right) \quad k = \ldots - 1, 0, 1, \ldots$$

This sequence is applied to the D-A converter. Because the D-A converter holds the signal constant over a sampling period, the output is the same as if the signal w were applied directly to a hold circuit. The discussion of the previous section can thus be applied: The output contains the fundamental component with frequency ω and sidebands $k\omega_s \pm \omega$. The signal transmission of the fundamental component may be described by the transfer function

$$K(i\omega) = \begin{cases} \dfrac{1}{h}\, H(e^{i\omega h})F(i\omega) & \omega \neq k\omega_N \\[2ex] \dfrac{2}{h}\, H(e^{i\omega h})F(i\omega)e^{i(\pi/2 - \varphi)}\sin\varphi & \omega = k\omega_N \end{cases}$$

where ω_N is the Nyquist frequency and

$$F(s) = \frac{1}{s}\left(1 - e^{-sh}\right)G(s)$$

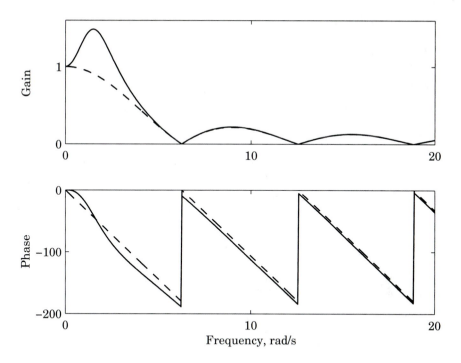

Figure 7.24 Magnitude and argument curves of the transfer function for first-order-hold (full) and zero-order-hold (dashed) circuits. The Nyquist frequency is $\omega_N = \pi$.

When ω is not a multiple of the Nyquist frequency, the signal transmission of the fundamental component can be characterized by a transfer function that is a product of four terms: the gain $1/h$ of the sampler, the transfer function $(1 - \exp(-sh))/s$ of the hold circuit, the pulse-transfer function $H(\exp(sh))$ of the algorithm in the computer, and the transfer function $G(s)$ of the process. Notice, however, that there are other frequencies in the output of the system because of the sampling. At the Nyquist frequency the fundamental component and the lowest sideband coincide.

It follows from the discussion that the hold circuit can be interpreted as a filter. The frequency functions of zero-order and first-order-hold circuits are shown in Fig. 7.24. It is clear from the figure that both the zero-order and the first- order hold permit significant signal transmission above the Nyquist frequency $\omega_N = \pi/h$. Notice that the phase curve is discontinuous at arguments $\omega_h = 2k\pi$, $k = 1, 2, \ldots$. Because the phase is defined modulo 2π, the discontinuities may be $\pm\pi$. In the figure they are shown as π for convenience only.

The following example illustrates the calculation and interpretation of the frequency response of a sampled system.

Example 7.9   Frequency response of a sampled-data system

Consider a system composed of a sampler and a zero-order hold, given by (7.27), followed by a linear system, with the transfer function

$$G(s) = \frac{1}{s+1}$$

The sampling period is $h = 0.05$ s. The Nyquist frequency is thus $\pi/0.05 = 62.8$ rad/s. Figure 7.25 shows the Bode diagram of the system. For comparison, the Bode diagram of the transfer function G is also shown in the figure. The curves are very close for frequencies that are much smaller than the Nyquist frequency. The deviations occur first in the phase curve. At $\omega = 0.1\omega_N$ the phase curves differ by about 10°. There is no signal transmission at frequencies that are multiples of the sampling frequency ω_s, because the transfer function of the zero-order hold is zero for these frequencies. The phase curve is also discontinuous at these frequencies. (Compare with Fig. 7.24.) Notice also that there are ambiguities of the transfer function at frequencies that are multiples of the Nyquist frequency that are not shown in Fig. 7.25. The value of ω_N is indicated by a vertical dashed line in Fig. 7.25.

The interpretation of the Bode diagram requires some care because of the modulation introduced by the sampling. If a sine wave of frequency ω is introduced, the output signal is the sum of the outputs of the sine wave and all its aliases.

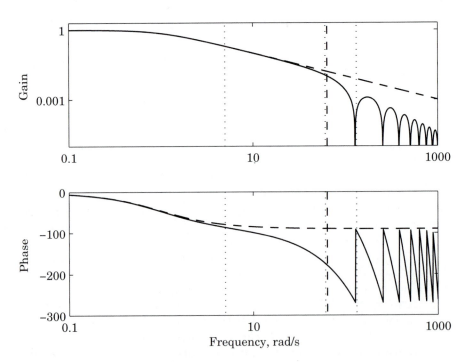

Figure 7.25 Bode diagrams for a zero-order sample-and-hold circuit followed by a first-order lag (solid). The sampling period is 0.05 s. The dashed line is the frequency curve for the continuous-time first-order lag. The vertical dotted lines indicate the frequencies $\omega = 5$, 60, and 130 rad/s, respectively. The vertical dashed line indicates the Nyquist frequency.

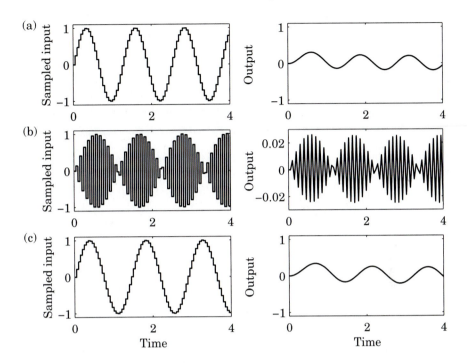

Figure 7.26 Steady-state responses to sinusoids with different frequencies for a zero-order hold followed by a first-order system with a unit time constant. The sampling period is 0.05 s. The frequencies are 5 rad/s in (a), 60 rad/s in (b), and 130 rad/s in (c). They are indicated by dotted lines in Fig. 7.25.

This is illustrated in Fig. 7.26, which shows the steady-state outputs for different frequencies. For frequencies smaller than the Nyquist frequency, the contribution from the fundamental frequency dominates. At frequencies close to the Nyquist frequency, there is a substantial interaction with the first alias, $\omega_s - \omega$. Typical beats are thus obtained. At the Nyquist frequency, the signal and its first alias have the same frequency and magnitude. The resulting signal then depends on the phase shift between the signals. For frequencies higher than the Nyquist frequency, the contribution from the alias in the frequency range $(0, \omega_N)$ dominates.

This clearly shows how important it is to filter a signal before the sampling, so that the signal transmission above the Nyquist frequency is negligible. Compare this conclusion with the discussion of aliasing in Sec. 7.4. ■

Frequency Response of an Internal-Combustion Engine

An internal-combustion engine is a typical example of a system that is inherently sampled. The sampling is caused by the ignition mechanism, and its frequency is the number of independently fired cylinders divided by the time required for a full cycle.

When an attempt was made to investigate the dynamic response of the engines, reproducible results were easily obtained for frequencies lower than the sampling frequency. For a long time, however, the results for higher frequencies were erratic: Different results were obtained at different measurements and results of experiments could not be verified when the experiments were repeated. This was due to the sampled nature of the process. For input signals with a frequency close to the Nyquist frequency, there is interference from the sidebands. At the Nyquist frequency, the results depend on how the sinusoid is synchronized to the ignition pulses.

When the source of the difficulty was finally understood, it was easy to find a solution. The sinusoid was simply synchronized to the ignition pulses; then it became possible to measure the frequency response to high frequencies. A typical result is shown in Fig.7.27. Notice, in particular, that the measurement is done in a range of frequencies that includes the Nyquist frequency.

The Idea of Lifting

The notion of *lifting* is an elegant way to deal with periodically sampled systems. The idea is to represent a finite-dimensional sampled system as a time-invariant infinite-dimensional discrete system. In this way it is possible to define a notion of frequency response properly. It is also possible to give a nice description of intersample behavior.

Consider a system described by Eq. (2.1). Assume that the system is sampled with a period h, and that the input signal and the states are in L_2. We introduce the discrete signal $u_k \in L_2(0, h)$ defined by

$$u_k(\tau) = u(kh + \tau) \qquad 0 < \tau < h \qquad (7.30)$$

and the signals x_k and y_k, which are defined analogously. Define the discrete signal x_k is the same way. It follows from Eq. (2.1) that

$$x_{k+1}(\tau) = \varphi(\tau)x_k(h) + \int_0^\tau \psi(\tau - s)Bu_k(s)\,ds \qquad (7.31)$$
$$y_k(\tau) = Cx_k(\tau)$$

where

$$\varphi(\tau) = e^{A\tau}$$
$$\psi(\tau) = e^{A(\tau)}B$$

This system is a time-invariant discrete-time system. Equation (7.31) gives a complete description of the intersample behavior because the function $y_k(\tau)$, which is defined for $0 \le \tau \le h$, is the output in the interval $kh \le t \le kh + h$. The description thus includes the phenomenon of aliasing. Notice, however, that u_k, x_k, and y_k are elements of function spaces. Because the system is linear and time-invariant, the frequency response can be defined as $H(e^{i\omega h})$, where H is

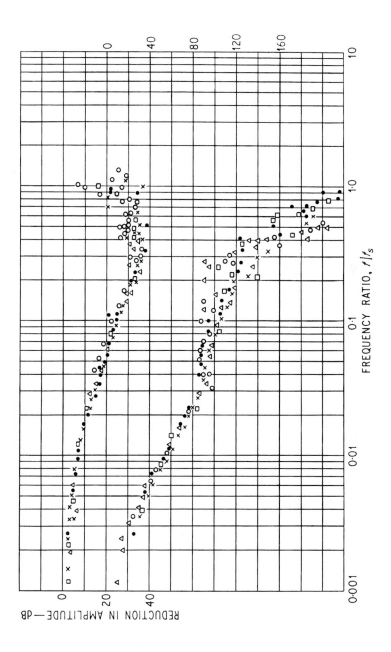

Figure 7.27 Measured frequency response for a diesel engine. The frequencies are normalized with respect to the sampling frequency. [Redrawn from D. E. Bowns, "The Dynamic Transfer Characteristics of Reciprocating Engines," *Proc. Mech. Eng.*, 185. (1971) with permission.]

the transfer function of the infinite-dimensional system (7.31). The transfer function H is, however, a nontrivial mathematical object. It can be computed numerically by a finite-dimensional approximation of the state. This can, for example, be obtained through the discrete-time system obtained by dividing the sampling interval h into N equal parts. A complete treatment requires functional analysis, which is outside the scope of this book. Details are given in the References.

Another way to deal with frequency response of sampled systems is to realize that the output generated by a sinusoid with frequency ω_0 contains the frequencies $\omega_n = n\omega_n \pm \omega_0$. The system can then be properly characterized by the transfer functions for all those frequencies.

Practical Consequences

The fact that sampled systems are time-varying means practically that some care must be exercised when interpreting frequency responses of sampled systems. Discrete frequency responses such as $H(e^{i\omega h})$, $L(e^{i\omega h})$, and $S(e^{i\omega h})$ give a correct description of what happens at the sampling instants, but they may give misleading results when intersample behavior is considered. For example, if we compute a phase margin based on $L(e^{i\omega h})$ it may happen that the system becomes unstable for a much smaller increase of the phase lag of the physical process. Realize that the reason for the difficulty is that a sinusoidal input, with frequency ω_0, to a sampled system gives outputs that contains many other frequencies. With an ideal antialiasing filter the signal components with frequencies different from ω_0 will not be present and the difficulty disappears. Ideal antialiasing filters cannot be implemented practically. There will not be much difficulties with plants with good attenuation of high frequency if the sampling period and the antialiasing filter are chosen properly. There may, however, be severe problems if there are resonant modes close to the Nyquist frequency. In such cases it is necessary to choose sampling rates and antialiasing filters very carefully. It is also advisable to use the theory of lifting to compute the proper frequency responses.

7.8 Pulse-Transfer-Function Formalism

Linear continuous-time systems can be conveniently described, analyzed, and synthesized using algebraic methods. When the theory of sampled-data systems was developed, it was natural to try to develop similar algebraic tools. Much of the early development of the theory of sampled-data systems went in this direction.

The approach is useful, simple, and successful if the system is viewed from the computer or if the process is observed at times that are synchronized with the computer clock, because the system is then time-invariant. (See Chapter 3 for the appropriate analysis.) However, when the system is analyzed from the process point of view, as is done in this chapter, the system is time-variable. The algebraic approach then loses some of its simplicity, because multiplication with time functions does not commute with differential and difference operators. For the case of completeness, a brief description of the algebraic system theory in the more complicated case is given. The main reason is historical. Much of the theory of sampled-data systems was originally developed using this approach, which is also used in many papers.

Goals

Before going into the details, it is useful to state the goals. The main purpose is to develop a formalism for manipulating the system descriptions. The formalism will have many properties in common with the transform methods for linear time-invariant systems. Each A-D and D-A converter is associated with a sampling operation. Because sampling can be described as an amplitude modulation, the time-varying parts will be associated with these operations. The system can then be separated into different parts: Some parts are ordinary linear time-invariant systems that can be handled by the ordinary transform methods; the other parts consist of the samplers that are intrinsically time-varying.

The z-Transform

Section 2.7 introduces z-transforms as mappings from sequences to functions of a complex variable. A different z-transform whose domain is continuous functions can be defined as follows:

DEFINITION 7.1   THE z-TRANSFORM   The z-transform of a *continuous-time function* is defined as

$$\tilde{F}(z) = \sum_{k=0}^{\infty} z^{-k} f(kh) \tag{7.32}$$

The inverse transform is given by

$$f(kh) = \frac{1}{2\pi i} \oint_{\Gamma} z^{k-1} \tilde{F}(z) \, dz$$

where the contour of integration Γ encloses all singularities of the integrand. ∎

The z-transform of a continuous-time signal is thus obtained by sampling the signal and then applying the z-transform to the sampled sequence. Because the transform depends only on the values at the sampling instants, all time functions that agree at the sampling instants have the same transform.

Notice that the transform is inherently related to the clock, which defines the sampling instants. Also notice that the inverse transform defines the function at the sampling instants only.

These properties of the z-transform of a continuous-time function are easily misunderstood and have led to much confusion and many mistakes.

Two Basic Theorems

To develop an algebra that allows formal manipulation of the systems, two theorems are needed. The first theorem tells how the z-transform of a continuous-time function is related to its Laplace transform.

THEOREM 7.2   Let the function f have the Laplace transform F and the z-transform $\tilde{F}$, and let F^* be the Laplace transform of the sampled representation f^* of f. Assume that for some $\varepsilon > 0, |F(s)| \leq |s|^{-1-\varepsilon}$ for large $|s|$ then

$$F^*(s) = \tilde{F}(e^{sh}) = \frac{1}{h} \sum_{k=-\infty}^{\infty} F(s + ik\omega_s) \tag{7.33}$$

where $\omega_s = 2\pi/h$ is the sampling frequency.

Proof.   The definition of F^* gives

$$F^*(s) = \int_0^\infty e^{-st} f^*(t)\, dt$$

$$= \int_0^\infty e^{-st} f(t) m(t)\, dt$$

$$= \int_0^\infty e^{-st} f(t) \left(\sum_{k=-\infty}^{\infty} \delta(t - kh) \right) dt$$

where the last equality is obtained from (7.26). Interchange the order of integration and summation gives

$$F^*(s) = \sum_{-\infty}^{\infty} \int_0^\infty e^{-st} f(t) \delta(t - kh)\, dt$$

$$= \sum_{k=0}^{\infty} (e^{sh})^{-k} f(kh)$$

$$= \tilde{F}(e^{sh})$$

The last equality follows from (7.32).

Because the Laplace transform of a product of two functions is a convolution of their transforms, it follows that

$$F^*(s) = F(s) * M(s) = \frac{1}{2\pi i} \int_{\gamma-i\infty}^{\gamma+i\infty} F(v) M(s-v)\, dv$$

$$= \frac{1}{2\pi i} \int_{\gamma-i\infty}^{\gamma+i\infty} F(v) \frac{1}{1 - e^{-h(s-v)}}\, dv \tag{7.34}$$

The integration path should be to the right of all poles of F and to the left of all poles of M (see Fig. 7.28). If F goes to zero faster than $|s|^{-1-\varepsilon}$ as $|s| \to \infty$,

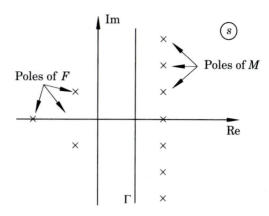

Figure 7.28   Singularities of F and M and the integration contour Γ.

the integral of FM on a large semicircle will vanish. Upon completion of the integration path by a large semicircle to the right, the integral can be evaluated with residue calculus. In the domain enclosed by the contour the integrand has simple poles at the zeros of

$$e^{h(s-v)} = 1$$

that is, at

$$v_k = s + \frac{2\pi ik}{h} = s + ik\omega_s \qquad k = \ldots - 1, 0, 1, \ldots$$

The residues at these poles are

$$\frac{1}{h} F\left(s + \frac{2\pi ik}{h}\right)$$

Summation of the residues now gives (7.33).   ■

Remark 1.   Notice that Eq. (7.33) can also be written as

$$F^*(s) = \frac{1}{h}\left(F(s) + F(s + i\omega_s) + F(s - i\omega_s) + \cdots\right)$$

Remark 2.   Notice that if F is analytic for Re $s < -\gamma_0$, the integration path in (7.34) may be closed by a large semicircle to the left. The following formula is obtained:

$$\tilde{F}(z) = \sum_{\text{Poles of } F} \text{Res}\left(F(s)\frac{z}{z - \exp(sh)}\right)$$

This gives a proof of formula (2.31).

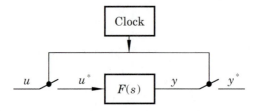

Figure 7.29 Block diagram of a system with two samplers.

Remark 3.   The theorem can be extended to the case in which the function F goes to zero as $1/|s|$ for large $|s|$. Equation (7.33) is then replaced by

$$F^*(s) = \frac{1}{h} \sum_{k=-\infty}^{\infty} F(s + ik\omega_s) + \frac{1}{2} f(0+)$$

Remark 4.   In the literature the same notation is sometimes used for the functions F^* and F. This is confusing and should be avoided.

Remark 5.   Notice that (7.33) is closely related to (7.3) for the Fourier transform of a sampled signal.

Pulse-transfer functions.   Section 7.6 shows that the input-output relationship of a sampler followed by a linear transfer function is given by Eq. (7.29). This equation cannot be described by a transfer function. If a *fictitious sampler* is added to the system output, the configuration shown in Fig. 7.29 is obtained. For this system it is possible to define a transfer function. The input-output relationship is given by

$$y^*(t) = \left(f(t) * u^*(t) \right)^*$$

The following theorem is useful for obtaining the corresponding transforms.

THEOREM 7.3   Let f and g be functions that have Laplace transforms and let m be the modulation function corresponding to an impulse train. Then

$$m(t)\left(f(t) * \left(m(t)g(t) \right) \right) = \left(m(t)f(t) \right) * \left(m(t)g(t) \right) \qquad (7.35)$$

or, equivalently,

$$\left(f(t) * g^*(t) \right)^* = f^*(t) * g^*(t) \qquad (7.36)$$

Proof.   Use of the definition of a convolution allows the left-hand side of (7.35) to be written as

$$(f * g^*)^*(t) = m(t) \int_{-\infty}^{\infty} f(t - \tau)g^*(\tau)\, d\tau = \int_{-\infty}^{\infty} m(t)f(t - \tau)m(\tau)g(\tau)\, d\tau$$

Similarly, the right-hand side of Eq. (7.36) can be written as

$$(f^* * g^*)(t) = \int_{-\infty}^{\infty} m(t-\tau)f(t-\tau)m(\tau)g(\tau)\,d\tau$$

$$= \int_{-\infty}^{\infty} m(t)f(t-\tau)m(\tau)g(\tau)\,d\tau$$

The last equality holds because $m(\tau) \neq 0$ only for $\tau = nh$ and $m(t-nh) = m(t)$.

∎

Remark 1.   The Laplace transformation of (7.35) gives

$$\left(F(s)G^*(s)\right)^* = F^*(s)G^*(s) \tag{7.37}$$

Remark 2.   Notice that the multiplication by m outside the brace in (7.35) can be interpreted as introduction of a fictitious sampler.

A Formalism

It is now straightforward to develop a formalism for dealing with sampled systems. First, a system is represented by a block diagram. Each A-D converter is represented as an ideal sampler. Each D-A converter is represented as a hold circuit having the transfer function (7.27). Linear continuous-time blocks are represented by their transfer functions, and linear calculations in the computer, by their pulse-transfer functions. The paths between the samplers can be reduced using ordinary rules for linear time-invariant systems. The equations describing the system are then written down. Theorems 7.2 and 7.3 are then used to rewrite the equations. The procedure is illustrated by two examples.

Example 7.10   Translation of a simple computer-controlled system

Consider the standard configuration of a computer-controlled system shown in Fig. 7.30(a). The process is characterized by a linear transfer function G, and the calculations performed in the computer are represented by a pulse-transfer function H. The analog and digital parts of the system are, as usual, connected via D-A and A-D converters. To apply the formalism, the A-D converter is represented by an ideal sampler. The computer is represented as a system that transforms an impulse-modulated signal to another impulse-modulated signal. The D-A converter is represented by a sampler, followed by a zero-order hold. It is assumed that the samplers are perfectly synchronized. The block diagram shown in Fig. 7.30(b) is then obtained. The analog parts are thus the hold and the process. Their combined transfer function is

$$F(s) = \frac{1}{s}\left(1 - e^{-sh}\right)G(s)$$

The Laplace transform Y of the output y is given by

$$Y(s) = F(s)U^*(s)$$

(a)

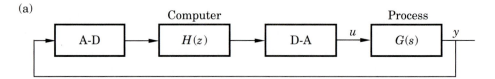

(b)

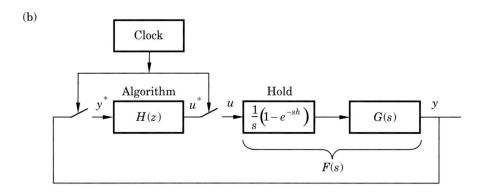

(c)

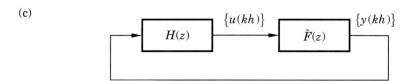

Figure 7.30   Standard configuration of a computer-controlled system.

The sampled output has the transform

$$Y^*(s) = \left(F(s)U^*(s)\right)^* = F^*(s)U^*(s)$$

where (7.37) is used to obtain the last equality. The relationship between y^* and u^* can thus be represented by the pulse-transfer function

$$\tilde{F}(z) = F^*(s)\,|_{s=(\ln z)/h}$$

The calculations in the computer can furthermore be represented by the pulse-transfer function $H(z)$. If the loop is cut in the computer the pulse-transfer function is thus

$$H(z)\tilde{F}(z)$$

A block diagram of the properties of the system that can be seen from the computer is shown in Fig. 7.30(c). By considering all signals as sequences like $\{y(kh),k = \dots -1,0,1,\dots\}$ and by introducing appropriate pulse-transfer functions for the algorithm and the process with the sample-and-hold, a representation that is equivalent to the ordinary block-diagram representation of continuous-time systems was thus obtained.   ∎

A further illustration is given by a slightly more complicated example.

Example 7.11   Translation of a computer-controlled system with two loops

The system illustrated in Fig. 7.31(a) has two measured analog signals, y_1 and y_2, and one analog command signal, u_c. The analog signals are scanned by a multiplexer and converted to digital form. The computer calculates the control signal, which is fed to the process via the D-A converter. Figure 7.31(b) is obtained by the procedure given in Example 7.10. We now introduce

$$F_1(s) = G_1(s)\frac{1}{s}(1 - e^{-sh})$$
$$F_2(s) = G_2(s)F_1(s)$$

The Laplace transforms of the output signals are then given by

$$Y_1(s) = F_1(s)U^*(s)$$
$$Y_2(s) = F_2(s)U^*(s)$$

Hence

$$Y_1^*(s) = (F_1(s)U^*(s))^* = F_1^*(s)U^*(s)$$
$$Y_2^*(s) = (F_2(s)U^*(s))^* = F_2^*(s)U^*(s)$$

It follows from (7.33) and (7.37) that

$$Y_1(z) = \tilde{F}_1(z)U(z)$$
$$Y_2(z) = \tilde{F}_2(z)U(z)$$

Let the calculations performed by the control computer be represented by

$$U(z) = H_c(z)U_c(z) - H_1(z)Y_1(z) - H_2(z)Y_2(z)$$

The relationship between the output, Y_2, and the sampled command signal, U_c, is

$$Y_2(z) = \frac{H_c(z)\tilde{F}_2(z)}{1 + H_1(z)\tilde{F}_1(z) + H_2(z)\tilde{F}_2(z)}U_c(z)$$

Notice, however, that the relationship between the analog signals y_1 and u_c cannot be represented by a simple pulse-transfer function because of the periodic nature of the sampled-data system.

With the introduction of the sampled signals as sequences and pulse-transfer functions, the system can be represented as in Fig. 7.31(c).   ∎

Modified z-Transforms

The problem of sampling a system with a delay can be handled by the modified z-transform defined in Definition 2.2. The modified z-transform is useful for many purposes—for example, the intersample behavior can easily be investigated using these transforms. There are extensive tables of modified z-transforms and many theorems about their properties (see the References).

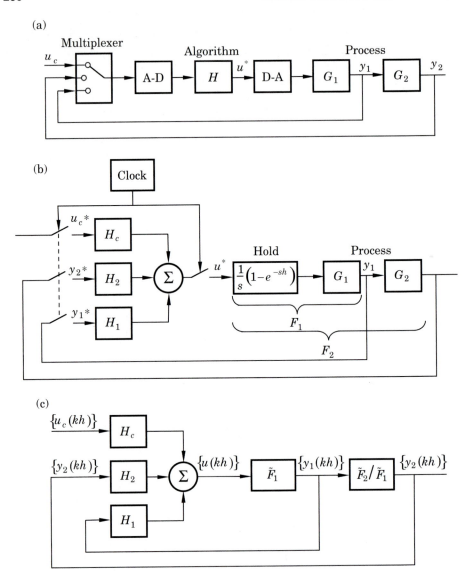

Figure 7.31 Computer-controlled system with multiplexer and two feedback loops and equivalent block diagram.

7.9 Multirate Sampling

So far only systems in which the A-D and the D-A conversions are made at the same rates have been discussed. In the discussion of postsampling filters in Sec. 7.4 it was indicated that it may be advantageous to make the D-A conversion more rapidly. There are also situations where the converse is true. It is, for example, difficult to implement antialiasing filters with long time constants using analog techniques. In such cases it is much easier to sample the signals rapidly with analog antialiasing filters and to do digital filtering afterward. In both cases the systems have two samplers that operate at different rates. This is called *multirate sampling*. Such sampling schemes may be necessary for systems with special data-transmission links or special sensors and actuators and are useful for improving the responses of systems in which measurements are obtained at slow rates, for example, when laboratory instruments are used. Multirate systems may allow better control of what happens between the sampling instants. In multivariable systems it may also be advantageous to have different sampling rates in different loops to reduce the computational load and to improve the numeric conditioning. Use of multirate sampling is also natural in multiprocessor systems.

A detailed treatment of multirate systems is outside the scope of this book; however, a short discussion of the major ideas will be given to show how the methods presented in the book can be extended to also cover multirate systems.

State-Space Descriptions

Consider a system composed of two subsystems that are continuous constant-coefficient dynamic systems. Assume that there are two periodic samplers with periods h_1, and h_2. Let the ratio of the periods be a rational number $h_1/h_2 = m_1/m_2$, where m_1 and m_2 have no common factor. Then there exists a smallest integer m and a real number h such that

$$m = m_1 m_2 \qquad h_1 = \frac{hm_1}{m} \qquad h_2 = \frac{hm_2}{m}$$

If the samplers are synchronized, it follows that the control signals will be constant over sampling periods of length h/m. Sampling with that period gives a discrete-time system that is periodic with period h. The system can then be described as a constant discrete-time system if the values of the system variables are considered only at integer multiples of h. The ordinary discrete-time theory can then be applied. An example illustrates the idea.

Example 7.12   Multirate systems

Consider the system shown in Fig.7.32, which has two subsystems and two samplers with periods 0.5 and 1. It is assumed that the samplers are synchronized. It is also assumed that the hold circuits are included in the subsystems. If the

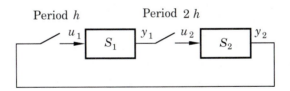

Figure 7.32 Block diagram of a simple multirate system.

subsystems are sampled with period 0.5 and 0.5 is chosen as a time unit, then

$$\begin{cases} x_1(k+1) = \Phi_1 x_1(k) + \Gamma_1 u_1(k) \\ \quad y_1(k) = C_1 x_1(k) \end{cases}$$

$$\begin{cases} x_2(k+1) = \Phi_2 x_2(k) + \Gamma_2 u_2(k) \\ \quad y_2(k) = C_2 x_2(k) \end{cases}$$

The interconnection are described by

$$\begin{aligned} u_1(k) &= y_2(k) & k &= \ldots -1, 0, 1, 2, \ldots \\ u_2(k) &= y_1(k) & k &= \ldots -1, 0, 1, 2, \ldots \end{aligned}$$

The system is periodic with a period of two sampling intervals. A time-invariant description can be obtained by considering the system variables at even sampling periods only. Straightforward calculations give

$$\begin{pmatrix} x_1(2k+2) \\ x_2(2k+2) \end{pmatrix} = \begin{pmatrix} \Phi_1^2 + \Gamma_1 C_2 \Gamma_2 C_1 & \Phi_1 \Gamma_1 C_2 + \Gamma_1 C_2 \Phi_2 \\ (\Phi_2 \Gamma_2 + \Gamma_2)C_1 & \Phi_2^2 \end{pmatrix} \begin{pmatrix} x_1(2k) \\ x_2(2k) \end{pmatrix} \qquad (7.38)$$

This equation can be used to analyze the response of the multirate system. For example, the stability condition is that the matrix on the right-hand side of (7.38) has all its eigenvalues inside the unit disc. The values of the state variables at odd sampling periods are given by

$$\begin{pmatrix} x_1(2k+1) \\ x_2(2k+1) \end{pmatrix} = \begin{pmatrix} \Phi_1 & \Gamma_1 C_2 \\ \Gamma_2 C_1 & \Phi_2 \end{pmatrix} \begin{pmatrix} x_1(2k) \\ x_2(2k) \end{pmatrix}$$

∎

The analysis illustrated by the example can be extended to an arbitrary number of samplers provided that the ratios of the sampling periods are rational numbers. Delayed sampling can also be handled by the methods described in Sec. 2.3.

Input-Output Methods

Multirate systems can also be investigated by input-output analysis. First, observe as before that the system is periodic with period h if the ratios of the sampling periods are rational numbers. The values of the system variables at times that are synchronized to the period can then be described as a time-invariant dynamic system. Ordinary operator or transfer-function methods for linear systems can then be used. The procedure for analyzing a system can be described as follows: A block diagram of the system including all subsystems and all samplers is first drawn. The period h is determined. All samplers appearing in the system then have periods h/m, where m is an integer. A trick called *switch decomposition* is then used to convert samplers with rate h/m to a combination of samplers with period h. The system can then be analyzed using the methods described in Sec. 7.8.

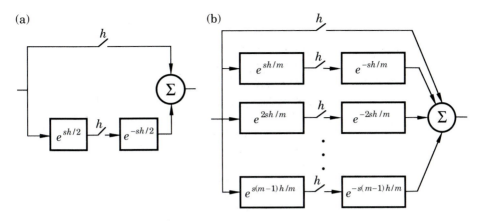

Figure 7.33   Representation of samplers with periods (a) $h/2$ and (b) h/m by switch decomposition.

Switch Decomposition

To understand the concept of switch decomposition, first consider a sampler with period $h/2$. Such a sampling can be obtained by combining a sampler with period h and another sampler with period h that is delayed $h/2$. The scheme is illustrated in Fig. 7.33(a). The idea can easily be extended to sampling at rate h/m, where m is an arbitrary integer [see Fig. 7.33(b)].

Multirate Systems with Nonrational Periods

The methods described so far will work only when the ratios of the sampling periods are rational numbers. If this is not the case, it is not possible to obtain a periodic system; different techniques must then be used. The multirate techniques also lead to complicated analysis if there are many samplers with a wide range of periods.

7.10 Problems

7.1   The signal

$$f(t) = a_1 \sin 2\pi t + a_2 \sin 20t$$

is the input to a zero-order sample-and-hold circuit. Which frequencies are there at the output if the sampling period is $h = 0.2$?

7.2   A signal that is going to be sampled has the spectrum shown in Fig. 7.34. Of interest are the frequencies in the range from 0 to f_1 Hz. A disturbance has a fixed known frequency with $f_2 = 5f_1$. Discuss choice of sampling interval and presampling filter.

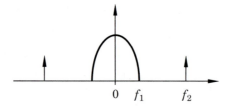

Figure 7.34

7.3 Show that the system in Fig. 7.35 is an implementation of a first-order hold and determine its response to a pulse of unit magnitude and a duration of one sampling interval.

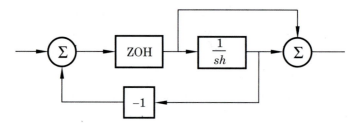

Figure 7.35

7.4 Sample a sinusoidal signal $u(t) = sin(t)$ using zero-order hold, first-order hold, and predictive first-order hold. Compare the different hold circuits when the sampling period is changed.

7.5 The magnitude of the spectrum of a signal is shown in Fig. 7.36. Sketch the magnitude of the spectrum when the signal has been sampled with (a) $h = 2\pi/10$ s, (b) $h = 2\pi/20$ s, and (c) $h = 2\pi/50$ s.

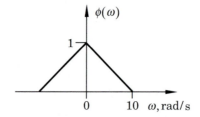

Figure 7.36

7.6 Consider the signal in Problem 7.5, but let the spectrum be centered around $\omega = 100$ rad/s and with (a) $\omega_s = 120$ rad/s and (b) $\omega_s = 240$ rad/s.

7.7 A camera is used to get a picture of a rotating wheel with a mark on it. The wheel rotates at r revolutions per second. The camera takes one frame each h seconds. Discuss how the picture will appear when shown on a screen. (Compare with what you see in western movies.)

7.8 The signal $y(t) = \sin 3\pi t$ is sampled with the sampling period h. Determine h such that the sampled signal is periodic.

7.9 An amplitude modulated signal

$$u(t) = \sin(4\omega_0 t)\cos(2\omega_0 t)$$

is sampled with $h = \pi/3\omega_0$. Determine the frequencies f, $0 \le f \le 3\omega_0/2\pi$ that are represented in the sampled signal.

7.10 Find Y^* for the systems in Fig. 7.37.

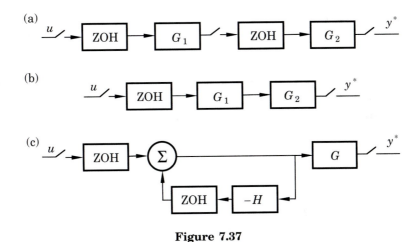

Figure 7.37

7.11 Write a program to compute the frequency response of a sampled-data system. Let the following be the input to the program:

(a) The polynomials in the pulse-transfer function $H(z)$.

(b) The sampling interval.

(c) The maximum and minimum frequencies.

Use the program to plot $H(\exp(i\omega h))$ for the normalized motor sampled with a zero-order hold and compare with the continuous-time system.

7.11 Notes and References

The fact that a sinusoid can be retrieved from its sampled values if it is sampled at least twice per period was stated in Nyquist (1928). The sampling theorem in the form presented in this chapter was introduced in Shannon (1949), where the implications for communication were emphasized. The results had, however, been known earlier as a theorem in mathematics. In the Soviet communication literature, the theorem was introduced by Kotelnikov (1933). A review of the sampling theorem with many references is given in Jerri (1977).

There are many ways of sampling. A review of different schemes is given in Jury (1961). Different types of hold circuits are discussed in more detail in Ragazzini and Franklin (1958). Selection of the sampling period for signal processing is discussed in Gardenhire (1964). The different trade-offs in the areas of control and signal processing may lead to very different rules for choosing the sampling rate. Predictive first-order hold is discussed in Bernhardsson (1990) an and application to motion control is described in Åström and Kanniah (1994).

The approach taken in this chapter corresponds to the classic treatment of sampled-data systems. The modulation model was proposed by MacColl (1945) and elaborated on by Linvill (1951). A more detailed treatment is given in the classic texts by Ragazzini and Franklin (1958) and Jury (1958). The ideal-sampler approximation is discussed in Li, Meiry, and Curry (1972).

Frequency response is important from the point of view of both analysis and design. A fuller treatment of this problem is given in Lindorff (1965). Practical applications of frequency-response analysis are discussed in Flower, Windett, and Forge (1971). New aspects of frequency analysis of sampled-data systems are found in Araki and Ito (1993), Yamamoto (1994), Yamamoto and Araki (1994), and Yamamoto and Khargonekar (1996).

More material on the z-transform is given in Jury (1982). The modified z-transform is discussed in Jury (1958). Tables of modified z-transforms are also given in that book.

Systems with multirate sampling were first analyzed in Kranc (1957). Additional results are given in Jury (1967a, 1967b), Konar and Mahesh (1978), Whitbeck (1980), and Crochieve and Rabiner (1983).

8

Approximating Continuous-Time Controllers

8.1 Introduction

There are situations when a continuous-time controller is already available. A typical case is when an analog-control system is replaced by a computer-control system. It is then natural to try to convert the continuous-time controller to a discrete-time controller directly. A straightforward approach is to use a short sampling interval and to make some discrete-time approximations of the continuous-time controller. This approach is illustrated in Example 1.2. See, for example, Fig. 1.6, which compares a continuous-time controller with an approximating discrete-time controller. In Sec. 8.2 we will present several methods for approximating a continuous-time controller given in terms of their transfer functions. Similar methods for controllers given in state-space form are presented in Sec. 8.3. In Sec. 8.5 the results are used to obtain digital PID controllers. Some practical aspects of implementing a digital PID controller will also be discussed in that section.

8.2 Approximations Based on Transfer Functions

This section assumes that a continuous-time controller is given as a transfer function, $G(s)$. It is desired to find an algorithm for a computer so that the digital system approximates the transfer function $G(s)$ (see Fig. 8.1). This problem is interesting for implementation of both analog controllers and digital filters. The approximation may be done in many different ways. Digital implementation includes a data reconstruction, which also can be made in different ways—for example, zero- or first-order hold.

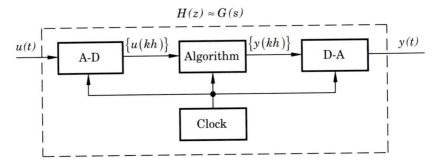

Figure 8.1   Approximating a continuous-time transfer function, $G(s)$, using a computer.

Differentiation and Tustin Approximations

A transfer function represents a differential equation. It is natural to obtain a difference equation by approximating the derivatives with a forward difference (Euler's method)

$$px(t) = \frac{dx(t)}{dt} \approx \frac{x(t+h) - x(t)}{h} = \frac{q-1}{h} x(t)$$

or a backward difference

$$px(t) = \frac{dx(t)}{dt} \approx \frac{x(t) - x(t-h)}{h} = \frac{q-1}{qh} x(t)$$

In the transform variables, this corresponds to replacing s by $(z-1)/h$ or $(z-1)/zh$. Section 2.8 shows that the variables z and s are related in some respects as $z = \exp(sh)$. The difference approximations correspond to the series expansions

$$z = e^{sh} \approx 1 + sh \qquad \text{(Euler's method)} \qquad (8.1)$$

$$z = e^{sh} \approx \frac{1}{1 - sh} \qquad \text{(Backward difference)} \qquad (8.2)$$

Another approximation, which corresponds to the trapezoidal method for numerical integration, is

$$z = e^{sh} \approx \frac{1 + sh/2}{1 - sh/2} \qquad \text{(Trapezoidal method)} \qquad (8.3)$$

In digital-control context, the approximation in (8.3) is often called *Tustin's approximation*, or *the bilinear transformation*. Using the approximation methods above, the pulse-transfer function $H(z)$ is obtained by simply replacing the

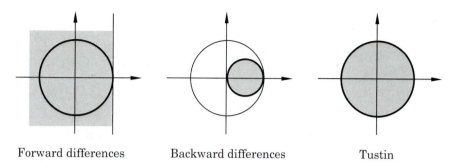

Forward differences   Backward differences   Tustin

Figure 8.2 Mapping of the stability region in the s-plane on the z-plane for the transformations (8.4), (8.5), and (8.6).

argument s in $G(s)$ by s', where

$$s' = \frac{z - 1}{h} \qquad \text{(Forward difference or Euler's method)} \qquad (8.4)$$

$$s' = \frac{z - 1}{zh} \qquad \text{(Backward difference)} \qquad (8.5)$$

$$s' = \frac{2}{h} \cdot \frac{z - 1}{z + 1} \qquad \text{(Tustin's approximation)} \qquad (8.6)$$

Hence

$$H(z) = G(s')$$

The methods are very easy to apply even for hand calculations. Figure 8.2 shows how the stability region $\mathrm{Re}\, s < 0$ in the s-plane is mapped on the z-plane for the mappings (8.4), (8.5), and (8.6).

With the forward-difference approximation it is thus possible that a stable continuous-time system is mapped into an unstable discrete-time system. When the backward approximation is used, a stable continuous-time system will always give a stable discrete-time system. There are, however, also unstable continuous-time systems that are transformed into stable discrete-time systems. Tustin's approximation has the advantage that the left half-s-plane is transformed into the unit disc. Stable continuous-time systems are therefore transformed into stable sampled systems, and unstable continuous-time systems are transformed into unstable discrete-time systems.

Frequency Prewarping

One problem with the approximations discussed earlier is that the frequency scale is distorted. For instance, if it is desired to design band-pass or notch filters, the digital filters obtained by the approximations may not give the correct frequencies for the band-pass or the notches. This effect is called *frequency warping*. Consider an approximation obtained by Tustin's approximation. The

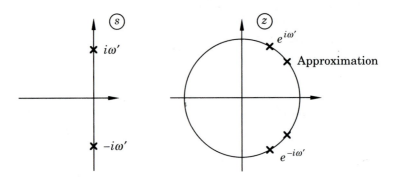

Figure 8.3 Frequency distortion (warping) obtained with approximation.

transmission of sinusoids for the digital filter is given by

$$H(e^{i\omega h}) = \frac{1}{i\omega h}\,(1 - e^{-i\omega h})\,G\left(\frac{2}{h}\cdot\frac{e^{i\omega h} - 1}{e^{i\omega h} + 1}\right)$$

The first two factors are due to the sample-and-hold operations; compare with (7.28). The argument of G is

$$\frac{2}{h}\frac{e^{i\omega h} - 1}{e^{i\omega h} + 1} = \frac{2}{h}\frac{e^{i\omega h/2} - e^{-i\omega h/2}}{e^{i\omega h/2} + e^{-i\omega h/2}} = \frac{2i}{h}\,\tan\left(\frac{\omega h}{2}\right)$$

The frequency scale is thus distorted. Assume, for example, that the continuous-time system blocks signals at the frequency ω'. Because of the frequency distortion, the sampled system will instead block signal transmission at the frequency ω, where

$$\omega' = \frac{2}{h}\,\tan\left(\frac{\omega h}{2}\right)$$

That is,

$$\omega = \frac{2}{h}\,\tan^{-1}\left(\frac{\omega' h}{2}\right) \approx \omega'\left(1 - \frac{(\omega' h)^2}{12}\right) \tag{8.7}$$

This expression gives the distortion of the frequency scale (see Fig. 8.3). It follows from (8.7) that there is no frequency distortion at $\omega = 0$ and that the distortion is small if ωh is small. It is easy to introduce a transformation that eliminates the scale distortion at a specific frequency ω_1 by modifying Tustin's transformation from (8.6) to the transformation

$$s' = \frac{\omega_1}{\tan(\omega_1 h/2)}\cdot\frac{z - 1}{z + 1} \qquad \text{(Tustin with prewarping)} \tag{8.8}$$

From (8.8), it follows that

$$H\left(e^{i\omega_1 h}\right) = G(i\omega_1)$$

that is, the continuous-time filter and its approximation have the same value at the frequency ω_1. There is, however, still a distortion at other frequencies.

Example 8.1   Frequency prewarping

Assume that the integrator

$$G(s) = \frac{1}{s}$$

should be implemented as a digital filter. Using the transformation of (8.6) without prewarping gives

$$H_T(z) = \frac{1}{\dfrac{2}{h} \cdot \dfrac{z-1}{z+1}} = \frac{h}{2} \cdot \frac{z+1}{z-1}$$

Prewarping gives

$$H_P(z) = \frac{\tan\left(\omega_1 h/2\right)}{\omega_1} \cdot \frac{z+1}{z-1}$$

The frequency function of H_P is

$$H_P(e^{i\omega h}) = \frac{\tan\left(\omega_1 h/2\right)}{\omega_1} \cdot \frac{e^{i\omega h}+1}{e^{i\omega h}-1} \frac{\tan\left(\omega_1 h/2\right)}{\omega_1} \cdot \frac{1}{\tan\left(\omega h/2\right)}$$

thus $G(i\omega)$ and $H_P\left(e^{i\omega h}\right)$ are equal for $\omega = \omega_1$.   ∎

Step Invariance

Another way to generate approximations is to use the ideas developed in Chapter 2. In this way it is possible to obtain approximations that give correct values at the sampling instants for special classes of input signals. For example, if the input signal is constant over the sampling intervals, Table 2.1 or Eq. (2.30) give an appropriate pulse-transfer function $H(z)$ for a given transfer function $G(s)$. Because this relation gives the correct values of the output when the input signal is a piecewise constant signal that changes at the sampling instants, it is called *step invariance*.

Ramp Invariance

The notion of step invariance is ideally suited to describe a system where the input signal is generated by a computer, because the input signal is then constant over the sampling period. The approximation is, however, not so good when dealing with input signals that are continuous. In this case it is much better to use an approximation where the input signal is assumed to vary linearly between the sampling instants. The approximation obtained is called *ramp invariance* because it gives the values of the output at the sampling instants exactly for ramp signals. It is identical to predictive first-order-hold sampling that was discussed in Sec. 7.5. Notice that because of the predictive nature of ramp invariance there must be one delay in the controller; see Sec. 7.5.

Comparison of Approximations

The step-invariant method is not suitable for approximation of continuous-time transfer functions. The reason is that the approximation of the phase curve is unnecessarily poor. Both Tustin's method and the ramp-invariant method give better approximations. Tustin's method is a little simpler than the ramp-invariant method. The ramp-invariant method does give correct sampled poles. This is not the case for Tustin's method. This difference is particularly important when implementing notch filters where Tustin's method gives a frequency distortion. Another drawback with Tustin's method is that very fast poles of the continuous-time system appear in sampled poles close to $z = -1$, which will give rise to ringing in the digital filter. The different approximations are illustrated by an example.

Example 8.2   Sampled approximations of transfer function

Consider a continuous-time system with the transfer function

$$G(s) = \frac{(s + 1)^2(s^2 + 2s + 400)}{(s + f)^2(s^2 + 2s + 100)(s^2 + 3s + 2500)}$$

Let $H(z)$ be the pulse-transfer function representing the algorithm in Fig. 8.1. The transmission properties of the digital filter in Fig. 8.1 depend on the nature of the D-A converter. If it is assumed that the converter keeps the output constant between the sampling periods, the transmission properties of the filter are described by

$$\hat{G}(s) = \frac{1}{sh}\ (1 - e^{-sh})H(e^{sh})$$

where the pulse-transfer function H depends on the approximation used. Figure 8.4 shows Bode diagrams of H for the different digital filters obtained by step equivalence, ramp equivalence, and Tustin's method. The sampling period is 0.03 s in all cases. This implies that the Nyquist frequency is 105 rad/s. All methods except Tustin's give a good approximation of the amplitude curve. The frequency distortion by Tustin's method is noticeable at the notch at 20 rad/s and very clear at the resonance at 50 rad/s.

 The step-equivalence method gives a noticeable phase error. This corresponds approximately to a time delay of half a sampling interval. Ramp equivalence gives a negligible phase error. The phase curve for Tustin's approximation also deviates because of the frequency warping. Notice that all approximations suffer from the time delay due to the sample and hold. Ramp equivalence gives the best approximation of both amplitude and phase.   ■

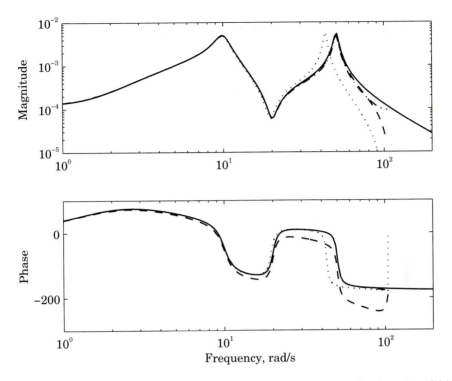

Figure 8.4 Bode diagrams of a continuous-time transfer function $G(s)$ and different sampled approximations $H(e^{sh})$, continuous-time transfer function (solid), ramp invariance (dashed-dotted), step invariance (dashed), and Tustin's approximation (dotted).

Antialiasing Filters

The consequences of aliasing and the importance of antialiasing filters were discussed in Sec. 7.4. Choice of sampling rate and antialiasing filters is important in digital systems that are based on translation of analog design. Some consequences of the selection of sampling rates have been discussed previously. The sampling rate must be so large that the errors due to the approximation are negligible.

The necessity of taking the antialiasing filter into account in the control design can be determined from the results of Sec. 7.4. In general, the antialiasing filter must be taken into consideration when making the design of the controller.

Selection of Sampling Interval

The choice of sampling period depends on many factors. One way to determine the sampling period is to use continuous-time arguments. The sampled system can be approximated by the hold circuit, followed by the continuous-time system. For small sampling periods, the transfer function of the hold circuit can be approximated as

$$\frac{1 - e^{-sh}}{sh} \approx \frac{1 - 1 + sh - (sh)^2/2 + \cdots}{sh} = 1 - \frac{sh}{2} + \cdots$$

The first two terms correspond to the series expansion of $\exp(-sh/2)$. That is, for small h, the hold can be approximated by a time delay of half a sampling interval. Assume that the phase margin can be decreased by 5° to 15°. This gives the following rule of thumb:

$$h\omega_c \approx 0.15 \text{ to } 0.5$$

where ω_c is the crossover frequency (in radians per second) of the continuous-time system. This rule gives quite short sampling periods. The Nyquist frequency will be about 5 to 20 times larger than the crossover frequency.

Example 8.3   Digital redesign of lead compensator

Consider the system in Example A.2, which is a normalized model of a motor. The closed-loop transfer function

$$G_c(s) = \frac{4}{s^2 + 2s + 4}$$

is obtained with the lead compensator

$$G_k(s) = 4\,\frac{s+1}{s+2} \tag{8.9}$$

The closed-loop system has a damping of $\zeta = 0.5$ and a natural frequency of $\omega_0 = 2$ rad/s. The objective is now to find $H(z)$ in Fig. 8.5, which approximates (8.9).
Euler's method gives the approximation

$$H_E(z) = 4\,\frac{z - 1 + h}{z - 1 + 2h} = 4\,\frac{z - (1 - h)}{z - (1 - 2h)} \tag{8.10}$$

while Tustin's approximation gives

$$H_T(z) = 4\,\frac{(2 + h)z - 2 + h}{(2 + 2h)z - 2 + 2h} = 4\,\frac{2 + h}{2 + 2h} \cdot \frac{z - (2 - h)/(2 + h)}{z - (1 - h)/(1 + h)}$$

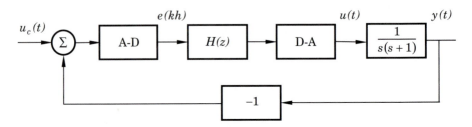

Figure 8.5   Digital control of the motor example.

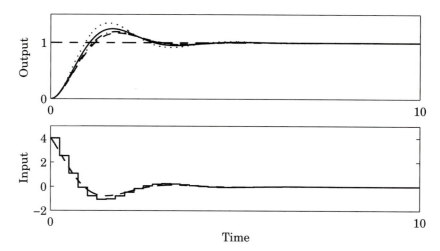

Figure 8.6 Process output, $y(t)$, when the motor is controlled using the compensator of (8.10) when $h = 0.1$ (dashed-dotted), 0.25 (solid), and 0.5 (dotted). The control signal is shown for $h = 0.25$. For comparison, the continuous-time signals are also shown (dashed).

Finally, zero-order-hold sampling of (8.9) gives

$$H_{\text{zoh}}(z) = \frac{4z - 2(1 + e^{-2h})}{z - e^{-2h}} = 4\,\frac{z - 0.5(1 + e^{-2h})}{z - e^{-2h}}$$

All approximations have the form

$$H(z) = \frac{b_0 z + b_1}{z + a_1}$$

The crossover frequency of the continuous-time process in cascade with the compensator (8.9) is $\omega_c = 1.6$ rad/s. The earlier rule of thumb gives a sampling period of about 0.1 to 0.3 s.

Figure 8.6 shows the control signal and the process output when Euler's approximation has been used for different sampling times. The other approximations give similar results. The closed-loop system has a satisfactory behavior for all compensators when the sampling time is short. The rule of thumb also gives reasonable values for the sampling period. The overshoot when $h = 0.5$ is about twice as large as for the continuous-time compensator. In the example, the change in u_c occurs at a sampling instant. This is not true in practice, and there may be a delay in the response of at most one sampling period.   ∎

8.3 Approximations Based on State Models

In this section we will make discrete-time approximations of controllers described by continuous-time state-space models. State-feedback controllers can be regarded as generalized P-controllers. The formulation of the problem assumes that the process is described by the equations

$$\frac{dx}{dt} = Ax + Bu$$

$$y = Cx$$

(8.11)

where all the states are assumed to be measurable. By using a controller of the form

$$u(t) = Mu_c(t) - Lx(t)$$

(8.12)

it is possible to place the poles of the closed-loop system arbitrarily if the system is controllable. The controller in (8.12) can be implemented in digital form by sampling the states and holding the control signal constant over the sampling intervals. This is how the control is done in Example 1.2. If the sampling period is increased, then the behavior of the closed-loop system starts to deteriorate. It is, however, possible to modify the controller in order to improve the performance of the closed-loop system. Assume that the discrete-time controller is

$$u(kh) = \tilde{M}u_c(kh) - \tilde{L}x(kh)$$

(8.13)

One way to solve the problem is to design the controller in (8.12) using sampled-data theory. This is done in Chapter 4. Here, an approximate method is used to translate the controller in (8.12) into discrete-time form.

Controlling (8.11) with the continuous-time controller in (8.12) gives the closed-loop system

$$\frac{dx}{dt} = (A - BL)x + BMu_c = A_c x + BMu_c$$

$$y = Cx$$

If $u_c(t)$ is constant over one sampling period, then this equation can be integrated; this gives

$$x(kh + h) = \Phi_c x(kh) + \Gamma_c Mu_c(kh)$$

(8.14)

where

$$\Phi = e^{A_c h}$$

$$\Gamma = \int_0^h e^{A_c s}\, ds\, B$$

If the discrete-time controller in (8.13) is used to control (8.11), then

$$x(kh + h) = (\Phi - \Gamma\tilde{L})x(kh) + \Gamma\tilde{M}u_c(kh)$$

(8.15)

where Φ and Γ are the system matrices obtained when (8.11) is sampled. It is in general not possible to choose $\tilde{L}$ such that

$$\Phi_c = \Phi - \Gamma\tilde{L}$$

However, we can make a series expansion and equate terms of different powers of h. Assume that

$$\tilde{L} = L_0 + L_1 h/2$$

then

$$\Phi_c \approx I + (A - BL)h + \left(A^2 - BLA - ABL + (BL)^2\right)h^2/2 + \cdots$$

and

$$\Phi - \Gamma\tilde{L} \approx I + (A - BL_0)h + \left(A^2 - ABL_0 - 2BL_1\right)h + \cdots$$

The systems (8.14) and (8.15) have the same poles up to and including order h^2 when

$$\tilde{L} = L\left(I + (A - BL)h/2\right) \tag{8.16}$$

Without modification of L the poles are the same up to and including order h.

The modification of M is determined by assuming that the steady-state values are the same for (8.14) and (8.15). Let the reference value be constant and assume that the steady-state value of the state is x^0. This gives the relations

$$(I - \Phi_c)x^0 = \Gamma_c M u_c$$

and

$$\left(I - (\Phi - \Gamma\tilde{L})\right)x^0 = \Gamma\tilde{M}u_c$$

The series expansions of the left-hand sides of these two relations are equal for powers of h up to and including h^2. Now determine $\tilde{M}$ such that the series expansions of the right-hand sides are the same for h and h^2. Assume that

$$\tilde{M} = M_0 + M_1 h/2$$

then

$$\Gamma_c M \approx BMh + (A - BL)BMh^2/2 + \cdots$$

and

$$\Gamma\tilde{M} \approx BM_0 h + (BM_1 + ABM_0)h^2/2 + \cdots$$

This gives

$$\tilde{M} = (I - LBh/2)M \tag{8.17}$$

The modifications (8.16) and (8.17) are easily computed using the continuous-time system and the continuous-time controller.

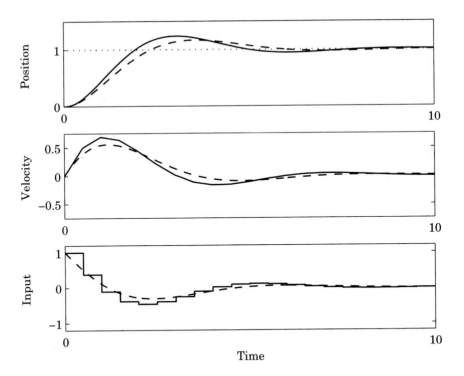

Figure 8.7 Digital control of the double integrator (solid) using the control law in (8.19) when $h = 0.5$. The continuous-time response when using (8.18) is shown by the dashed curves.

Example 8.4   Modification of a state-feedback controller

The system in Example A.1 is the double integrator; that is, the system is defined by the matrices

$$A = \begin{pmatrix} 0 & 1 \\ 0 & 0 \end{pmatrix} \qquad B = \begin{pmatrix} 0 \\ 1 \end{pmatrix} \qquad \text{and} \qquad C = \begin{pmatrix} 1 & 0 \end{pmatrix}$$

Let the continuous-time controller be

$$u(t) = u_c(t) - \begin{pmatrix} 1 & 1 \end{pmatrix} x(t) \tag{8.18}$$

Figure 8.7 shows the behavior when the sampled controller

$$u(kh) = u_c(kh) - \begin{pmatrix} 1 & 1 \end{pmatrix} x(kh) \tag{8.19}$$

is used when $h = 0.5$. Using the modifications in (8.16) and (8.17), we get

$$\tilde{L} = \begin{pmatrix} 1 - 0.5h & 1 \end{pmatrix}$$
$$\tilde{M} = 1 - 0.5h \tag{8.20}$$

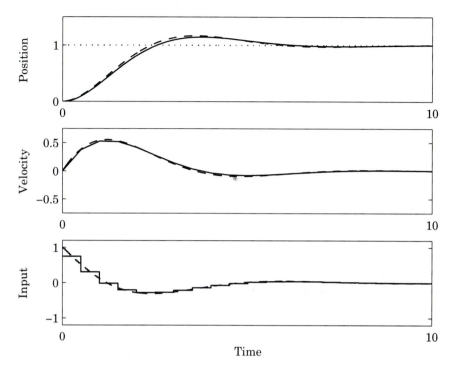

Figure 8.8 Control of the double integrator using the modified controller in (8.20) when $h = 0.5$ (solid). The continuous-time response when using (8.18) is also shown (dashed).

Figure 8.8 shows the behavior when the modified controller is used for $h = 0.5$; there is an improvement compared with the unmodified controller. However, the sampling period cannot be increased much further before the closed-loop behavior starts to deteriorate, even when the modified controller is used. The example shows that a simple modification can have a large influence on the performance. ∎

8.4 Frequency-Response Design Methods

This chapter has so far shown how continuous-time controllers can be translated into discrete-time forms. This section discusses how continuous-time frequency-design methods can be used to design discrete-time controllers.

Frequency-design methods based on Bode and Nichols plots are useful for designing compensators for systems described by transfer functions. The usefulness of the methods depends on the simplicity of drawing the Bode plots and on rules of thumb for choosing the compensators. The Bode plots are easy to draw because the transfer functions are in general rational functions in $i\omega$, except for pure time delays. Frequency curves for discrete-time systems are more difficult to draw because the pulse-transfer functions are not rational functions in $i\omega$, but in $\exp(i\omega h)$. The *w-transform method* is one way to circumvent this difficulty. The method can be summarized into the following steps:

1. Sample the continuous-time system that should be controlled using a zero-order-hold circuit. This gives $H(z)$.

2. Introduce the variable

$$w = \frac{2}{h}\frac{z-1}{z+1}$$

   [compare (8.6)]. Transform the pulse-transfer function of the process into the w-plane giving

$$H'(w) = H(z)\Big|_{z=\frac{1+wh/2}{1-wh/2}}$$

   For $z = \exp(i\omega h)$ then

$$w = i\frac{2}{h}\tan(\omega h/2) = iv$$

   (compare frequency prewarping in Sec.8.2). The transformed transfer function $H'(iv)$ is a rational function in iv.

3. Draw the Bode plot of $H'(iv)$ and use conventional methods to design a compensator $H'_c(iv)$ that gives desired frequency domain properties. The distortion of the frequency scale between v and ω must be taken into account when deciding, for instance, crossover frequency and bandwidth.

4. Transform the compensator back into the z-plane and implement $H_c(z)$ as a discrete-time system.

The advantage with the w-transform method is that conventional Bode diagram techniques can be used to make the design. One difficulty is to handle the distortion of the frequency scale and to choose the sampling interval.

8.5 Digital PID-Controllers

Many practical control problems are solved by PID-controllers. The "textbook" version of the PID-controller can be described by the equation

$$u(t) = K\left(e(t) + \frac{1}{T_i}\int^t e(s)\,ds + T_d\frac{de(t)}{dt}\right) \tag{8.21}$$

where error e is the difference between command signals u_c (the set point) and process output y (the measured variable). K is the *gain* or *proportional gain* of the controller, T_i the *integration time* or *reset time*, and T_d the *derivative time*. The PID-controller was originally implemented using analog technology that went through several development stages, that is, pneumatic valves, relays and motors, transistors, and integrated circuits. In this development much know-how was accumulated that was embedded into the analog design. Today virtually all PID-controllers are implemented digitally. Early implementations were often a pure translation of (8.21), which left out many of the extra features that were incorporated in the analog design. In this section we will discuss the digital PID-controller in some detail. This is a good demonstration that a good controller is not just an implementation of a "textbook" algorithm. It is also a good way to introduce some of the implementation issues that will be discussed in depth in Chapter 9.

Modification of Linear Response

A pure derivative can, and should not be, implemented, because it will give a very large amplification of measurement noise. The gain of the derivative must thus be limited. This can be done by approximating the transfer function sT_d as follows:

$$sT_d \approx \frac{sT_d}{1 + sT_d/N}$$

The transfer function on the right approximates the derivative well at low frequencies but the gain is limited to N at high frequencies. N is typically in the range of 3 to 20.

In the work with analog controllers it was also found advantageous not to let the derivative act on the command signal. Later it was also found suitable to let only a fraction b of the command signal act on the proportional part. The PID-algorithm then becomes

$$U(s) = K \left(bU_c(s) - Y(s) + \frac{1}{sT_i} \left(U_c(s) - Y(s) \right) - \frac{sT_d}{1 + sT_d/N} Y(s) \right) \quad (8.22)$$

where U, U_c, and Y denote the Laplace transforms of u, u_c, and y. The idea of providing different signal paths for the process output and the command signal is a good way to separate command signal response from the response to disturbances. Alternatively it may be viewed as a way to position the closed-loop zeros. There are also several other variations of the PID-algorithm that are used in commercial systems. An extra first-order lag may be used in series with the controller to obtain a high-frequency roll-off. In some applications it has also been useful to include nonlinearities. The proportional term Ke can be replaced by $Ke|e|$ and a dead zone can also be included.

Discretization

The controller described by (8.22) can be discretized using any of the standard methods such as Tustin's approximation or ramp equivalence. Because the PID-controller is so simple, there are some special methods that are used. The following is a popular approximation that is very easy to derive. The proportional part

$$P(t) = K \left(bu_c(t) - y(t) \right)$$

requires no approximation because it is a purely static part. The integral term

$$I(t) = \frac{K}{T_i} \int^t e(s) \, ds$$

is approximated by a forward approximation, that is,

$$I(kh + h) = I(kh) + \frac{Kh}{T_i} e(kh) \tag{8.23}$$

The derivative part given by

$$\frac{T_d}{N} \frac{dD}{dt} + D = -KT_d \frac{dy}{dt}$$

is approximated by taking backward differences. This gives

$$D(kh) = \frac{T_d}{T_d + Nh} D(kh - h) - \frac{KT_dN}{T_d + Nh} \left(y(kh) - y(kh - h) \right)$$

This approximation has the advantage that it is always stable and that the sampled pole goes to zero when T_d goes to zero. Tustin's approximation gives an approximation such that the pole instead goes to $= -1$ as T_d goes to zero. The control signal is given as

$$u(kh) = P(kh) + I(kh) + D(kh) \tag{8.24}$$

This approximation has the pedagogical advantage that the proportional, integral, and derivative terms are obtained separately. The other approximations give similar results. They can all be represented as

$$R(q)u(kh) = T(q)u_c(kh) - S(q)y(kh) \tag{8.25}$$

where the polynomials R, S, and T are of second order. The polynomial R has the form

$$R(q) = (q - 1)(q - a_d) \tag{8.26}$$

The number a_d and the coefficients of the polynomials S and T obtained for different approximation methods are given in Table 8.1.

Table 8.1 Coefficients in different approximations of the continuous-time PID-controller.

	Special	Tustin	Ramp Equivalence
s_0	$K(1 + b_d)$	$K(1 + b_i + b_d)$	
s_1	$-K(1 + a_d + 2b_d - b_i)$	$-K\left(1 + a_d + 2b_d - b_i(1 - a_d)\right)$	
s_2	$K(a_d + b_d - b_i a_d)$	$K(a_d + b_d - b_i a_d)$	
t_0	Kb	$K(b + b_i)$	
t_1	$-K(b(1 + a_d) - b_i)$	$-K\left(b(1 + a_d) - b_i(1 - a_d)\right)$	
t_2	$Ka_d(b - b_i)$	$Ka_d(b - b_i)$	
a_d	$\dfrac{T_d}{Nh + T_d}$	$\dfrac{2T_d - Nh}{2T_d + Nh}$	$\exp\left(-\dfrac{Nh}{T_d}\right)$
b_d	Na_d	$\dfrac{2NT_d}{2T_d + Nh}$	$\dfrac{T_d}{h}(1 - a_d)$
b_i	$\dfrac{h}{T_i}$	$\dfrac{h}{2T_i}$	$\dfrac{h}{2T_i}$

Incremental Algorithms

Equation (8.24) is called a *position algorithm* or an *absolute algorithm*. The reason is that the output of the controller is the absolut value of the control signal, for instance, a valve position. In some cases it is advantageous to move the integral action outside the control algorithm. This is natural when a stepper motor is used. The output of the controller should then represent the increments of the control signal, and the motor implements the integrator. Another case is when an actuator with pulse-width control is used.

To obtain an incremental algorithm the control algorithm is rewritten so that its output is the increment of the control signal. Because it follows from (8.26) that the polynomial R in (8.25) always has a factor $(q - 1)$ this is easy to do. Introducing

$$\Delta u(kh) = u(kh) - u(kh - h)$$

we get

$$(q - a_d)\Delta u(kh + h) = T(q)u_c(kh) - S(q)y(kh)$$

This is called the *incremental* form of the controller. A drawback with the incremental algorithm is that it cannot be used for P- or PD-controllers. If this is attempted the controller will be unable to keep the reference value, because an unstable mode $z - 1$ is canceled.

Integrator Windup

A controller with integral action combined with an actuator that becomes saturated can give some undesirable effects. If the control error is so large that the integrator saturates the actuator, the feedback path will be broken, because the actuator will remain saturated even if the process output changes. The integrator, being an unstable system, may then integrate up to a very large value. When the error is finally reduced, the integral may be so large that it takes considerable time until the integral assumes a normal value again. This effect is called *integrator windup*. The effect is illustrated in Fig. 8.9.

There are several ways to avoid integrator windup. One possibility is to stop updating the integral when the actuator is saturated. Another method is illustrated by the block diagram in Fig. 8.10(a). In this system an extra feedback path is provided by measuring the actuator output and forming an error signal (e_s) as the difference between the actuator output (u_c) and the controller output (v) and feeding this error back to the integrator through the gain $1/T_t$. The error signal e_s is zero when the actuator is not saturated. When the actuator is saturated the extra feedback path tries to make the error signal e_s equal zero. This means that the integrator is reset, so that the controller output is at the saturation limit. The integrator is thus reset to an appropriate value with the time constant T_t, which is called the tracking-time constant. The advantage with this scheme for *antiwindup* is that it can be applied to any actuator, that is, not only a saturated actuator but also an actuator with arbitrary characteristics, such as a dead zone or an hysteresis, as long as the actuator output is measured. If the actuator output is not measured, the actuator can be modeled and an

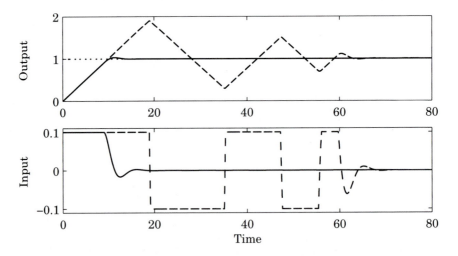

Figure 8.9 Illustration of integrator windup. The dashed lines show the response with an ordinary PID-controller. The solid lines show the improvement with a controller having antiwindup.

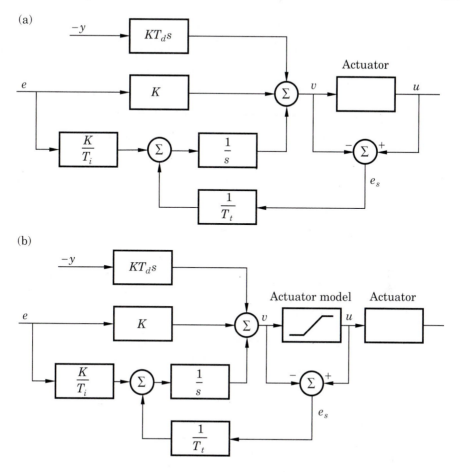

Figure 8.10 Controller with antiwindup. A system in which the actuator output is measured is shown in (a) and a system in which the actuator output is estimated from a mathematical model is shown in (b).

equivalent signal can be generated from the model, as shown in Fig. 8.10(b). Figure 8.9 shows the improved behavior with controllers having an anti-windup scheme. Antiwindup is further discussed in Sec. 9.4.

Operational Aspects

Practically all PID-controllers can run in two modes: manual and automatic. In manual mode the controller output is manipulated directly by the operator, typically by push buttons that increase or decrease the controller output. The controllers may also operate in combination with other controllers, as in a cascade or ratio connection, or with nonlinear elements such as multipliers and selectors. This gives rise to more operational modes. The controllers also have parameters that can be adjusted in operation. When there are changes of

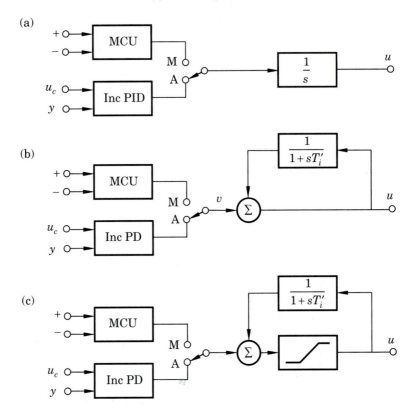

Figure 8.11 Controllers with bumpless transfer from manual to automatic mode. The controller in (a) is incremental. The controllers in (b) and (c) are special forms of position algorithms. The controller in (c) has antiwindup (MCU = Manual Control Unit).

modes and parameters, it is essential to avoid switching transients. The way mode switchings and parameter changes are made depends on the structure chosen for the controller.

Bumpless transfer. Because the controller is a dynamic system it is necessary to make sure that the state of the system is correct when switching the controller between manual and automatic mode. When the system is in manual mode, the controller produces a control signal that may be different from the manually generated control signal. It is necessary to make sure that the value of the integrator is correct at the time of switching. This is called *bumpless transfer*. Bumpless transfer is easy to obtain for a controller in incremental form. This is shown in Fig. 8.11(a). The integrator is provided with a switch so that the signals are either chosen from the manual or the automatic increments. Because the switching only influences the increments, there will not be any large transients. A related scheme for a position algorithm is shown in Fig. 8.11(b). In this case the integral action is realized as positive feedback around a first-order

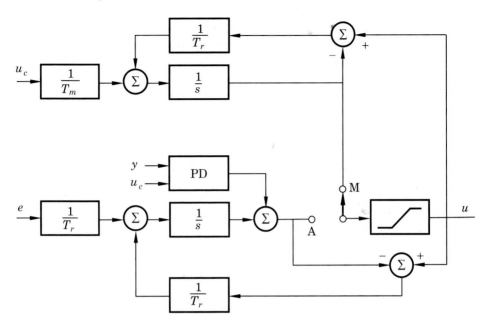

Figure 8.12 PID-controller with bumpless switching between manual and automatic control.

system. The transfer function from v to u is

$$\frac{1}{1 - \dfrac{1}{1 + sT_i'}} = \frac{1 + sT_i'}{sT_i'}$$

For simplicity the filters are shown in continuous-time form. In a digital system they are of course realized as sampled systems. The system can also be provided with an antiwindup protection, as shown in Fig. 8.11(c). A drawback with this scheme is that the PID-controller must be of the form

$$G(s) = K' \, \frac{(1 + sT_i')(1 + sT_d')}{sT_i'} \tag{8.27}$$

which is less general than (8.22). Moreover the reset-time constant is equal to T_i'. More elaborate schemes have to be used for general PID-algorithms on position form. Such a controller is built up of a manual control module and a PID-module, each having an integrator. See Fig. 8.12.

Bumpless Parameter Changes

A controller is a dynamic system. A change of the parameters of a dynamic system will naturally result in changes of its output even if the input is kept constant. Changes in the output can in some cases be avoided by a simultaneous change of the state of the system. The changes in the output will also depend on the chosen realization. With a PID-controller it is natural to require that there be no drastic changes in the output if the parameters are changed when the error is zero. This will obviously hold for all incremental algorithms, because the output of an incremental algorithm is zero when the input is zero irrespective of the parameter values. It also holds for a position algorithm with the structure shown in Figs. 8.11(b) and (c). For a position algorithm it depends, however, on the implementation. Assume, for example, that the state is chosen as

$$x_I = \int^t e(s)\, ds$$

when implementing the algorithm. The integral term is then

$$I = \frac{K}{T_i} x_I$$

Any change of K or T_i will then result in a change of I. To avoid bumps when the parameters are changed it is therefore essential that the state be chosen as

$$x_I = \int^t \frac{K(s)}{T_i(s)} e(s)\, ds$$

when implementing the integral term.

Tuning

A PID-controller has parameters K, T_i, T_d, T_t, b, N, u_{low}, and u_{high} that must be chosen. The primary parameters are K, T_i, and T_d. Parameter N can often be given a fixed default value, for example, $N = 10$. The tracking-time constant (T_t) is often related to the integration time (T_i). In some implementations it has to be equal to T_i; in other cases it can be chosen as 0.1 to 0.5 times T_i. The parameters u_{low} and u_{high} should be chosen close to the true saturation limits.

If the process dynamics and the disturbances are known parameters, then K, T_i, and T_d can be computed using the design methods of Chapters 4, 5, 11, and 12. Some special methods have, however, been developed to tune the PID-parameters experimentally. The behavior of the discrete-time PID-controller is very close to the analog PID-controller if the sampling interval is short. The traditional tuning rules for continuous-time controllers can thus be used. There are two classical heuristic rules due to Ziegler and Nichols (1942) that can be used to determine the controller parameters: the step-response method and the ultimate-sensitivity method.

The step-response method.   In this method the unit step response of the open-loop process is determined experimentally. The technique can be applied to processes whose step response is monotone or essentially monotone apart from an initial nonminimum phase characteristic. To use the method the tangent to

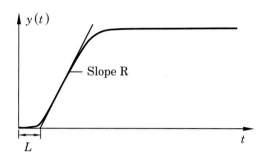

Figure 8.13   Determination of parameters $a = RL$ and L from the unit step response to be used in Ziegler-Nichols step-response method.

the step response that has the steepest slope is drawn and the intersections of the tangent with the axes are determined. See Fig. 8.13. The controller parameters are then obtained from Table 8.2. The Ziegler-Nichols rule was designed to give good response to load disturbances. It does, however, give fairly low damping of the dominant poles.

Parameter L is called the apparent deadtime. For stable processes parameter T, which is called the apparent-time constant, can also be determined from a step response of the open-loop system..

The ultimate-sensitivity method   In this method the key idea is to determine the point where the Nyquist curve of the open-loop system intersects the negative real axis. This is done by connecting the controller to the process and setting the parameters so that pure proportional control is obtained. The gain of the controller is then increased until the closed-loop system reaches the stability limit. The gain (K_u) when this occurs and the period of the oscillation (T_u) are determined. These parameters are called ultimate gain and ultimate period. The controller parameters are then given by Table 8.3.

Assessment   The Ziegler-Nichols tuning rules are conceptually attractive. Process dynamics is characterized by two parameters that are easy to determine experimentally and the controller parameters are then obtained from

Table 8.2   PID parameters obtained from the Ziegler-Nichols step-response method.

Controller Type	K	T_i	T_d
P	$1/a$		
PI	$0.9/a$	$3L$	
PID	$1.2/a$	$2L$	$0.5L$

Table 8.3 PID parameters obtained from Ziegler-Nichols ultimate-sensitivity method.

Controller Type	K	T_i	T_d
P	$0.5K_u$		
PI	$0.45K_u$	$T_u/1.2$	
PID	$0.6K_u$	$T_u/2$	$T_u/8$

simple tables. Because of this the rules have been very popular; they are the basis for much of practical controller tuning. The Ziegler-Nichols rules do however have some very serious drawbacks. The closed-loop systems obtained with the rules have very low relative damping typically about $\zeta \approx 0.2$. The tuning rules do not give all controller parameters, and the integrations time is always four times the derivative time. The damping can be improved by modifying the numerical values in the tables. To characterize process dynamics by two parameters is quite crude. More parameters are required for improved tuning. Much can be gained by also including the static gain K_p of a process. Tuning rules of the Ziegler-Nichols type should therefore be used with care. They can give a rough approximation but the tuning can often be improved.

Selection of Sampling Interval

When DDC-control was first introduced, computers were not as powerful as they are today. Long sampling intervals were needed to handle many loops. The following recommendations for the most common process variables are given for DDC-control.

Type of variable	Sampling time, s
Flow	1–3
Level	5–10
Pressure	1–5
Temperature	10–20

Commercial digital controllers for few loops often have a short fixed-sampling interval on the order of 200 ms. This implies that the controllers can be regarded as continuous-time controllers, and the continuous-time tuning rules may be used. Several rules of thumb for choosing the sampling period for a digital PID-controller are given in the literature. There is a significant difference between PI- and PID-controllers. For PI-controllers the sampling period is related to the integration time. A typical rule of thumb is

$$\frac{h}{T_i} \approx 0.1 \text{ to } 0.3$$

When Ziegler-Nichols tuning is used this implies

$$\frac{h}{L} \approx 0.3 \text{ to } 1$$

or

$$\frac{h}{T_u} \approx 0.1 \text{ to } 0.3$$

With PID-control the critical issue is that the sampling period must be so short that the phase lead is not adversely affected by the sampling. This implies that the sampling period should be chosen so that the number hN/T_d is in the range of 0.2 to 0.6. With $N = 10$ this means that for Ziegler-Nichols tuning the ratio h/L is between 0.01 and 0.06. This gives

$$\frac{hN}{T_d} \approx 0.2 \text{ to } 0.6$$

Significantly shorter sampling periods are thus required for controllers with derivative action. If computer time is at a premium, it is advantageous to use the sampled-data design methods used in this book.

Computer Code

A typical computer code for a discrete PID-controller is given in Listing 8.1 on page 318. The discretization of the integral term is made using a forward difference. The derivative term is approximated using a backward difference. The calculation PID_Init is made initially only. This saves computing time. In a real system these calculations have to be made each time the parameters are changed. The code given admits bumpless parameter changes if $b = 1$. When $b \neq 1$ the proportional term (P) is different from zero in steady state. To ensure bumpless parameter changes it is necessary that the quantity P + I is invariant to parameter changes. This means that the state I has to be changed as follows:

$$I_{\text{new}} = I_{\text{old}} + K_{\text{old}}(b_{\text{old}}u_c - y) - K_{\text{new}}(b_{\text{new}}u_c - y) \tag{8.28}$$

Word length and integration offset.   The integral part in digital PID-controllers is approximated as a sum. Computational problems, such as *integration offset*, may then occur due to the finite precision in the representation in the computer. Assume that there is an error, $e(kh)$. The integrator term is then increased at each sampling time with [see (8.23)]

$$\frac{Kh}{T_i} e(kh)$$

Listing 8.1 C code for PID-controller based on Tustin discretization.

```c
#include<Kernel.h>          /* Import real-time primitives */
/* PID controller based on Tustin discretization */
struct PID_Data {
  struct {
    double uc;              /* Input  : Set point               */
    double y;               /* Input  : Measured variable       */
    double u;               /* Output : Controller output       */
    double v;               /* Output : Limited controller output */
  } Signals;
  struct {
    double I;               /* Integral part                    */
    double D;               /* Derivative part                  */
    double yold;            /* Delayed measured variable        */
  } States;
  struct {
    double K;               /* Controller gain                  */
    double Ti;              /* Integral time                    */
    double Td;              /* Derivative time                  */
    double Tt;              /* Reset time                       */
    double N;               /* Maximum derivative gain          */
    double b;               /* Fraction of setpoint in prop. term */
    double ulow;            /* Low output limit                 */
    double uhigh;           /* High output limit                */
    double h;               /* Sampling period                  */
    double bi, ar, bd, ad;
  } Par;
} pid_data;

void PID_Init(struct PID_Data *data)
{
  data->States.I    = 0;
  data->States.D    = 0;
  data->States.yold = 0;
  data->Par.K  = 4.4;
  data->Par.Ti = 0.4;
  data->Par.Td = 0.2;
  data->Par.Tt = 10;
  data->Par.N  = 10;
  data->Par.b  = 1;
  data->Par.ulow  = -1;
  data->Par.uhigh =  1;
  data->Par.h  = 0.03;
  data->Par.bi = data->Par.K*data->Par.h/data->Par.Ti;
  data->Par.ar = data->Par.h/data->Par.Tt;
  data->Par.bd = data->Par.K*data->Par.N*data->Par.Td/
    (data->Par.Td+data->Par.N*data->Par.h);
  data->Par.ad = data->Par.Td/(data->Par.Td+data->Par.N*data->Par.h);
}
```

Listing 8.1   (Continued)

```c
void PID_CalculateOutput(struct PID_Data *data) {

  /* Proportional part */
  double P = data->Par.K*(data->Par.b*data->Signals.uc -
                          data->Signals.y);

  /* Derivative part */
  data->States.D = data->Par.ad * data->States.D -
    data->Par.bd * (data->Signals.y - data->States.yold);

  /* Calculate control signal */
  data->Signals.v = P + data->States.I + data->States.D;

  /* Handle actuator limitations */
  if ( data->Signals.v < data->Par.ulow ) {
    data->Signals.u = data->Par.ulow;
  } else if ( data->Signals.v > data->Par.uhigh ) {
    data->Signals.u = data->Par.uhigh;
  } else {
    data->Signals.u = data->Signals.v;
  }
}

void PID_UpdateStates(struct PID_Data *data) {
  /* Integral part */
  data->States.I = data->States.I +
    data->Par.bi*(data->Signals.uc - data->Signals.y) +
    data->Par.ar*(data->Signals.u  - data->Signals.v);

  data->States.yold = data->Signals.y;
}

void PID_Main() {
  Kernel_Time time;

  PID_Init(&pid_data);
  Kernel_CurrentTime(&time);      /* Get current time */
  for (;;) {
    Kernel_IncTime(&time, 1000 * pid_data.Par.h);
                                  /* Increment "time" with h*/
    read_y(&(pid_data.Signals.y));
    read_uc(&(pid_data.Signals.uc));
    PID_CalculateOutput(&pid_data);
    write_u(pid_data.Signals.u);
    PID_UpdateStates(&pid_data);
    Kernel_WaitUntil(time);       /* Wait until "time" */
  }
}
```

Assume that the gain is small and that the reset time is large compared to the sampling time. The change in the output may then be smaller than the quantization step in the D-A converter. For instance, a 12-bit D-A converter (that is, a resolution of 1/4096) should give sufficiently good resolution for control purposes. Yet if $K = h = 1$ and $T_i = 3600$, then any error less than 90% of the span of the A-D converter gives a calculated change in the integral part less than the quantization step. If the integral part is stored in the same precision as that of the D-A converter, then there will be an offset in the output. One way to avoid this is to use higher precision in the internal calculations. The results then have an error that is less than the quantization level of the output. Frequently at least 24 bits are used to implement the integral part in a computer, in order to avoid integration offset.

8.6 Conclusions

Different ways of translating a continuous-time controller to a digital controller have been presented. The problem is of substantial interest if an analog design is available, and a digital solution is needed. Several methods to compute a pulse-transfer function that corresponds to the continuous-time transfer function have been discussed, based on step invariance, ramp invariance, and Tustin's approximation. Tustin's method is commonly used because of its simplicity. It does, however, distort the frequency scale of the filter. The method based on ramp invariance gives very good results and is only moderately more complicated than Tustin's method. Digital systems designed in this way are always (slightly) inferior to analog systems because of the inherent time delay caused by the hold circuit. This time delay is approximately $h/2$.

The translation methods work well if the sampling period is short. A good way to choose the sampling period is to observe that the extra time delay decreases the phase margin by $\omega_c h/2$ radians or by $180\omega_c/\omega_s$ degrees, where ω_c is the crossover frequency. There are possibilities of creating better designs than those discussed in this chapter, as discussed in the following chapters.

8.7 Problems

8.1 Find how the left half-s-plane is transformed into the z-plane when using the mappings in (8.4) to (8.6).

8.2 Use different methods to make an approximation of the transfer function

$$G(s) = \frac{a}{s + a}$$

(a) Euler's method

(b) Tustin's approximation

(c) Tustin's approximation with prewarping if the warping frequency is $\omega_1 = a$ rad/s

8.3 The lead network given in (8.9) gives about $20°$ phase advance at $\omega_c = 1.6$ rad/s. Approximate the network for $h = 0.25$ using

(a) Euler's method

(b) Backward differences

(c) Tustin's approximation

(d) Tustin's approximation with prewarping using $\omega_1 = \omega_c$ as the warping frequency

(e) Zero-order-hold sampling

Compute the phase of the approximated networks at $z = \exp(i\omega_c h)$.

8.4 Verify the calculations leading to the rule of thumb for the choice of the sampling interval given in Sec. 8.2.

8.5 Show that (8.24) is obtained from (8.22) by approximating the integral part using Euler's method and backward difference for the derivative part. Discuss advantages and disadvantages for each of the following cases.

(a) The integral part is approximated using backward difference.

(b) The derivative part is approximated using Euler's method. (*Hint:* Consider the case when T_d is small.)

8.6 A continuous-time PI-controller is given by the transfer function

$$K \left(1 + \frac{1}{T_i s} \right)$$

Use the bilinear approximation to find a discrete-time controller. Find the relation between the continuous-time parameters K and T_i and the corresponding discrete-time parameters in (8.24).

8.7 Consider the tank system in Problem 2.10. Assume the following specifications for the closed-loop system:
1. The steady-state error after a step in the reference value is zero.
2. The crossover frequency of the compensated system is 0.025 rad/s.
3. The phase margin is about $50°$.

(a) Design a PI-controller such that the specifications are fulfilled.

(b) Determine the poles and the zero of the closed-loop system. What is the damping corresponding to the complex poles?

(c) Choose a suitable sampling interval and approximate the continuous-time controller using Tustin's method with warping. Use the crossover frequency as the warping frequency.

(d) Simulate the system when the sampled-data controller is used. Compare with the desired response, that is, when the continuous-time controller is used.

8.8 Make an approximation, analogous to (8.16) and (8.17), such that the modifications are valid for terms up to and including h^3.

8.9 The normalized motor has a state-space representation given by (A.5). The control law

$$u(t) = M u_c(t) - L x(t)$$

with $M = 4$ and $L = \begin{pmatrix} 2 & 4 \end{pmatrix}$ gives the continuous-time transfer function

$$\frac{4}{s^2 + 3s + 4}$$

from u_c to y. This corresponds to $\zeta = 0.75$ and $\omega_0 = 2$.

(a) Make a sampled-data implementation of the controller.

(b) Modify the control law using (8.16) and (8.17).

(c) Simulate the controllers in (a) and (b) for different sampling periods and compare with the continuous-time controller.

8.10 Given the continuous-time system

$$\frac{dx}{dt} = \begin{pmatrix} -3 & 1 \\ 0 & -2 \end{pmatrix} x + \begin{pmatrix} 0 \\ 1 \end{pmatrix} u$$

$$y = \begin{pmatrix} 1 & 0 \end{pmatrix} x$$

(a) Determine a continuous-time state-feedback controller

$$u(t) = -L x(t)$$

such that the characteristic polynomial of the closed-loop system is

$$s^2 + 8s + 32$$

A computer is then used to implement the controller as

$$u(kh) = -L x(kh)$$

(b) Modify the controller using (8.16).

(c) Simulate the controllers in (a) and (b) and decide suitable sampling intervals. Assume that $x(0) = [1\,0]$.

8.11 Use the w-plane method to design a compensator for the motor in Example 8.3 when $h = 0.25$. Design the compensator such that the transformed system has a crossover frequency corresponding to 1.4 rad/s and a phase margin of $50°$. Compare with the continuous-time design and the discrete-time approximations in Example 8.3. Investigate how long a sampling interval can be used for the w-plane method.

8.12 Consider the continuous-time double integrator described by (A.2). Assume that a time-continuous design has been made giving the controller

$$u(t) = 2 u_c(t) - \begin{pmatrix} 1 & 2 \end{pmatrix} \hat{x}(t)$$

$$\frac{d\hat{x}(t)}{dt} = A \hat{x}(t) + B u(t) + K(y(t) - C \hat{x}(t))$$

with $K^T = \begin{pmatrix} 1 & 1 \end{pmatrix}$.

(a) Assume that the controller should be implemented using a computer. Modify the controller (not the observer part) for the sampling interval $h = 0.2$ using (8.16) and (8.17).

(b) Approximate the observer using backward-difference approximation.

(c) Simulate the continuous-time controller and the discrete-time approximation. Let the initial values be $x(0) = \begin{pmatrix} 1 & 1 \end{pmatrix}^T$ and $\hat{x}(0) = \begin{pmatrix} 0 & 0 \end{pmatrix}^T$.

8.13 Derive ramp-invariant approximations of the transfer function

$$G(s) = \frac{1}{s + a}$$

and

$$G(s) = \frac{s}{s + a}$$

8.14 Derive the ramp-invariant equivalent of the PID-controller.

8.15 There are many different ways to sample a continuous-time system. The key difference is the assumption made on the behavior of the control signal over the sampling interval. So far we have discussed step invariance and ramp invariance. Derive formula for impulse invariant sampling of the system (8.11) when the continuous-time signal is assumed to be a sequence of impulses that occur just after the sampling instants .

8.16 Derive the impulse-invariant approximations of the transfer functions in Problem 8.13.

8.17 The frequency prewarping in Sec. 8.2 gives the correct transformation at one frequency along the imaginary axis. Derive the necessary warping transformation such that one point at an arbitrary ray through the origin is transformed correctly.

8.8 Notes and References

The problem of designing digital filters that implement analog-transfer functions approximately is discussed in the digital-filtering literature: Rabiner and Gold (1975), Antoniou (1979), and Oppenheim and Schafer (1989). Interesting views on similarities and differences between digital signal processing and control theory are presented in Willsky (1979). A more control-oriented presentation of different approximations is found in Franklin and Powell (1989). Redesign of state feedback is discussed in more detail in Kuo (1980).

Digital PID-controllers and their operational aspects are thoroughly discussed in Goff (1966), Bristol (1977), Shinskey (1988), and Åström and Hägglund (1995). The classical reference for tuning PID-controllers is Ziegler and Nichols (1942). A modification of the rules by Ziegler and Nichols which takes the length of the sampling interval into account is given in Takahashi, Chan, and Auslander (1971).

9

Implementation of Digital Controllers

9.1 Introduction

Design of algorithms for computer control is discussed in the previous chapters. The problem of implementing a control algorithm on a digital computer is discussed in this chapter. The control algorithms obtained in the previous chapters are discrete-time dynamic systems. The key problem is to implement a discrete-time dynamic system using a digital computer. An overview of this problem is given in Sec. 9.2, which shows that it is straightforward to obtain a computer code from the discrete-time algorithm. There are, however, several issues that must be considered. It is necessary to take the interfaces to the sensors, the actuators, and the human operators into account. It is also necessary to consider the numerical precision required.

The sensor interface is discussed in Sec. 9.3. This covers prefiltering and computational delays and shows that the computational delay depends critically on the details of the algorithm. Different ways to shorten the computational delay by reorganizing the code are discussed. Methods of filtering the signals effectively by introducing nonlinearities, which may reduce the influence of unreliable sensors, are shown. This is one of the major advantages of computer control. Most theory in this book deals with linear theory. There are, however, a few nonlinearities such as actuator saturation that must be taken into account. Different ways of handling these are discussed in Sec. 9.4. This leads to extensions of the methods for antireset windup used in classical process control.

The operator interface is important factor; it is discussed in Sec. 9.5. This includes treatment of operational modes and different ways to avoid switching transients. The information that should be displayed and different ways of influencing the control loop are also discussed. Digital computers offer many interesting possibilities; so far they have been used only to a very modest degree. There are many opportunities for innovations in this field. It is important to have sound numerics in the control algorithm, which is the topic of Sec. 9.6. Effects of a finite word length are also discussed. Realization of digital controllers is treated in Sec. 9.7. Programming of control algorithms is discussed in Sec. 9.8. For more ambitious systems in which parameters and control algorithms are changed on-line, it is necessary to understand concurrent programming.

9.2 An Overview

This section gives an overview of implementation of digital control laws. Different representations of the control laws obtained from the design methods in Chapters 4, 5, and 8 are first given; the algorithms are then implemented. A list of some important problems is given. These problems are discussed in greater detail in the following sections.

Different Representations of the Controller

The design methods of the previous chapters give control laws in the form of a discrete-time dynamic system. Different representations are obtained, depending on the approaches used. The design methods based on pole placement by state feedback in Sec. 4.5 give a controller of the form

$$\hat{x}(k|k) = \hat{x}(k|k-1) + K\left(y(k) - \hat{y}(k|k-1)\right)$$
$$u(k) = L\left(x_m(k) - \hat{x}(k|k)\right) + L_c u_c(k)$$
$$\hat{x}(k+1|k) = \Phi\hat{x}(k|k) + \Gamma u(k) \tag{9.1}$$
$$x_m(k+1) = f\left(x_m(k), u_c(k)\right)$$
$$\hat{y}(k+1|1) = C\hat{x}(k+1|k)$$

In this representation the state of the controller is $\hat{x}$ and x_m, where $\hat{x}$ is an estimate of the process state, and x_m is the state of the model that generates the desired response to command signals u_c. The form in (9.1) is called a *state representation with an explicit observer* because of the physical interpretation of the controller state. It is easy to include a nonlinear model for the desired state in this representation.

If the function f in (9.1) is linear, the controller is a linear system with the inputs y and u_c and the output u. Such a controller may be represented as

$$x(k+1) = Fx(k) + Gy(k) + G_c u_c(k)$$
$$u(k) = Cx(k) + Dy(k) + D_c u_c(k) \tag{9.2}$$

where x is the state of the controller (see Problem 4.7). Equation (9.2) is a *general-state representation* of a discrete-time dynamic system. This form is more compact than (9.1). The state does, however, not necessarily have a simple physical interpretation.

The design methods for single-input–single-output systems discussed in Chapter 5, which are based on external models, give a controller in the form of a general *input-output representation*

$$R(q)u(k) = T(q)u_c(k) - S(q)y(k) \tag{9.3}$$

where $R(q)$, $S(q)$, and $T(q)$ are polynomials in the forward-shift operator q. There are simple transformations between the different representations (compare with Chapter 2).

Realization

A control law is a dynamic system. Different realizations can be obtained by different choices of the state variables. The different representations are equivalent from an input-output point of view if we assume that the calculations are done with infinite precision. With finite precision in the calculations, the choice of the state-space representation is very important. Quantization and roundoff introduce nonlinearities. Linear and nonlinear operations do not commute. For instance, $Q(a + b) \neq Q(a) + Q(b)$, where $Q(\cdot)$ represents the quantization of a signal. It is thus important in which order different operations are done when an algorithm is implemented. A bad choice of the representation may give a controller that is sensitive to errors in the computations.

It is very important that the controller is transformed into a robust form before the controller is implemented as a computer program. Suitable forms are serial and parallel realizations of first- and second-order blocks. It is also important to organize the computations in a numerically good way. For instance, it should be avoided to take the difference of large numbers. These aspects are further discussed in Sec. 9.7.

Implementing a Computer-Controlled System

The implementation of a discrete-time system described by (9.1), (9.2), or (9.3) using a digital computer is straightforward. The details depend on the hardware and software available. To show the principles, it is assumed that the system described by (9.2) should be implemented using a digital computer with A-D and D-A converters and a real-time clock. A graphical representation of the program is shown in Fig. 9.1. The execution of the program is controlled by the clock. The horizontal bar indicates that execution is halted until an interrupt comes from the clock. The clock is set so that an interrupt is obtained at each sampling instant. The code in the box is executed after each interrupt.

The body of the code is given in Listing 9.1. Analog-to-digital conversion is commanded in the first line. The appropriate values are stored in the arrays y and uc. The control signal u is computed in the second line using matrix-vector multiplication and vector addition. The state vector x is updated in the third line, and the digital-to-analog conversion is performed in the fourth line. To

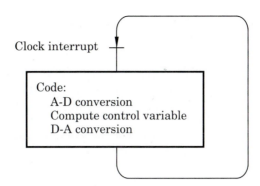

Figure 9.1 Graphical representations of a program used to implement a discrete-time system.

obtain a complete code, it is also necessary to have type declarations for the vectors u, uc, x, and y and the matrices F, G, Gc, C, D, and Dc. It is also necessary to assign values to the matrices and the initial value for the state x. When using computer languages that do not have matrix operations, it is necessary to write appropriate procedures for generating matrix operations using operations on scalars. Notice that the second and third lines of the code correspond exactly to the algorithm in (9.2).

To obtain a good control system, it is also necessary to consider

- Prefiltering and computational delay

- Actuator nonlinearities

- Operational aspects

- Numerics

- Realization

- Programming aspects

These issues are discussed in the following sections.

Listing 9.1 Computer code skeleton for the control law of (9.2). Line numbers are introduced only for purposes of referencing.

```
        Procedure Regulate
        begin
1          Adin y uc
2          u:=C*x+D*y+Dc*uc
3          x:=F*x+G*y+Gc*uc
4          Daout u
        end
```

9.3 Prefiltering and Computational Delay

The interactions between the computer and its environment are important when implementing a control system. Compare with Chapter 7. The sensor interface is discussed in this section. The consequences of disturbances are discussed in Chapters 4 and 7. The importance of using an analog prefilter to avoid aliasing is treated in Chapter 7.

It is also clear from Sec. 9.2 that there is always a time delay associated with the computations. The prefilter and the computational delay give rise to additional dynamics, which may be important when implementing a digital controller. These effects are now discussed.

Analog Prefiltering

To avoid aliasing (see Sec. 7.4), it is necessary to use an analog prefilter for elimination of disturbances with frequencies higher than the Nyquist frequency associated with the sampling rate. Different prefilters are discussed from the signal-processing point of view in Sec. 7.4. The discussion is based on knowledge of the frequency content of the signal. In a control problem there is normally much more information available about the signals in terms of differential equations for the process models and possibly also for the disturbances.

It is often useful to sample the analog signals at a comparatively high rate and to avoid aliasing by an ordinary analog prefilter designed from the signal-processing point of view. The precise choice depends on the order of the filter and the character of the measured signal. The dynamics of the prefilter should be taken into account when designing the system. Compare the discussion in Secs. 5.8 and 5.9. If the sampling rate is changed, the prefilter must also be changed. With reasonable component values, it is possible to construct analog prefilters for sampling periods shorter than a few seconds. For slower sampling rates, it is often simpler to sample faster with an appropriate analog prefilter and apply digital filtering to the sampled signal. This approach also makes it possible to change the sampling period of the control calculations by software only.

Because the analog prefilter has dynamics, it is necessary to include the filter dynamics in the process model. If the prefilter or the sampling rate is changed, the coefficients of the control law must be recomputed. With normal sampling rates—that is, 15 to 45 times per period—it is necessary to consider the prefilter dynamics in the controller design (compare with Secs. 5.9 and 7.4).

Computational Delay

Because A-D and D-A conversions and computations take time, there will always be a delay when a control law is implemented using a computer. The delay, which is called the *computational delay*, depends on how the control algorithm is implemented. There are basically two different ways to do this (see Fig. 9.2). In case A, the measured variables read at time t_k may be used to compute the control signal to be applied at time t_{k+1}. Another possibility, case B, is to read

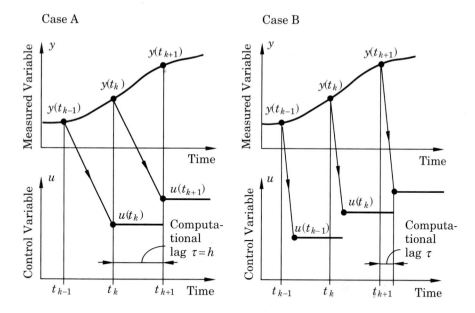

Figure 9.2   Two ways of synchronizing inputs and outputs. In case A the signals measured at time t_k are used to compute the control signal to be applied at time t_{k+1}. In case B the control signals are changed as soon as they are computed.

the measured variables at time t_k and to make the D-A conversion as soon as possible.

The disadvantage of case A is that the control actions are delayed unnecessarily; the disadvantage of case B is that the delay will be variable, depending on the programming. In both cases it is necessary to take the computational delay into account when computing the control law. This is easily done by including a time delay of h (case A) or τ (case B) in the process model. A good rule is to read the inputs before the outputs are set out. If this is not done, there is always the risk of electrical cross-coupling.

In case B it is desirable to make the computational delay as small as possible. This can be done by making as few operations as possible between the A-D and D-A conversions. Consider the program in Listing 9.1. Because the control signal u is available after executing the second line of code, the D-A conversion can be done before the state is updated. The delay may be reduced further by calculating the product C*x after the D-A conversion. The algorithm in Listing 9.1 is then modified to Listing 9.2.

To judge the consequences of computational delays, it is also useful to know the sensitivity of the closed-loop system to a time delay. This may be evaluated from a root locus with respect to a time delay. A simpler way is to evaluate how much the closed-loop poles change when a time delay of one sampling period is introduced.

Listing 9.2 Computer code skeleton that implements the control algorithm (9.2). This code has a smaller computational delay than the code in Listing 9.1.

```
        Procedure Regulate
        begin
1           Adin y uc
2           u:=u1+D*y+Dc*uc
3           Daout u
4           x:=F*x+G*y+Gc*uc
5           u1:=C*x
        end
```

Outliers and Measurement Malfunctions

The linear filtering theory that will be discussed in Chapter 11 is very useful in reducing the influence of measurement noise. However, there may also be other types of errors, such as instrument malfunction and conversion errors. These are typically characterized by large deviations, which occur with low probabilities. It is very important to try to eliminate such errors so that they do not enter into the control-law calculations. There are many good ways to achieve this when using computer control.

The errors may be detected at the source. In systems with high-reliability requirements, this is done by duplication of the sensors. Two sensors are then combined with a simple logic, which gives an alarm if the difference between the sensor signals is larger than a threshold. A pair of redundant sensors may be regarded as one sensor that gives either a reliable measurement or a signal that it does not work.

Three sensors may be used in more extreme cases. A measurement is then accepted as long as two out of the three sensors agree (two-out-of-three logic). It is also possible to use even more elaborate combinations of sensors and filters.

An observer can also be used for error detection. For example, consider the control algorithm of (9.1) with an explicit observer. Notice that the one-step prediction error

$$\varepsilon(k) = y(k) - \hat{y}(k|k-1) = y(k) - C\hat{x}(k|k-1) \tag{9.4}$$

appears explicitly in the algorithm. This error can be used for diagnosis and to detect if the measurements are reasonable. This will be further discussed in connection with the Kalman filter in Chapter 11.

In computer control there are also many other possibilities for detecting different types of hardware and software errors. A few extra channels in the A-D converter, which are connected to fixed voltages, may be used for testing and calibration. By connecting a D-A channel to an A-D channel, the D-A converter may also be tested and calibrated.

9.4 Nonlinear Actuators

The design methods of Chapters 4, 5, and 8 are all based on the assumption that the process can be described by a linear model. Although linear theory has a wide applicability, there are often some nonlinearities that must be taken into account. For example, it frequently happens that the actuators are nonlinear, as is shown in Fig. 9.3. Valves are commonly used as actuators in process-control systems. This corresponds to a nonlinearity of the saturation type where the limits correspond to a fully open or closed valve. The system shown in Fig. 9.3 can be described linearly when the valve does not saturate. The nonlinearity is thus important when large changes are made. There may be difficulties with the control system during startup and shutdown, as well as during large changes, if the nonlinearities are not considered. A typical example is *integrator windup*. Other typical nonlinearities in practical systems are rate limitations, hysteresis, and backlash.

The rational way to deal with the saturation is to develop a design theory that takes the nonlinearity into account. This can be done using optimal-control theory. However, such a design method is quite complicated. The corresponding control law is also complex. Therefore, it is practical to use simple heuristic methods.

Difficulties occur because the controller is a dynamic system. When the control variable saturates, it is necessary to make sure that the state of the controller behaves properly. Different ways of achieving this are discussed in what follows.

Antiwindup for State-Space Controllers with an Explicit Observer

Consider first the case when the control law is described as an observer combined with a state feedback (9.1). The controller is a dynamic system, whose state is represented by the estimated state $\hat{x}$ in (9.1). In this case it is straightforward to see how the difficulties with the saturation may be avoided.

The estimator of (9.1) gives the correct estimate if the variable u in (9.1) is the actual control variable u_p in Fig. 9.3. If the variable u_p is measured, the estimate given by (9.1) and the state of the controller will be correct even if the control variable saturates. If the actuator output is not measured, it can be estimated—provided that the nonlinear characteristics are known. For the case

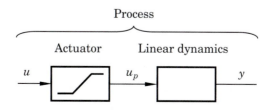

Figure 9.3 Block diagram of a process with a nonlinear actuator having saturation characteristics.

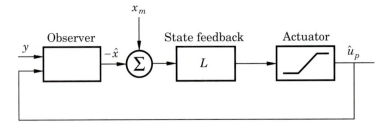

Figure 9.4 Controller based on an observer and state feedback with anti-windup compensation.

of a simple saturation, the control law can be written as

$$\hat{x}(k|k) = \hat{x}(k|k-1) + K\Big(y(k) - C\hat{x}(k|k-1)\Big)$$
$$= (I - KC)\Phi\hat{x}(k-1|k-1) + Ky(k) + (I - KC)\Gamma\hat{u}_p(k-1)$$
$$\hat{u}_p(k) = \mathrm{sat}\Big(L\big(x_m(k) - \hat{x}(k|k)\big) + Du_c(k)\Big) \tag{9.5}$$
$$\hat{x}(k+1|k) = \Phi\hat{x}(k|k) + \Gamma\hat{u}_p(k)$$

where the function sat is defined as

$$\mathrm{sat}\,u = \begin{cases} u_{\text{low}} & u \le u_{\text{low}} \\ u & u_{\text{low}} < u < u_{\text{high}} \\ u_{\text{high}} & u \ge u_{\text{high}} \end{cases} \tag{9.6}$$

for a scalar and

$$\mathrm{sat}\,u = \begin{pmatrix} \mathrm{sat}\,u_1 \\ \mathrm{sat}\,u_2 \\ \vdots \\ \mathrm{sat}\,u_n \end{pmatrix} \tag{9.7}$$

for a vector. The values u_{low} and u_{high} are chosen to correspond to the actuator limitations. A block diagram of a controller with a model for the actuator non-linearity is shown in Fig. 9.4. Observe that even if the transfer function from y to u for (9.1) is unstable, the state of the system in (9.5) will always be bounded if the matrix $(I - KC)\Phi$ is stable. It is also clear that x will be a good estimate of the process state even if the valve saturates, provided that u_{low} and u_{high} are chosen properly.

Antiwindup for the General State-Space Model

The controller may also be specified as a state-space model of the form in (9.2):

$$x(k+1) = Fx(k) + Gy(k) \tag{9.8}$$
$$u(k) = Cx(k) + Dy(k) \tag{9.9}$$

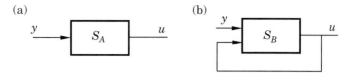

Figure 9.5   Different representations of the control law.

which does not include an explicit observer. The command signals have been neglected for simplicity. If the matrix F has eigenvalues outside the unit disc and the control variable saturates, it is clear that windup may occur. Assume, for example, that the output is at its limit and there is a control error y. The state and the control signal will then continue to grow, although the influence on the process is restricted because of the saturation.

To avoid this difficulty, it is desirable to make sure that the state of (9.8) assumes a proper value when the control variable saturates. In conventional process controllers, this is accomplished by introducing a special *tracking mode*, which makes sure that the state of the controller corresponds to the input-output sequence $\{u_p(k), y(k)\}$. The design of a tracking mode may be formulated as an observer problem. In the case of state feedback with an explicit observer, the tracking is done automatically by providing the observer with the actuator output u_p or its estimate $\hat{u}_p$. In the controller of (9.8) and (9.9), there is no explicit observer. To get a controller that avoids the windup problem, the solution for the controller with an explicit observer will be imitated. The control law is first rewritten as indicated in Fig. 9.5. The systems in (a) and (b) have the same input-output relation. The system S_B is also stable. By introducing a saturation in the feedback loop in (b), the state of the system S_B is always bounded if y and u are bounded. This argument may formally be expressed as follows. Multiply (9.9) by K and add to (9.8). This gives

$$
\begin{aligned}
x(k+1) &= Fx(k) + Gy(k) + K\Big(u(k) - Cx(k) - Dy(k)\Big) \\
&= (F - KC)x(k) + (G - KD)y(k) + Ku(k) \\
&= F_0x(k) + G_0y(k) + Ku(k)
\end{aligned}
$$

If the system of (9.8), and (9.9) is observable, the matrix K can always be chosen so that $F_0 = F - KC$ has prescribed eigenvalues inside the unit disc. Notice that this equation is analogous to (9.5). By applying the same arguments as for the controller with an explicit observer, the control law becomes

$$
\begin{aligned}
x(k+1) &= F_0x(k) + G_0y(k) + Ku(k) \\
u(k) &= \text{sat}(Cx(k) + Dy(k))
\end{aligned}
\tag{9.10}
$$

The saturation function is chosen to correspond to the actual saturation in the actuator. A comparison with the case of an explicit observer shows that (9.10) corresponds to an observer with dynamics given by the matrix F_0. The system

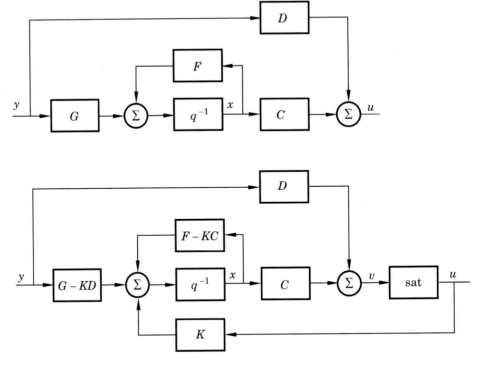

Figure 9.6 Block diagram of the controller (9.2) and the modification in (9.10) that avoids windup.

of (9.10) is also equivalent to (9.2) for small signals. A block diagram of the controller with antireset windup compensation is shown in Fig. 9.6.

Antiwindup for the Input-Output Form

The corresponding construction can also be carried out for controllers characterized by input-output models. Consider a controller described by

$$R(q)u(k) = T(q)u_c(k) - S(q)y(k) \tag{9.11}$$

where R, S, and T are polynomials in the forward-shift operator. The problem is to rewrite the equation so that it looks like a dynamic system with the observer dynamics driven by three inputs, the command signal u_c, the process output y, and the control signal u. This is accomplished as follows.

Let $A_{aw}(q)$ be the desired characteristic polynomial of the antiwindup observer. Adding $A_{aw}(q)u(k)$ to both sides of (9.11) gives

$$A_{aw}u = Tu_c - Sy + (A_{aw} - R)u$$

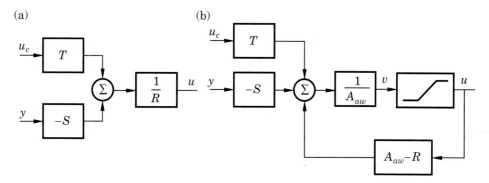

Figure 9.7   Block diagram of the controller of (9.11) and the modification in (9.12) that avoids windup.

A controller with antiwindup compensation is then given by

$$A_{aw}v = Tu_c - Sy + (A_{aw} - R)u$$
$$u = \text{sat}\, v \tag{9.12}$$

This controller is equivalent to (9.11) when it does not saturate. When the control variable saturates, it can be interpreted as an observer with dynamics given by polynomial A_{aw}.

A block diagram of the linear controller of (9.11) and the nonlinear modification of (9.12) that avoids windup is shown in Fig. 9.7. A particularly simple case is that of a deadbeat observer, that is, $A_{aw}^* = 1$. The controller can then be written as

$$u(k) = \text{sat}\left(T^*(q^{-1})u_c(k) - S^*(q^{-1})y(k) + \left(1 - R^*(q^{-1})\right)u(k)\right) \tag{9.13}$$

An example illustrates the implementation.

Example 9.1   Double integrator with antireset windup

A controller with integral action for the double integrator was designed in Sec. 5.7. In this example we use the same design procedure with parameters $\omega = 0.4$ and $\omega h = 0.2$. The result when using the antireset windup procedure in Fig. 9.7 with $A_{aw} = (q-0.5)^2$ is shown in Fig. 9.8. The antireset windup gives less overshoot and the control signal is only saturating at the maximum value. The response with the antireset windup is similar but somewhat slower then for the unsaturated case.   ∎

A generalization of the antireset windup in Fig. 9.7 is given in Fig. 9.9. An extra degree of freedom is introduced through the polynomial A_n. This polynomial, as well as A_{aw}, should be monic and stable. The case in Fig. 9.7 is obtained for $A_n = 1$. The polynomial A_n can be used to shape the response from errors due to the saturation.

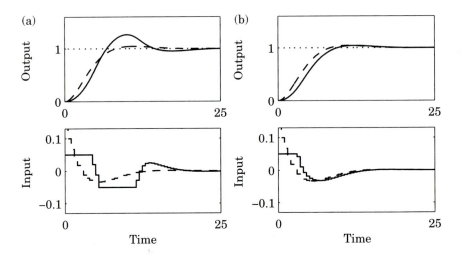

Figure 9.8 Output and input when the double integrator is controlled with a controller (a) without and (b) with antirest windup given by (9.12). The dashed lines show the behavior when there is no saturation.

9.5 Operational Aspects

The interface between the controller and the operator is discussed in this section. This includes an evaluation of the information displayed to the operator and the mechanisms for the operator to change the parameters of the controller. In conventional analog controllers it is customary to display the set point, the measured output, and the control signal. The controller may also be switched from manual to automatic control. The operator may change the gain (or proportional band), the integration time, and the derivative time. This organization was motivated by properties of early analog hardware. When computers are used to implement the controllers, there are many other possibilities. So far the

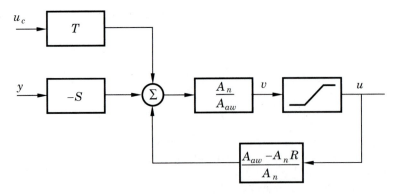

Figure 9.9 A generalization of the antiwindup scheme in Fig.9.7.

potentials of the computer have been used only to a very modest degree.

To discuss the operator interface, it is necessary to consider how the system will be used operationally. This is mentioned in Sec. 6.2 and a few additional comments are given here. First, it is important to realize the wide variety of applications of control systems. There is no way to give a comprehensive treatment, so a few examples are given. For instance, the demands are very different for an autopilot, a process-control room, or a pilot plant.

Example 9.2   Importance of operational aspects

To illustrate that the operational aspects and security are important we take two examples from practical implementations.

The first example is a control system for a steel rolling mill. In this application the control, signal conditioning and logic took about 30% of the code and the rest was related to operator interface and security measures.

The second example is the implementation of an autotuner based on relay feedback. A straightforward implementation of the tuning algorithm could be done in 1.5 pages of C code. The commercial algorithm with all bells and whistles needed for operator communication and security required 15 pages of code.   ■

Operating Modes

It is often desirable to have the possibility of running a system under manual control. A simple way to do this is to have the arrangement shown in Fig. 9.10, where the control variable may be adjusted manually. Manual control is often done with push buttons for increasing or decreasing the control variable.

Because the controller is a dynamic system, the state of the controller must have the correct value when the mode is switched from manual to automatic. If this is not the case, there will be a switching transient. A smooth transition is called *bumpless transfer*, or *bumpless transition*.

In conventional analog controllers, it is customary to handle bumpless transition by introducing a *tracking mode*, which adjusts the controller state so that it is compatible with the given inputs and outputs of the controller. A tracking mode may be viewed as an implementation of an observer.

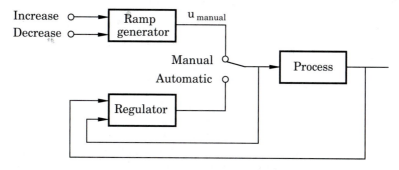

Figure 9.10   Control system with manual and automatic control modes.

A tracking mode is obtained automatically in the controllers of (9.5), (9.10), and (9.12) because they have an observer built into them. To run them in a tracking mode, simply put

$$u_{\text{low}} = u_{\text{high}} = u_{\text{manual}}$$

This implies that the control signal is always equal to the manual input signal. The state of the controller will be reset automatically because of the internal feedback in the controller. The saturation introduced in the controller to handle actuator saturation will automatically give bumpless transfer. There are also other ways to have modes for semiautomatic control by keeping some feedback paths for stabilization.

With computer control, it is possible to have many other operating modes. Parameter estimation and control-design algorithms can be included in the controller. An estimation mode, in which a model of the process is estimated, may be introduced. The estimated model may be used in the design algorithm to give an update of the parameters of the controller in a tuning mode. Adaptive control modes, in which the parameters are updated continuously, may also be added. Computer control offers many other interesting possibilities. The performance of the control loop may be displayed instead of set point and error.

Initialization

Because a controller is a dynamic system, it is important to set the controller state appropriately when the controller is switched on. If this is not done, there may be large switching transients. In conventional PI-controllers, the controller has one state only—namely, the integrator. It is customary to initialize such a controller by operating it in manual control until the process output comes close to its desired value. For an algorithm with an explicit observer, the controller state may be initialized by keeping the control signal fixed for the time required for the observer to settle. A controller with antiwindup may also be initialized by running it in manual mode during a period that corresponds to the settling time of the observer.

Parameterization and Parameter Changes

In conventional process controllers, the operator can manipulate the set point and the parameters in the control law (gain, integration time, and derivative time). With computer control, there are many other interesting alternatives. Because of the simplicity of computing, it is possible to use one parameterization in the control algorithm and another in the operator communication. The parameters displayed to the operator may then be related to the performance of the system rather than to the details of the control algorithm. The conversion between the parameters is made by an algorithm in the computer.

To illustrate the idea of performance-related parameters, consider design of a servo using the pole-placement method described in Chapter 5. The closed-loop properties may be specified in terms of the relative damping ζ and the bandwidth ω_B. To perform the design, it is also necessary to have a model of the open-loop dynamics. One possibility is to have the process engineer enter the desired bandwidth and damping and a continuous-time model. The computer can then make the necessary conversions in order to obtain the control law. If the computer also has a recursive estimation algorithm, it is not necessary to introduce the model. Clearly, there are many interesting possibilities if estimation and design algorithms are included in the controller.

There are two operational problems with on-line parameter changes. One problem is related to real-time programming. Data representing parameters are shared among different programs. It is then necessary to make sure that one program is not using data that are being changed by another program. This is discussed in Sec. 9.8.

The other problem is algorithmic. There may be switching transients when the parameters are changed in a control algorithm. To get some insight into what can happen, consider the simple PI-algorithm

```
e:=uc-y
u:=k*(e+i/ti)
i:=i+e*h
```

It is clear that a change of the integration time ti will cause a step in the control signal unless the integral part, i, is zero. The problem can be avoided by changing the state from i to i*ti/ti', where ti' is the new value of the integration time. Another simpler way is to write the algorithm as

```
e:=uc-y
u:=k*e+i
i:=i+k*e*h/ti
```

Compare with Sec. 8.5. The need for changing the state when parameters are changed is dictated by the fact that the state of the controller depends on its parameters. One way to obtain bumpless parameter changes is to store a set of past input-output data and to run an observer when the controller parameters are changed. However, it is often possible to use a simpler solution.

To see what should be done, consider the algorithm of (9.5) with an explicit observer and state feedback. First, a realization should be chosen so that the matrices C and D do not depend on the adjustable parameters. If the state x represents an estimate of physical state variables, there are very few difficulties because the estimated state will not change drastically when model parameters are changed. Transients due to changes in the feedback gain cannot be avoided if there is a nonzero error $e = x - \hat{x}$. Similarly, there will not be any switching transients with the algorithm in (9.10), provided that there is a representation in which the matrices C and D do not contain any parameters that are modified. It is more complicated to see what should be done with the algorithm of (9.12). In the representation of (9.12), the state is delayed inputs and outputs. This state is not minimal. Although the state does not depend on the coefficients of the polynomials R, S, and T, there is no guarantee that the given R, S, T, u, and y are compatible with Eq. (9.12). With the representation of (9.12), there will be switching transients when the parameters are changed.

Security

It is very important to make sure that a computer-control system operates safely. Ideally, this means that the system should either give the correct result or an alarm if it is not functioning properly. Systems with extremely high requirements may be tripled (or quadrupled) and the output accepted if two subsystems give the same result. For simpler systems, it may be sufficient to rely on self-checking. There are many ways to do this using computer control. Arithmetic units may be checked by computing functions with known results. Memory and data transmission may be checked through checksums. A-D and D-A converters may be checked by using a few extra connected channels. A D-A conversion is commanded and the result of an A-D conversion of the same channel is checked. Timing may similarly be investigated by connecting a network with a known time constant between a D-A and an A-D converter.

9.6 Numerics

When implementing a computer-control system it is necessary to answer questions such as: How accurate should the converters be? What precision is required in the computations? Should computations be made in fixed-point or floating-point arithmetic? To answer these questions, it is necessary to understand the effects of the limitations and to estimate their consequences for the closed-loop system. This is not a trivial question, because the answer depends on a complex interaction of the feedback, the algorithm, and the sampling rate. Fortunately, only crude estimates have to be done. For instance, should the resolution be 10 or 12 bits and should the word length be 24 or 32 bits? Such questions may be answered using simplified analysis.

Error Sources

The major sources of error are the following:

- Quantization in A-D converters

- Quantization of parameters

- Roundoff, overflow, and underflow in addition, subtraction, multiplication, division, function evaluation, and other operations

- Quantization in D-A converters

Common types of A-D converters have accuracies of 8, 10, 12, and 14 bits, which correspond to a resolution of 0.4%, 0.1%, 0.025%, and 0.006%, respectively. The percentages are in relation to full scale. The D-A converters also have a limited precision. An accuracy of 10 bits is typical. The error due to the quantization of the parameters depends critically on the sampling period and on the chosen realization of the control law.

Word Length

Digital-control algorithms are typically implemented on microcomputers and minicomputers, which have word lengths of 8, 16, or 32 bits. The essential numerical difficulty with short word length is illustrated by the example.

Example 9.3   Scalar-product calculations

Consider the vectors

$$a = \begin{pmatrix} 100 & 1 & 100 \end{pmatrix}$$

$$b = \begin{pmatrix} 100 & 1 & -100 \end{pmatrix}$$

The scalar product is $(a, b) = 1$. If the scalar product is computed in floating-point representation with a precision corresponding to three decimal places, the result will be zero because $100 \cdot 100 + 1 \cdot 1$ is rounded to 10,000. Notice that the result obtained depends on the order of the operations. Finite-word-length operations are neither associative nor distributive.   ∎

The difficulty may be avoided without using complete double-precision calculation by adding the terms in double precision and rounding to single precision afterwards. This method can be applied to fixed-point and floating-point calculations. Notice that the multiply instruction for many computers is implemented so that the product is available in double precision. Many high-level languages also have constructions that support this type of calculation. Generally speaking, roundoff and quantization will give rise to small errors, whereas the effects of overflow will be disastrous.

Digital-signal processors (DSPs) are now commonly used to implement computer-controlled systems when short sampling periods are required. The low-cost signal processors are all using fixed-point calculations. For the TMS family from Texas Instruments, the standard word length is 16 bits but the accumulator is 32 bits wide. A 16-bit DSP from AT&T has a 36-bit accumulator, and Motorola has a DSP with 24-bit word length and a 56-bit accumulator. The architecture with a long accumulator is ideal for computing scalar products, which is the key operation when implementing linear filters, because the products of the terms can be accumulated in double precision. The signal processors are very fast. The operation of multiply and accumulate (MAP) typically takes 100 nanoseconds. There are also more expensive signal processors with floating-point hardware.

There is also an increased use of computer-controlled systems implemented using special-purpose VLSI circuits. In these applications the word length is a design parameter that can be chosen freely. Such a choice naturally requires a more detailed investigation than a simple choice between single or double precision. There are applications of custom VLSI both in the aerospace industry and for mass-produced consumer goods like VCRs and CD players. For these applications it is of major concern to minimize chip area. A typical example is a CD player in which both audio and servo functions are implemented on one chip. For a stationary CD player there are fewer demands on the servo than for a CD player for a car. The chip area for the control system can thus be smaller for the stationary player.

There are many number representations used in digital computers. Integers are typically 16, 32, or 48 bits. For a long time there were many representations of floating-point numbers. The IEEE did, however, take the initiative to standardize them, and a standard ANSI-IEEE 754 was published in 1985. In this standard the numbers are represented as

$$\pm a \cdot 2^b$$

where $0 \leq a < 2$ is the *significand*, also called the mantissa, and b is the *exponent*. In the standard there are three types of floating-point numbers:

short real (32 bits)	1 sign	8 exponent	23 significand
long real (64 bits)	1 sign	11 exponent	52 significand
short temporary real (80 bits)	1 sign	15 exponent	64 significand

The IEEE standard has gained widespread acceptance, and the floating-point chips from Intel and Motorola are based on it.

Overview of Effects of Roundoff and Quantization

The consequences of roundoff and quantization depend on the feedback system and on the details of the algorithm. The properties may be influenced considerably by changing the representation of the control law or the details of the algorithm. Thus it is important to understand the phenomenon.

A detailed description of roundoff and quantization leads to a complicated nonlinear model, which is very difficult to analyze. Investigation of simple cases shows, however, that roundoff and quantization may lead to limit-cycle oscillations. Such examples are presented later, together with approximative analysis. Limit-cycle oscillations have also been observed in more complex cases.

Some properties of roundoff and quantization in a feedback system may also be captured by linear analysis. Roundoff and quantization are then modeled as ideal operations with additive or multiplicative disturbances. The disturbances may be either deterministic or stochastic. This type of analysis is particularly useful for order-of-magnitude estimation. It allows investigation of complex systems and it is useful when comparing different algorithms.

Techniques from sensitivity analysis and numerical analysis are also useful in finding the sensitivity of algorithms to changes of parameters. Such methods may be used to compare and screen different algorithms. However, the methods are limited to comparison of the open-loop performances of the algorithms. It is also necessary to compare the effects of roundoff and quantization with the other disturbances in the system.

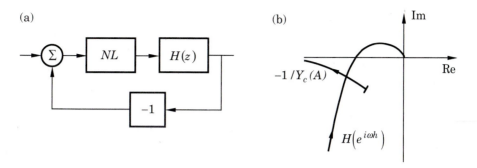

Figure 9.11 (a) Discrete-time system with one nonlinearity NL. (b) Using the method of describing function.

Nonlinear Analysis Using Describing Functions

If there is only one nonlinearity in the loop, it is possible to use the method of describing function to determine limit cycles approximately.

Consider the system in Fig. 9.11(a). The method of describing function can be regarded as a generalization of the Nyquist criterion. The critical point -1 is replaced by $-1/Y_c(A)$, where $Y_c(A)$ is the describing function of the nonlinearity. The describing function characterizes the transmission of a sinusoidal signal with amplitude A through the nonlinearity. The method predicts a limit cycle if

$$H(e^{i\omega h}) = -\frac{1}{Y_c(A)}$$

[compare with Fig. 9.11(b)]. The frequency, ω_1 (from the Nyquist curve), and the amplitude, A_1 (from the describing function), at the intersection are the estimated frequency and the estimated amplitude of the limit cycle. The describing function of a roundoff quantizer is

$$Y_c(A) = \begin{cases} 0 & 0 < A < \dfrac{\delta}{2} \\[2ex] \dfrac{4\delta}{\pi A}\displaystyle\sum_{i=1}^{n}\sqrt{1-\left(\dfrac{2i-1}{2A}\delta\right)^2} & \dfrac{2n-1}{2}\delta < A < \dfrac{2n+1}{2}\delta \end{cases}$$

The function Y_c only takes real values. Its smallest value is zero and its largest value is $4/\pi \approx 1.27$. The function is graphed in Fig. 9.12. This means that the critical part for quantization consists of the part of the negative real axis from $-\infty$ to -0.78. Describing function analysis thus predicts oscillations due to quantization if the Nyquist curve of the loop gain intersects this line segment. For stable systems this means that quantization will not give rise to oscillations if the amplitude margin is larger than 1.27. Describing function analysis predicts oscillations for all systems that are open-loop unstable.

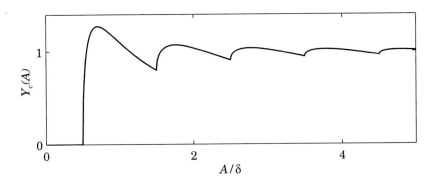

Figure 9.12   The describing function of roundoff.

Example 9.4   Roundoff effects

Consider the system in Example 3.4 with the pulse-transfer function

$$H(z) = \frac{0.25K}{(z-1)(z-0.5)}$$

In Example 3.4 it is shown that the closed-loop system without roundoff is asymptotically stable if $K < 2$. With roundoff of the error signal in Fig. 3.5, the method of describing function predicts that there will be a limit cycle if K is greater than about 1.3. Figure 9.13 shows the behavior of the system for a roundoff level of $\delta = 0.2$. The limit cycle is clearly noticeable for $K = 1.6$.   ∎

Linear Analysis

The effects of roundoff and quantization may also be estimated by linear analysis. The idea is to represent the operations by their ideal models and an additive disturbance e_a. The D-A and A-D converters are then simply represented as linear gains with a disturbance that models the quantization. With fixed-point calculations, the additions are exact. There will, however, be errors in multiplications. These are represented as exact multiplications, with an additive error, which represent the roundoff. This is illustrated in Fig. 9.14.

The errors may be modeled as deterministic or stochastic signals. In a deterministic model, the error is modeled as constants having the sizes of quantization errors and with the resolution in the arithmetic calculations. In the stochastic model, the error introduced by rounding or quantization is then described as additive white noise with a rectangular distribution. The errors at different sampling times are thus assumed to be uncorrelated. If the quantization is done as rounding, then the error is equally distributed over the interval $(-\delta/2, \delta/2)$, where δ is the quantization step. If the quantization is done as truncation, the error is equally distributed over $(0, \delta)$. A rectangular noise distributed over an interval of length δ has a variance of $\delta^2/12$.

By using the linear models of roundoff and quantization, it is possible to reduce the problem of estimating the effect of the roundoff and quantization

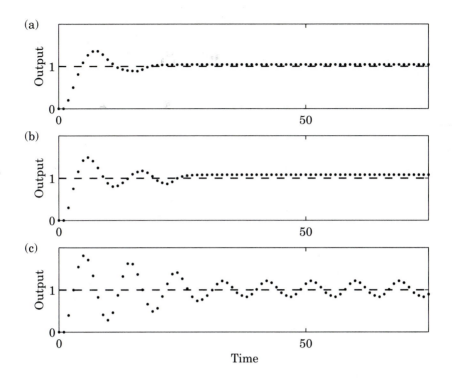

Figure 9.13   The output of the system in Example 9.4 when δ = 0.2 and (a) K = 0.8, (b) K = 1.2, and (c) K = 1.6.

to the problem of calculating responses of a linear system to deterministic or stochastic inputs. By using the linear model, it is possible to assess the effects of quantization qualitatively without going into detailed calculation. It is also easy to compare roundoff with other disturbances in the system. In the linear model, the effect of roundoff in the A-D converter is the same as the effect of measurement noise. The effect on the control signal may be substantial for those frequencies where the controller has high gain. The effect of roundoff in the D-A converter is the same as a disturbance in the process input. Because the process normally attenuates high frequencies, the effect on the process output is normally small. Remember, however, that the linear model does not capture all aspects of roundoff.

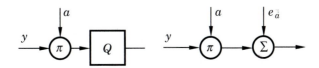

Figure 9.14   Linear models for multiplication with roundoff.

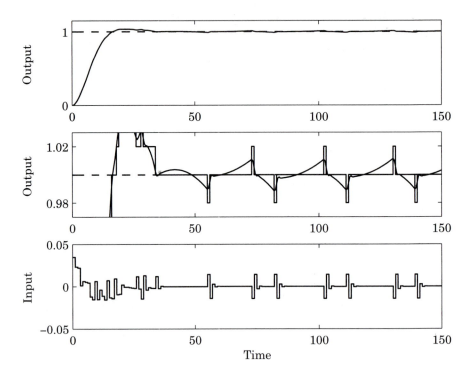

Figure 9.15   Control of a double integrator with a quantized A-D converter. The sampling period is 1 s and the quantization level is 0.02. The middle curve shows the quantized as well as the unquantized output.

Selection of Resolution of A-D and D-A Converters

Today the resolution of the D-A converters are about 10 bits and the A-D converters have normally 14 bits. With double-precision arithmetic the computations are typically done with a resolution of 64 bits. This implies that it is the accuracy of the converters that have the largest influence on the performance. A couple of examples illustrate the influence of the quantization in the converters.

Example 9.5   Effects of A-D quantization for the double integrator

Figure 9.15 shows a simulation of digital control of a double integrator where the A-D converter is quantized with the level 0.02. The controller is the same as the one used in Sec. 5.7. It is given by

$$R(q)u(k) = T(q)u_c(k) - S(q)y(k)$$

where

$$
\begin{aligned}
R(q) &= (q - 1)(q + 0.188) \\
S(q) &= 0.715q^2 - 1.281q + 0.580 \\
T(q) &= (3.473q^2 - 2.555q + 4.700) \cdot 10^{-2}
\end{aligned}
\qquad (9.14)
$$

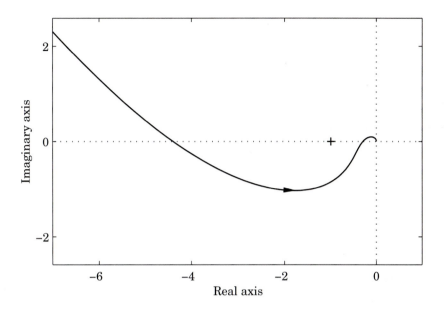

Figure 9.16 Nyquist curve for the sampled loop gain for the double integrator, when using the controller defined by (9.14).

and the sampling period 1 s. The simulation clearly shows that there is a limit-cycle oscillation where the output changes one quantization level. The period is 28 s. The describing functions analysis predicts a limit cycle with period 39 s. See Fig. 9.16, which shows the Nyquist curve of the sampled loop gain. The describing function method predicts that the amplitude of the oscillation is $\delta/2$, which agrees well with the behavior of the output of the process. Describing function analysis gives correct qualitative results in this case, but the prediction of the period is poor because the signals deviate significantly from sinusoids.

We will also use another method to estimate the amplitudes of the fluctuations caused by the quantization. To do this, first observe that the output signal oscillates with one quantization level up or down at widely spaced intervals. This means that the controller output is given by the pulse response of the controller, that is,

$$-\frac{S(z)}{R(z)} = -0.710 + 0.700z^{-1} - 0.145z^{-2} + 0.0134z^{-3} \cdots \qquad (9.15)$$

multiplied by the quantization level. This gives an excellent prediction of the fluctuations in the control signal. Compare with Fig. 9.15. Notice that the first coefficient in the expansion of $-S/R$ is equal to $-s_0$. ∎

This example shows that the periodic ripple in the control signal due to quantization in the A-D converter can be estimated from a simple pulse-response calculation. In the next example we will analyze the effect of quantization in the D-A converter.

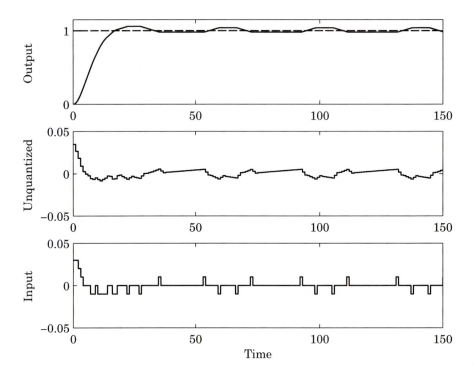

Figure 9.17 Digital control of the double integrator with a quantized D-A converter.

Example 9.6   Effects of D-A quantization for the double integrator

Figure 9.17 shows a simulation of the double integrator with a D-A converter with quantization level 0.01. The quantization causes a limit-cycle oscillation. The process output is, however, much more sinusoidal than with A-D quantization. The reason for this is that the nonlinearity is just before the process that attenuates high frequencies. As before, the describing function predicts an oscillation with the period 39 s whereas the actual period is 39 s. The amplitude of the oscillation in the process output can be estimated by evaluating the magnitude of the pulse-transfer function of the controller at the period of oscillation. The controller gain is approximately 0.12. With $\delta = 0.01$, the amplitude of the output can be estimated to 0.04 and the measured amplitude is about 0.03. See Fig. 9.17.

Notice that the oscillations due to the quantization in the D-A converter can be avoided if the output of the D-A converter is fed back into the control law in the same way as was done to avoid windup.   ∎

By using the insight obtained from the examples, some recommendation on the selection of resolution of the converters can now be given.

The resolution of the A-D converter must be chosen so that it gives the desired precision in the process output. One should investigate whether quantization can give rise to limit-cycle oscillations. The magnitude of the ripple in the control signal caused by the A-D quantization should be investigated. This can be estimated simply from Eq. (9.15). If the ripple in the control signal is too large, better resolution of the A-D converter is required.

To determine the required resolution of the D-A converter, the frequency of a possible limit cycle is first determined. If there is a limit-cycle oscillation, the amplitude can be estimated crudely from the process gain at the oscillation frequency, or more accurately using the theory of relay oscillations quoted in the References. The estimates obtained in this way will typically give the order of magnitude. It is recommended to use simulation to get more accurate results. The procedure is illustrated by an example.

Example 9.7   Choosing resolution in D-A and A-D converters

Consider the double integrator that we have investigated. Assume that the process output is in the range $[-1, 1]$, that the range of the control signal is $[-0.04, 0.04]$, and that it is desired to control the output with a precision of 1%. If we let each converter contribute 0.5%, the A-D converter must have a resolution of at least 0.005, which is equivalent to 9 bits. Because the gain of the process at the limit cycle is about 15, the resolution of the D-A converter must be better than 0.00033. With the given signal range, this corresponds to 1 part in 240, or 8 bits.   ∎

9.7 Realization of Digital Controllers

The previous section illustrated how roundoff and quantization in A-D and D-A converters influence the behavior of the system. Roundoff errors in the computations of the control law also cause quantization, which can be modeled and analyzed in the same way as converter quantization. The quantization arising from the computations depends critically on how the computations are organized, for example, on how the sampled-data controller is realized. This section discusses different realizations. Some advantages and disadvantages of different methods are given.

Assume that we want to realize the controller

$$y(k) = H(q^{-1})u(k) = \frac{b_0 + b_1 q^{-1} + \cdots + b_m q^{-m}}{1 + a_1 q^{-1} + a_2 q^{-2} + \cdots + a_n q^{-n}} u(k) \qquad (9.16)$$

Some different realizations are

- Direct form

- Companion form

- Series (Jordan) form

- Parallel (diagonal) form

- Ladder form

- δ-operator form

Coefficient-Pole Sensitivity

Finite precision in the representation of the coefficients of the controller gives a distortion of the poles and zeros of the controller. The following analysis gives quantitative results for the sensitivity of the roots of a polynomial with respect to changes in the coefficients. Consider a linear filter with distinct poles in p_i and the characteristic polynomial

$$A(z, a_i) = z^n + a_1 z^{n-1} + \cdots + a_n = (z - p_1) \cdots (z - p_n)$$

The characteristic polynomial A can be regarded as a function of z and a_i. When the parameter a_i is changed to $a_i + \delta a_i$, the poles are changed from p_k to $p_k + \delta p_k$. Hence

$$0 = A(p_k + \delta p_k, a_i + \delta a_i) \approx A(p_k, a_i) + \left.\frac{\delta A}{\delta z}\right|_{p_k} \delta p_k + \left.\frac{\delta A}{\delta a_i}\right|_{p_k} \delta a_i + \cdots$$

The first term on the right-hand side is zero. If terms of second order and higher are neglected, it follows that

$$\delta p_k \approx - \left.\frac{\delta A/\delta a_i}{\delta A/\delta z}\right|_{z = p_k} \cdot \delta a_i$$

Because

$$\left.\frac{\delta A}{\delta a_i}\right|_{z = p_k} = p_k^{n-i} \quad \text{and} \quad \left.\frac{\delta A}{\delta z}\right|_{z = p_k} = \prod_{j \neq k}(p_k - p_j)$$

the following estimate is obtained:

$$\delta p_k \approx - \frac{p_k^{n-i}}{\prod_{j \neq k}(p_k - p_j)} \delta a_i \tag{9.17}$$

If the polynomial has a root p_k with multiplicity m, Eq. (9.17) becomes

$$\delta p_k \approx - \frac{p_k^{n-i}}{\prod_{j \neq k}(p_k - p_j)} (\delta a_i)^{1/m} \tag{9.18}$$

If the filter is stable, then $|p_k| < 1$ and the numerator of (9.17) has its largest magnitude for $i = n$. The coefficient a_n is thus the most sensitive parameter. Furthermore, the denominator will be small if the poles are close, which then makes the system sensitive to changes in the coefficients. Equation (9.18) shows that the sensitivity is even higher if the polynomial has multiple roots. Equations (9.17) and (9.18) may be used to determine the conditioning numbers for the transformation from the diagonal form to the companion form. It follows from the equations that the computation of companion forms may be poorly conditioned.

Direct- and Companion-Form Realizations

The most straightforward way to realize (9.16) is to write it in the *direct form*

$$y(k) = \sum_{i=0}^{m} b_i u(k - i) - \sum_{i=1}^{m} a_i y(k - i)$$

It is then necessary to store $y(k - 1)$, $y(k - 2), \ldots, y(k - n)$, and $u(k - 1), \ldots,$ $u(k - m)$, that is, $n + m$ variables. The direct realization has $n + m$ states and thus is not a minimal realization. The controllable or observable canonical forms, see Sec. 3.4, have n states. The direct form has the advantage that the state variables are simply delayed versions of the input and output signals. This means that the state does not have to be recomputed when the parameters are changed. Both the direct form and the companion form have the disadvantage that the coefficients in the characteristic polynomial are the coefficients in the realizations. This makes the realizations extremely sensitive to computational errors if the system is of high order and if the poles or zeros are close to one as discussed before.

Well-Conditioned Realizations

The difficulty associated with the companion form can be avoided simply by representing the system as a combination of first- and second-order systems. If the dynamic system representing the controller has n_r distinct real poles and n_c complex-pole pairs, the control algorithm may be transformed to the modal form

$$z_i(k + 1) = \lambda_i z_i(k) + \beta_i y(k) \qquad\qquad i = 1, \ldots, n_r$$

$$v_i(k + 1) = \begin{pmatrix} \sigma_i & \omega_i \\ -\omega_i & \sigma_i \end{pmatrix} v_i(k) + \begin{pmatrix} \gamma_{i1} \\ \gamma_{i2} \end{pmatrix} y(k) \qquad i = 1, \ldots, n_c \qquad (9.19)$$

$$u(k) = D y(k) + \sum_{i=1}^{n_r} \gamma_i z_i(k) + \sum_{i=1}^{n_c} \delta_i^T v_i(k)$$

where the complex poles are represented using real variables. Notice that z_i are scalars and v_i are vectors with two elements. To avoid numerical difficulties, the control law should be transformed to the form of (9.19), which is then implemented in the control computer. The transformation may easily be done in a package for computer-aided design. It is easy to use fixed-point calculations and scaling for equations in the form of (9.19). If the control law has multiple eigenvalues, a Jordan canonical form replaces (9.19). An eigenvalue λ of multiplicity 3 thus corresponds to a block

$$z(k + 1) = \begin{pmatrix} \lambda & 1 & 0 \\ 0 & \lambda & 1 \\ 0 & 0 & \lambda \end{pmatrix} z(k) + \begin{pmatrix} \beta_1 \\ \beta_2 \\ \beta_3 \end{pmatrix} y(k)$$

Ladder Realizations

Ladder realizations are other representations that avoid the coefficient sensitivity in the implementations. One representation of the ladder network is obtained by making a continued-fraction expansion in the pulse-transfer operator in the following way:

$$H(z) = \alpha_0 + \cfrac{1}{\beta_1 z + \cfrac{1}{\alpha_1 + \cfrac{1}{\beta_2 z + \cfrac{1}{\ddots \cfrac{1}{\alpha_{n-1} + \cfrac{1}{\beta_n z + \cfrac{1}{\alpha_n}}}}}}} \tag{9.20}$$

Another realization is obtained by making the continued-fraction expansion in z^{-1}. The ladder forms have low sensitivity against coefficient errors and round-off errors. If

$$H(z) = \frac{B(z)}{A(z)}$$

where $\deg A(z) = \deg B(z) = n$, the coefficients α_i and β_i can be computed in the following way: Compute $\alpha_0 = B(z) \operatorname{div} A(z)$, $A_1(z) = A(z)$, and $B_1(z) = B(z) \operatorname{mod} A(z)$ and repeat for $i = 1$ to n

$$\beta_i = A_i \operatorname{div} z B_i \qquad\qquad A_{i+1} = A_i \operatorname{mod} z B_i$$
$$\alpha_i = B_i \operatorname{div} A_{i+1} \qquad\qquad B_{i+1} = B_i \operatorname{mod} A_{i+1}$$

The ladder-network representation can be expressed by the following state equations:

$$\beta_1 x_1(k+1) = \frac{1}{\alpha_1}(x_2(k) - x_1(k)) + u(k)$$

$$\beta_2 x_2(k+1) = \frac{1}{\alpha_1}(x_1(k) - x_2(k)) + \frac{1}{\alpha_2}(x_3(k) - x_2(k))$$

$$\vdots$$

$$\beta_i x_i(k+1) = \frac{1}{\alpha_{i-1}}(x_{i-1}(k) - x_i(k)) + \frac{1}{\alpha_i}(x_{i+1}(k) - x_i(k))$$

$$\vdots$$

$$\beta_n x_n(k+1) = \frac{1}{\alpha_{n-1}}(x_{n-1}(k) - x_n(k)) - \frac{1}{\alpha_n} x_n(k)$$

$$y(k) = x_1(k) + \alpha_0 u(k)$$

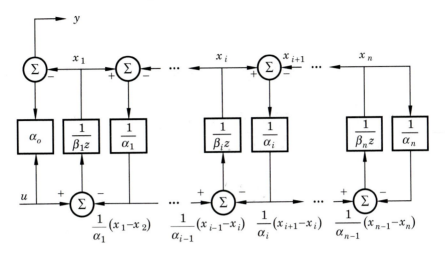

Figure 9.18   Block diagram of a ladder network representation of the transfer function (9.20).

A block diagram of the representation is shown in Fig. 9.18. The name ladder network derives from the shape of the graph.

Short-Sampling-Interval Modification

We have shown that it is useful to transform the system to a well-conditioned form before it is implemented on a digital computer. This reduces the coefficient sensitivity of the realization. An additional modification that is useful when the sampling period is short will now be discussed. Consider the compensator described by the general-state model (9.8) and (9.9). For short sampling periods the matrix F is close to the unit matrix, that is, all the eigenvalues are close to one. Further, the matrix G is proportional to the sampling period. With a short sampling period, the matrices F and G may therefore differ by several orders of magnitude. By rewriting the state equation it is possible to obtain a representation that is better conditioned. It is convenient to rewrite equation (9.8) as

$$x(k + 1) = x(k) + (F - I)x(k) + Gy(k) \qquad (9.21)$$

where the matrix $F - I$ is also proportional to the sampling period h. The numerical representation of $F - I$ requires fewer decimals than the representation of F itself. The term $(F - I)x(k) + Gy(k)$ represents a correction to the state, which will be small if the sampling period is short. This representation is particularly useful in fixed-word-length computations. The state is stored in double precision. The change in the control is calculated using single-precision multiplication, and the product is then added to the state. Compare with the discussion of the scalar-product computation in Example 9.3.

The δ-Operator

The system (9.21) can be written as

$$x(k + 1) = x(k) + h\left(\bar{F}x(k) + \bar{G}y(k)\right) \tag{9.22}$$

where

$$h\bar{F} = F - 1$$
$$h\bar{G} = G$$

Instead of the shift operator we can now introduce the δ-operator that is defined by

$$\delta = \frac{q - 1}{h} \tag{9.23}$$

Equation (9.22) can now be written as

$$\delta x(k) = \bar{F}x(k) + \bar{G}y(k)$$

A general pulse-transfer operator can be transformed from shift form to δ-form as

$$H(q) = \frac{B(q)}{A(q)} = \frac{B(\delta h + 1)}{A(\delta h + 1)} = \frac{\bar{B}(\delta)}{\bar{A}(\delta)} = \bar{H}(\delta)$$

The δ-operator is thus equivalent to the shift operator. All the analysis done for the shift operator can be translated into δ-form. The δ-operator has the property

$$\delta f(kh) = \frac{f(kh + h) - f(kh)}{h}$$

that is, it can be interpreted as the forward-difference approximation of the differential operator $p = d/dt$. In this respect the δ-operator is "closer" to the continuous-time domain than the shift operator. For instance, the stability region in the δ-form is a circle with radius $1/h$ and with the origin in $-1/h$. When $h \to 0$ the stability region becomes the left half-plane.

The δ-operator representation has the property that it translates into the corresponding continuous-time system when the sampling interval approaches zero. Hence

$$\lim_{h \to 0} \bar{H}(\delta) = G(\delta)$$

where G is the continuous-time transfer function. This implies, for instance, that the zeros of the transfer function in the δ-form approach the zeros (finite as well as infinite) of the continuous-time transfer function.

Example 9.8   Double integrator in δ-form

Consider the double integrator $G(s) = 1/s^2$. Then

$$H(q) = \frac{h^2(q+1)}{2(q-1)^2} = \frac{h^2(\delta h + 2)}{2h^2\delta^2} = \frac{1 + \delta h/2}{\delta^2} = \bar{H}(\delta)$$

When h goes to zero we get

$$\lim_{h\to 0} \bar{H}(\delta) = \frac{1}{\delta^2} = G(\delta)$$

Notice that the δ-form also has "sampling zeros," $\delta = -2/h$. This zero will approach $-\infty$ when $h \to 0$.   ∎

Heuristically we can interpret the δ-operator as a shift of origin and scaling. This is a common trick in numerical analysis and has the consequence that the δ-form can obtain better numerical properties than the shift operator. A controller in δ-form can be described by the state equations

$$\delta x(kh) = \bar{F}x(kh) + \bar{G}y(kh) = d(kh)$$
$$u(kh) = Cx(kh) + Dy(kh) \tag{9.24}$$

The shift operator and its inverse are implemented exactly using an assignment statement. To make a realization in δ-form we must implement the operator δ^{-1}. Solving (9.24) for $x(kh)$ gives

$$x(kh) = \delta^{-1}d(kh) = x(kh - h) + hd(kh)$$

The extra amount of computations compared to the shift form is marginal. One extra vector addition is necessary. Notice that because $hd(kh)$ is normally much smaller than $x(kh - h)$, it is necessary to represent $x(kh - h)$ with a sufficient word length.

An Example

An example illustrates the properties of the different realizations. Consider a system with the pulse-transfer function

$$H(z) = \frac{b^4}{(z+a)^4} \tag{9.25}$$

where $b = 1 + a$. The system has multiple poles that are close to one when a is close to -1. The previous discussion then shows that the system is very sensitive to coefficient perturbations.

    To obtain the computer program, a state-space realization of the pulse-transfer function is first determined. The computer code is then obtained as a direct implementation of the difference equations. There are many possible choices of the coordinate system in the state-space realization. A controllable

Listing 9.3 Computer code for implementation of (9.25) based on the shift-controllable canonical form.

```
begin
  y:=b*b*b*b*x4
  s:=-a1*x1-a2*x2-a3*x3-a4*x4+u
  x4:=x3
  x3:=x2
  x2:=x1
  x1:=s
end
```

canonical form in shift and δ-operator and a Jordan canonical form are chosen to demonstrate that the numerical properties may differ considerably. For the shift-operator controllable form we implement

$$\frac{b^4}{(z+a)^4} = \frac{b^4}{z^4 + 4az^3 + 6a^2z^2 + 4a^3z + a^4}$$

The code is given in Listing 9.3. The numerical values of the parameters for the controllable canonical form when $a = -0.99$ are given by

$$a_1 = 4a = -3.96 \qquad\qquad a_2 = 6a^2 = 5.8806$$
$$a_3 = 4a^3 = -3.881196 \qquad\qquad a_4 = a^4 = 0.96059601$$

Listing 9.4 gives an implementation based on the Jordan canonical form. By rewriting the Jordan form as (9.21) Listing 9.5 is obtained. Notice that this slight modification gives a significant improvement over the form in Listing 9.4, because the state is now obtained by adding a small correction to the previous state.

Listing 9.4 Computer code for implementing (9.25) based on the Jordan canonical form.

```
begin
  x4:=-a*x4+b*u
  x3:=-a*x3+b*x4
  x2:=-a*x2+b*x3
  x1:=-a*x1+b*x2
  y:=x1
end
```

Listing 9.5 Rearrangement to short-sampling-interval modification of the code in Listing 9.4.

```
begin
  x4:=x4+b*(u-x4)
  x3:=x3+b*(x4-x3)
  x2:=x2+b*(x3-x2)
  x1:=x1+b*(x2 - x1)
  y:=x1
end
```

For the δ-form we implement (in controllable canonical form)

$$\frac{b^4}{(\delta + 1 + a)^4} = \frac{b^4}{(\delta + b)^4} = \frac{b^4}{\delta^4 + 4b\delta^3 + 6b^2\delta^2 + 4b^3\delta + b^4}$$

$$= \frac{b^4}{\delta^4 + b_1\delta^3 + b_2\delta^2 + b_3\delta + b_4}$$

where $b = 1 + a$. Notice that b is a small number when a is close to -1. The system is implemented using (9.21), where $\bar{F}$ and $\bar{G}$ have the same form as for the shift-operator companion form in Listing 9.3. The implementation in δ-companion form is given in Listing 9.6, where b_i are the coefficients in the characteristic polynomial.

The implementation of the discrete-time system also includes a monitor system that runs the program each sampling period. Notice that the code contains only addition, multiplication, and assignment statements; thus it can easily be implemented using many computer languages. Because assignment statements only transfer data, they will not introduce any numerical errors. This means that 0 and 1 in a standard matrix representation are represented exactly.

Listing 9.6 Computer code for implementing (9.25) based on δ-operator controllable canonical form.

```
begin
  y:=b*b*b*b*x4
  s:=-b1*x1-b2*x2-b3*x3-b4*x4+u
  x4:=x4+x3
  x3:=x3+x2
  x2:=x2+x1
  x1:=x1+s
end
```

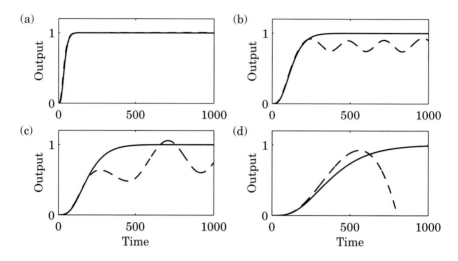

Figure 9.19 Step responses for the system (9.25) for different implementations and different values of a: shift-operator controllable canonical form (Listing 9.3) (dashed), Jordan form (Listing 9.4) and δ-operator controllable canonical form (Listing 9.6) are both the full line. (a) $a = -0.9$, (b) $a = -0.97$, (c) $a = -0.98$, and (d) $a = -0.99$.

Figure 9.19 shows simulations using MATLAB®. The simulation is simply an iteration of the state equations in Listings 9.3, 9.4, and 9.6. The results are obtained when chopping the result of all operations to seven digits. The figure shows the results when different values of a are used. For $a = -0.9$ all the implementations give compatible results, as shown in Fig. 9.19(a). When a is decreased, the shift-operator controllable canonical form is very sensitive and the solution is inaccurate. The other two implementations give approximately the same results. They will, however, differ when even lower numerical precision is used. The modified Jordan form is better than the δ-operator controllable canonical form when a is decreased further.

The sensitivity of the shift-operator controllable form with respect to parameter changes is given by (9.18). Perturbing the characteristic equation with a constant term ε,

$$(z - 0.99)^4 + \varepsilon = 0$$

gives the roots

$$z = 0.99 + (-\varepsilon)^{1/4}$$

The roots are moved from 0.99 to a circle with origin at 0.99 and the radius $r = |\varepsilon|^{1/4}$. If $\varepsilon = 10^{-8}$ then $r = 10^{-2}$; that is, the system can be unstable even if the perturbation is very small.

Making a similar calculation for the δ companion form we get

$$(\delta + 0.01)^4 + \varepsilon = 0$$

which gives the roots

$$\delta = -0.01 + (-\varepsilon)^{1/4}$$

The roots are moved from -0.01 to a circle with origin at -0.01 and the radius $r = |\varepsilon|^{1/4}$. If $\varepsilon = 10^{-8}$ then $r = 10^{-2}$, which is the same as for the shift-operator case. Notice, however, that the relative variation in parameters required to make the system unstable is two orders of magnitude larger with the δ-operator. The δ companion form is thus less sensitive to parameter perturbations than the shift companion form. Notice, however, that the Jordan realizations in shift or δ-forms are superior.

Effects of the Sampling Period

The sampling period also has a considerable influence on the conditioning, as shown by the following example.

Example 9.9   Numerical precision required for PI-control

Consider the formula for updating the integral in a PI-controller:

$$i(kh + h) = i(kh) + e(kh) \cdot h/T_i$$

If the sampling period is 0.03 s and the integration time is 15 min = 900 s, the ratio h/T_i becomes $3 \cdot 10^{-5}$, which corresponds to about 15 bits. To avoid that the quantity $e(kh)h/T_i$ is rounded it is thus necessary to make the computations with a longer word length. This is the reason why the integral term is often implemented in 24-bit representation in dedicated PI-controllers.   ∎

The examples show that a rapid sampling requires a high precision in the coefficients.

How to Choose Representations

The selection of representations is crucial when implementing a control law using a digital signal processor or with custom VLSI. It is less crucial for implementations using microcomputers with floating-point hardware. The companion forms should be avoided, because so much is gained by using series or parallel forms. Each block should be implemented on Jordan form. This is particularly important for high-order compensators and short sampling periods. For low-order controllers implemented with floating-point hardware and with poles well inside the stability area, the choice of realization is less crucial. It is, however, good practice to hedge against possible numerical problems.

9.8 Programming

Practically all discrete-time controllers are implemented in a real-time operating system. In some systems the different parts of the algorithms may be distributed among different processors. The communication can then introduce time-varying delays (jitter) in the sampling period. Programming is an important aspect of the implementation of a control system, both with respect to the efficiency of the system and the time required for the implementation.

The effort required and the approaches used depend on the available software and the nature of the control problem. The code is typically written in C or C++. Ada, which was developed by the U.S. Department of Defense for computer-control applications, is the first language designed and developed for real-time applications. The character and the difficulty of the programming depend very much on the application. The requirements on operator communications are critical. The code required for operator communication is often much larger than the pure control code. A few examples illustrate this.

A Simple Dedicated Control System

Consider a simple control loop that has a few measured signals, a few outputs, and limited operator communication. Information may be displayed to the operator, and the operator may have a few buttons and a few dials. The programming of such a system is very simple. If a real-time clock is available, the code is in essence given by Listing 9.7.

The first line is simply a procedure that halts execution until a clock-interrupt occurs. The procedure *Regulate* is the code required to implement the desired control algorithm.

The procedure *Display* in Listing 9.7 computes some variables and displays them in analog or digital form. Notice that it is straightforward to introduce facilities for the operator to change parameters simply by introducing them as analog inputs.

The program in Listing 9.7 is fairly easy to debug. The procedures *Regulate* and *Display* are simple sequential procedures that can be tested off-line. It is also easy to check that the wait procedure gives an interrupt every sampling period. It will fail only if the time required to execute the procedures is longer than one sampling period. This may be tested by timing.

Listing 9.7   Computer code skeleton for a simple control loop.

```
repeat
  Wait for clock interrupt
  Regulate
  Display
forever
```

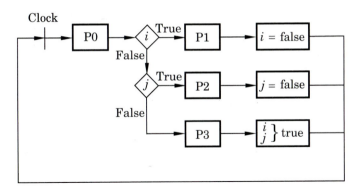

Figure 9.20   Flow chart for a multiloop control law with two sampling rates.

More Complicated Control Loops

The principles used in the program in Listing 9.7 may be extended to more complicated control systems with several loops having different sampling periods. A computer code, which may be represented by Fig. 9.20, is then obtained. The program P0 in Fig. 9.20 runs at the sampling rate given by the clock. The programs P1, P2, and P3 run each every third clock pulse. In order to obtain the representation in Fig. 9.20, it is necessary that the time required to execute each path be shorter than the shortest sampling period in the system. This is easy to do for systems with long sampling periods. For systems with fast sampling, it may be necessary to split up the computations in a tedious, unnatural, and error-prone fashion.

It is comparatively easy to debug the program shown in Fig. 9.20 if there are few paths and if the procedures are simple. The difficulty in debugging grows rapidly with increasing system complexity. New ideas and concepts are needed to handle such problems in a convenient way.

Concurrent, or Real-Time, Programming

It is natural to think about control loops as concurrent activities that are running in parallel. However, the digital computer operates sequentially in time. Ordinary programming languages can represent only sequential activities. Thus a key problem is to map a number of parallel concurrent activities into a sequential program. This may be done manually, as shown in Fig. 9.20. There are also special-purpose software—*real-time operating systems*—that make it possible to schedule tasks without making a strict sequential program. It is outside the scope of this book to discuss concurrent programs in detail. The basic ideas are given, together with a few examples.

The notions of *process* and *task* are fundamental concepts in real-time programming. They represent activities that may be thought of as running in parallel in time. Using these notions, it is possible to think of the computer running several activities in parallel. Hence, a real-time activity may be structured in the same way as a sequential activity is structured, using the notion of procedure or subroutine. The real-time operating system will organize the execution of the processes so that the desired result is obtained. To do this, a priority is associated with each process. Processes may also be scheduled to run periodically or in response to events such as interrupts or completion of other tasks.

The problem of *shared variables* and resources is one of the key problems in real-time programming. If different processes are using the same data, it is necessary to make sure that one process does not try to use data being modified by another process. If two processes may use the same resource, it is necessary to make sure that the system does not *deadlock* in a situation where both processes are waiting for each other.

Timing is a third problem. Computing power must be sufficient to allow all activities to be completed in the required time.

A Controller with Operator Interaction

A single control loop with operator interaction is one of the simplest examples of real-time programming. The task could be to run a control loop like the one in Listing 9.7 with sampling rate of 20 ms and to provide an interface so that the operator may change parameters from a keyboard or a terminal. Assume that the operator changes parameters by typing in a character string on the terminal. Because the time required for this is considerably longer than one sampling period, it is necessary to break down the operation into many small pieces in order to use the solution shown in Fig. 9.20. This is both tedious and unnatural. It is much more natural to think of the problem in the form of two concurrent processes. One process, *control*, should be run once every sampling period. The other process, *operator communication*, may run whenever the process *control* is idle. To ensure that control actions are taken at regular sampling periods, it is necessary to impose the rule that the process control has priority over *operator communication* and that it may interrupt the operator communication at any time. For convenience, the rule that the process *control* runs to completion once it has started is also introduced. In a case like this, *control* is called the *foreground task*, or foreground process, and *operator communication* is called a *background task*, or background process.

Real-Time Operating Systems

For problems with only two processes, it is not difficult to write an operating system that administrates the processes. Such a system may typically be written in less than 100 lines of assembly code.

The simple operating system may be extended to several processes. It is, however, a major task to make a program that can handle more complex situations. Such an operating system, which is also called a *real-time operating system*, may occupy anything from a few kilobytes to 20 kilobytes of code.

The real-time operating systems allow definition of tasks, or processes, in a high-level language such as Pascal, Modula 2, C, or C++. It is also possible to run processes at regular intervals or in given relationships to other tasks. Processes may also be introduced, started, and removed on-line. Priorities between different tasks may be introduced and modified. The introduction of real-time operating systems was one of the major innovations when process-control computers were introduced in the mid-1960s. Examples of such operating systems are RSX from Digital Equipment Corporation, VxWorks® from Wind River Systems, pSOS® from Integrated Systems, and QNX® from QNX Software Systems. Processes, or tasks, have also been introduced into simple languages such as BASIC.

Real-time operating systems are large general-purpose programs, which are often written in assembly code. They are difficult to maintain and modify. There has been a need to have real-time operating systems that can be tailored to specific applications. Computer languages with facilities for real-time programming have therefore been developed. Concurrent Pascal and Modula-2 are such languages. The language Ada is a standard tool for implementing computer-control systems in military systems.

DDC-Packages

Special techniques are used to program control systems consisting of a large number of identical control loops. The code is often structured as follows:

Read all analog inputs and store in a table.

Convert all signals to engineering units and store results in a table.

Apply the control algorithm sequentially to all values in the table using controller parameters stored in a parameter table.

Perform D-A conversion to all variables stored in the output table.

Programs of this type are called DDC-packages. The control algorithms are typically of the PID-type. Modules for gain scheduling, logic, supervision, and adaptation may also be available. The packages are easy to use because all programming is reduced to entering the appropriate data in the tables. Programs of this type are called *table-driven*.

DDC-packages usually also contain modules, that make it possible to make startup, shutdown, and alarm handling. Today these descriptions are often based on the standard IEC 1131-3 for function-block languages.

9.9 Conclusions

Implementation of control laws using a computer is discussed in this chapter. The key problem is to implement a discrete-time system. The principles for doing this have been covered in detail. It is straightforward to generate the code from the control algorithm. The importance of prefiltering to avoid aliasing has been mentioned. Sophisticated nonlinear digital filtering for removing outliers has also been discussed. The computational delay is influenced considerably by the organization of the computer code. Difficulties that arise from saturation in actuators and ways to avoid these difficulties are discussed. This also automatically gives a solution to mode switching and initialization.

Numerical problems and consequences of finite word length are also discussed. It is found to be very beneficial to transform the equations describing the control law to a form that is numerically well conditioned. Operational issues like mode switching and operator-machine interaction are discussed. There are many new possibilities in this area. Finally programming of control algorithms is discussed. Although the presentation is fairly short, the information given should be sufficient to implement control algorithms on minicomputers and microcomputers using high-level languages.

9.10 Problems

9.1 Consider control of a double integrator with a sampling period of 1 s. Calculate the deadbeat control for the system obtained using an antialiasing filter with the transfer function

$$G(s) = \frac{1}{s^2 + 1.4s + 1}$$

Compare the deadbeat strategy obtained with the deadbeat strategy for the pure double-integrator using simulation.

9.2 Write a program for computing the scalar product of two arrays

```
begin
  s:=0
  for i:=1 to n do
  s:=s+a[i]*b[i]
end
```

in each case.

(a) s: integer, a, b: arrays of integers.

(b) s: double-precision integer, a, b: arrays of integers.

(c) s: real, a, b: arrays of reals.

(d) s: double-precision real, a, b: arrays of reals.

Compare the computing times and precision. Try to find computers that have floating-point calculations in software, as well as in hardware.

9.3 Consider Example 9.9. Discuss the possibilities of using two loops with different sampling periods in order to improve the precision in the calculation.

9.4 Write a code for a digital PI-controller where the antiwindup is implemented as an observer with the time constant T_0.

9.5 Write a code for a digital PID-controller where the antiwindup is implemented as a deadbeat observer.

9.6 Write a code in your favorite high-level language for a digital PID-algorithm where antiwindup is implemented as an observer with time constant T_0. Determine the number of operations required for one iteration. Compile the program. Determine how many memory calls it requires. Time the program. How do the measured computing times relate to the number of operations and the computing times given in the computer manual?

9.7 Consider the control algorithm of (9.1), where x_m is considered to be an input. Assume that the state, the control variable, and the process output have dimensions n_x, n_u, and n_y, respectively, and that the matrices are full. Determine the number of additions, multiplications, and divisions required for one iteration.

9.8 Consider the control algorithm of (9.2). Write a code that implements the control algorithm in your favorite high-level language. Compile the code, determine how much memory space the code occupies, and determine the execution time. Try to find a good simple formula for determining the execution time.

9.9 Repeat Problem 9.8 but now use a subroutine to perform a scalar product. Discuss how computing time and storage requirements are influenced by the restructuring of the program.

9.10 Consider the control algorithm with rejection of outliers given by Eqs. (9.1) and (9.4). Make an estimate of the number of computations required for one iteration. (*Hint:* A matrix multiplication of an $n \times p$ matrix by a $p \times r$ matrix requires $N = npr$ operations, where an operation corresponds to one addition and one multiplication. Solution of the equation

$$Ax = B$$

where A is $n \times n$ and B is $n \times p$, requires approximately

$$N = \frac{1}{3} n^3 + \frac{1}{2} n^2 p$$

operations, where the major part of the calculations is the triangulation of the matrix A.)

9.11 Consider a discrete-time system characterized by the pulse-transfer function

$$H(z) = \frac{1}{(z - a)^n}$$

Calculate the sensitivity of the poles with respect to the parameters using Eq. (9.18) in each case.

(a) The filter is in companion form.

(b) The filter is in Jordan canonical form.

9.12 Make a flow chart similar to Fig. 9.20 for a system with loops having sampling periods of 1, 2, 5, and 60 s.

9.13 Consider a system with the transfer function

$$H(z) = \frac{1}{z^n + a_1 z^{n-1} + \cdots + a_n}$$

Assume that the system is realized with fixed-point arithmetic. Let the roundoff be described as normal rounding to integers. Show that the condition for a steady-state error k with no inputs is given by

$$k + \sum_{i=1}^{n} Q(a_i k) = 0$$

Furthermore, show that the condition for a limit-cycle oscillation with a period of two sampling periods is

$$k + \sum_{i=1}^{n}(-1)^i Q(a_i k) = 0$$

9.14 Consider the following algorithm for a PI-controller:

```
Adin uc y
e:=uc-y
v:=k*e+i
u:=max(min(512,v),0)
Daout u
i:=u+k*h*e/ti
```

Assume that the A-D and D-A converters have a resolution of 8 bits and that all calculations are made using integers. What is the word length required to represent variable i if overflow should be avoided? Use k = 50 and (a) h = 1, ti = 300 or (b) h = 0.01, ti = 1500.

Discuss how the result is influenced by the sampling period.

9.15 Three different algorithms for a PI-controller are listed. Use the linear model for roundoff to analyze the sensitivity of the algorithms to quantization in A-D and D-A converters and roundoff in the multiplications. Assume fixed-point calculations. Also, discuss the word lengths necessary for the algorithms.

Algorithm 1:

```
e:=uc-y
u:=k*(e+h*i/ti)
i:=i+e*h
```

Algorithm 2:

```
e:=uc-y
u:=k*(e+i)
i:=i+e*h/ti
```

Algorithm 3:

```
e:=uc-y
u:=i+k*e
i:=i+k*h*e/ti
```

9.16 Consider a dynamic system with the pulse-transfer function

$$H(z) = \frac{b_0 z^n + b_1 z^{n-1} + \cdots + b_n}{z^n + a_1 z^{n-1} + \cdots a_n}$$

There are many ways to introduce the states in a state-space representation. Show that the system has the following state descriptions:

(a)

$$x(k+1) = \begin{pmatrix} -a_1 & 1 & 0 & \cdots & 0 \\ -a_2 & 0 & 1 & \cdots & 0 \\ \vdots & \vdots & \vdots & \ddots & \vdots \\ -a_{n-1} & 0 & 0 & \cdots & 1 \\ -a_n & 0 & 0 & \cdots & 0 \end{pmatrix} x(k) + \begin{pmatrix} b_1 - b_0 a_1 \\ b_2 - b_0 a_2 \\ \vdots \\ b_{n-1} - b_0 a_{n-1} \\ b_n - b_0 a_n \end{pmatrix} u(k)$$

$$y(k) = \begin{pmatrix} 1 & 0 & 0 & \cdots & 0 \end{pmatrix} x(k) + b_0 u(k)$$

(b)

$$x(k+1) = \begin{pmatrix} -a_1 & 1 & 0 & \cdots & 0 \\ -a_2 & 0 & 1 & \cdots & 0 \\ \vdots & \vdots & \vdots & \ddots & \vdots \\ -a_n & 0 & 0 & \cdots & 1 \\ 0 & 0 & 0 & \cdots & 0 \end{pmatrix} x(k) + \begin{pmatrix} b_0 \\ b_1 \\ \vdots \\ b_{n-1} \\ b_n \end{pmatrix} u(k+1)$$

$$y(k) = \begin{pmatrix} 1 & 0 & 0 & \cdots & 0 \end{pmatrix} x(k)$$

(c)

$$x(k+1) = \begin{pmatrix} -a_1 & -a_2 & \cdots & -a_{n-1} & -a_n & b_1 & b_2 & \cdots & b_{n-1} & b_n \\ 1 & 0 & \cdots & 0 & 0 & 0 & 0 & \cdots & 0 & 0 \\ \vdots & \vdots & \ddots & \vdots & \vdots & \vdots & \vdots & \ddots & \vdots & \vdots \\ 0 & 0 & \cdots & 1 & 0 & 0 & 0 & \cdots & 0 & 0 \\ 0 & 0 & \cdots & 0 & 0 & 0 & 0 & \cdots & 0 & 0 \\ 0 & 0 & \cdots & 0 & 0 & 1 & 0 & \cdots & 0 & 0 \\ \vdots & \vdots & \ddots & \vdots & \vdots & \vdots & \vdots & \ddots & \vdots & \vdots \\ 0 & 0 & \cdots & 0 & 0 & 0 & 0 & \cdots & 1 & 0 \end{pmatrix} x(k)$$

$$+ \begin{pmatrix} b_0 \\ 0 \\ \vdots \\ 0 \\ 1 \\ 0 \\ \vdots \\ 0 \end{pmatrix} u(k+1)$$

$$y(k) = \begin{pmatrix} 1 & 0 & 0 & \cdots & 0 & 0 \end{pmatrix} x(k)$$

Assume that $H(z)$ represents a controller. Discuss the advantages and disadvantages with the different realizations of the controller.

9.17 A digital controller with the sampling period $h = 0.2$ has the pulse-transfer function

$$H(z) = \frac{6.25z^2 - 11.5z + 5.5}{z^2 - 1.5z + 0.5}$$

The controller is used to control a process with the transfer function

$$G(s) = \frac{1}{s^2 + 1}$$

Discuss the effect of roundoff and quantization noise when different realizations are used to implement the controller on a computer having fixed word length.

9.18 Show that the continued-fraction representation (9.20) can be obtained recursively as

$$H_i(z) = \cfrac{1}{\beta_i z + \cfrac{1}{\alpha_i + H_{i+1}(z)}}$$

where $H_{n+1}(z) = 0$.

9.19 Determine the δ-operator representations of the following continuous-time transfer functions:

(a) $1/(s^2 + 1)$

(b) $K/(1 + Ts)$

(c) $1/(s + a)^3$

Compute the poles and the zeros and investigate what happens when $h \to 0$.

9.20 Let $H(z)$ be the pulse-transfer function obtained from step-invariant sampling of the rational transfer function $G(s)$. Define

$$\bar{H}(\delta) = H(1 + \delta h)$$

Prove that

$$\lim_{h \to 0} \bar{H}(\delta) = G(\delta)$$

Show that this is true also for ramp-invariant and impulse-invariant sampling.

9.11 Notes and References

Design of filters is covered in standard texts on networks. Kuo (1980) ar.d Williams (1981) are good sources. Useful practical advice is also found in the handbooks published by manufacturers of operational amplifiers. Such handbooks are also useful for information about A-D and D-A converters. Make sure to get a new version of whatever handbook you use because the technology changes rapidly.

The problems associated with windup of PID-controllers is discussed in trade journals for the process industry. The general approach given in Sec. 9.4 was introduced in earlier editions of this book, see Åström and Hägglund (1995). The generalized form in Fig. 9.9 is described in Rönnbäck, Walgama, and Sternby (1992). Other references on antireset windup are Hanus (1988) and Græbe and Ahlén (1996). Possibilities for error detection and rejection of outliers are discussed in depth in Willsky (1979), which also contains many references.

A comprehensive text on the effects of quantization and roundoff in digital control systems is Moroney (1983), which contains many references. The following papers are classics in the area, Bertram (1958), Slaughter (1964), Knowles and Edwards (1965), Curry (1967), and Rink and Chong (1979). Describing function analysis is discussed in Atherton (1975, 1982). Limit cycles due to roundoff can be determined using the theory of relay oscillations. This is described in Tsypkin (1984).

A review of digital signal processors is given in Lee (1988). Design of special-purpose signal processors in VLSI is described in Catthoor et al. (1988) and in a series of books titled *VLSI Signal Processing I, II, and III*, published by IEEE. The books are based on presentations given at IEEE ASSIP workshops. The 1988 volume is Brown and Campbell (1948). A readable account of the IEEE standard and its impact on different high-level languages is found in Fateman (1982). Problems associated with quantization, roundoff, and overflow are also discussed in the signal-processing literature. Overviews are found in Oppenheim and Schafer (1989). Specialized issues are discussed in Jackson (1970a, 1970b, 1979), Parker and Hess (1971), Willson (1972a, 1972b), and Buttner (1977).

There are many standard texts on numerical analysis, Björk, Dahlqvist, and Andersson (1974) and Golub and Van Loan (1989) are good sources. Accuracy aspects in connection with control are found, for example, in Williamson (1991) and Gevers and Li (1993).

Concurrent programming is discussed in Brinch-Hansen (1973), Barnes (1982), Burns and Wellings (1990), and Burns and Davies (1993). Much useful information is also given in material from vendors of computer-control systems. The δ-operator is an old idea. See Tschauner (1963a, 1963b). The δ-operator has been given much attention because of its numerical properties. See Gawthrop (1980), Middleton and Goodwin (1987, 1989), and Gevers and Li (1993).

10

Disturbance Models

10.1 Introduction

The presence of disturbances is one of the main reasons for using control. Without disturbances there is no need for feedback control. The character of the disturbances imposes fundamental limitations on the performance of a control system. Measurement noise in a servo system limits the achievable bandwidth of the closed-loop system. The nature of the disturbances determines the quality of regulation in a process-control system. Disturbances also convey important information about the properties of the system. By investigating the characteristics of the disturbances it is thus possible to detect the status of the process, including beginning process malfunctions. We have already taken disturbances into account in the pole-placement design in Chapters 4 and 5. In this chapter we will give a systematic treatment of disturbances.

Different ways to describe disturbances and to analyze their effect on a system are discussed in this chapter. An overview of different ways to eliminate disturbances is first given. This includes use of feedback, feedforward, and prediction. The discussion gives a reason for the different ways of describing disturbances.

The classic disturbance models, impulse, step, ramp, and sinusoid, were discussed in Sec. 3.5. All these disturbances can be thought of as generated by linear systems with suitable initial conditions. The problem of analyzing the effect of disturbances on a linear system can then be reduced to an initial-value problem. From the input-output point of view, a disturbance may also be modeled as an impulse response of a linear filter. The disturbance analysis is then reduced to a response calculation. This is particularly useful for disturbances that are steps or sinusoids. In all cases the disturbance analysis can be done with the tools developed in Chapters 3 to 5, and 8. When the response of a system to a specific disturbance needs to be known, it is often necessary to resort to simulation. This is easily done with a simulation program because the disturbance analysis is again reduced to an initial-value problem.

When disturbances can be neither eliminated at the source nor measured, it is necessary to resort to prediction. To do so it is necessary to have models of disturbances that lead to a reasonable formulation of a prediction problem. For this purpose the concept of piecewise deterministic disturbances is introduced in Sec. 10.3.

Another way to arrive at a prediction problem is to describe disturbances as random processes. This formulation is presented in Sec. 10.4. A simple version of the famous Wiener-Kolmogorov-Kalman prediction theory is also presented. As seen in Chapters 11 and 12, the prediction error expresses a fundamental limitation on regulation performance. Continuous-time stochastic processes are discussed briefly in Sec. 10.5. Such models are required because of the desire to formulate models and specifications in continuous time. Sampling of continuous-time stochastic-state models is treated in Sec. 10.6.

10.2 Reduction of Effects of Disturbances

Before going into details of models for disturbances, it is useful to discuss how their effects on a system can be reduced. Disturbances may be reduced at their source. The effects of disturbances can also be reduced by local feedback or by feedforward from measurable disturbances. Prediction may also be used to estimate unmeasurable disturbances. The predictable part of the disturbance can then be reduced by feedforward. These different approaches will be discussed in more detail.

Reduction at the Source

The most obvious way to reduce the effects of disturbances is to attempt to reduce the source of the disturbances. This approach is closely related to process design. The following are typical examples:

Reduce variations in composition by a tank with efficient mixing.

Reduce friction forces in a servo by using better bearings.

Move a sensor to a position where there are smaller disturbances.

Modify sensor electronics so that less noise is obtained.

Replace a sensor with another having less noise.

Change the sampling procedure by spacing the samples better in time or space to obtain a better representation of the characteristics of the process.

These are just a few examples, but it is very important to keep these possibilities in mind. Compare with the integrated process and control design discussed in Chapter 6.

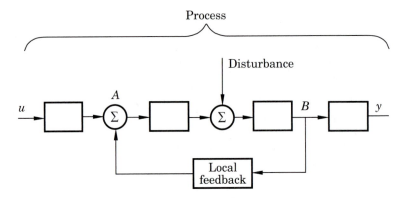

Figure 10.1   Reduction of disturbances by local feedback. The disturbance should enter the system between points A and B. The dynamics between A and B should be such that a high gain can be used in the loop.

Reduction by Local Feedback

If the disturbances cannot be reduced at the source, an attempt can be made to reduce them by local feedback. The generic principle of this approach is illustrated in Fig. 10.1. For this approach it is necessary that the disturbances enter the system locally in a well-defined way. It is necessary to have access to a measured variable that is influenced by the disturbance and to have access to a control variable that enters the system in the neighborhood of the disturbance. The effect of the disturbance can then be reduced by using local feedback. The dynamics relating the measured variable to the control variable should be such that a high-gain control loop can be used. This use of feedback is often very simple and effective because it is not necessary to have detailed information about the characteristics of the process, provided that a high gain can be used in the loop. However, an extra feedback loop is required. The following are typical examples of local feedback:

Reduce variations in supply pressure to valves, instruments, and regulators by introducing a pressure regulator.

Reduce variations in temperature control by stabilizing the supply voltage.

Reduction by Feedforward

Measurable disturbances can also be reduced to feedforward. The generic principle is illustrated in Fig. 6.3. The disturbance is measured, and a control signal that attempts to counteract the disturbance is generated and applied to the process. Feedforward is particularly useful for disturbances generated by changes in the command or reference signals or for cascaded processes when disturbances downstream are generated by variations in processes upstream.

Reduction by Prediction

Reduction by prediction is an extension of the feedforward principle that may be used when the disturbance cannot be measured. The principle is very simple; the disturbance is predicted using measurable signals, and the feedforward signal is generated from the prediction. It is important to observe that it is not necessary to predict the disturbance itself; it is sufficient to model a signal that represents the effect of the disturbance on the important process variables.

Goals for Modeling

To evaluate the needs for reduction of disturbances it is necessary to be able to estimate the influences of disturbances on important system variables, which is basically a problem of analyzing the response of a system to a given input. The models used for disturbances can be fairly simple, as long as they represent the major characteristics of true disturbances. Similarly simple models can also be used to estimate possible improvements obtained by local feedback and feedforward. More accurate models of disturbances are needed if prediction is applied. In this case the performance obtained depends critically on the character of the disturbances. There are also some fundamental difficulties in formulating disturbance models that give a sensible prediction problem.

10.3 Piecewise Deterministic Disturbances

The classical disturbance models discussed in Sec. 3.5 are useful for analyzing the effects of disturbances on a system. Possible improvements by using local feedback and feedforward can also be investigated using these models. The disturbance models discussed are, however, not suitable for investigating disturbance reduction by prediction. Fundamentally different models are required to formulate a sensible prediction problem. This leads to introduction of the piecewise deterministic disturbances. Alternative models, which also permit formulation of a prediction problem, are discussed in Secs. 10.4 and 10.5.

A Fundamental Problem

It is not trivial to construct models for disturbances that permit a sensible formulation of a prediction problem.

Example 10.1   Predictor for a step signal

To predict the future value of a step signal, it seems natural to use the current value of the signal. For discrete-time signals, the predictor then becomes

$$\hat{y}\big((k+m)h \mid kh\big) = y(kh)$$

The notation $\hat{y}(t \mid s)$ means the prediction of $y(t)$ based on data available at time s. This predictor has a prediction error at times $t = 0, h, 2h, \ldots, (m-1)h$, that is, m steps after the step change in y. It then predicts the signal without error.   ∎

Example 10.2 Predictor for a ramp signal

A predictor for a ramp can be constructed by calculating the slope from the past and the current observations and making a linear extrapolation, which can be expressed by the formula

$$\hat{y}\big((k+m)h \mid kh\big) \quad = \quad y(kh) + m\Big(y(kh) - y(kh - h)\Big)$$
$$= \quad (1+m)y(kh) - my(kh - h)$$

This predictor has an initial error for $t = h$, $2h$, ... , mh. After that it predicts the signal without error. ∎

The Basic Idea

These examples indicate that the prediction error will be zero except at a few points. This observation is not in close agreement with the practical experience that disturbances are hard to predict. The explanation is that the step and the ramp are not good models for prediction problems. Analytic signals are useless because an analytic function is uniquely given by its values in an arbitrarily short interval. The step and the ramp are analytic everywhere except at the origin.

One possibility of constructing signals that are less regular is to introduce more points of irregularity. Thus signals can be introduced that are generated by linear dynamic systems with irregular inputs. Instead of having a pulse at the origin, inputs that are different from zero at several points can be introduced. An interesting class of signals is obtained if the pulses are assumed to be isolated and spread by at least n samples, where n is the order of the system. It is assumed that it is not known a priori when the pulses occur. The amplitudes of the pulses are also unknown. Such signals are called *piecewise deterministic signals*. The name comes because the signals are deterministic except at isolated points, where they change in an unpredictable way. An example of a piecewise deterministic signal is shown in Fig. 10.2.

State-Space Models

Let a signal be generated by the dynamic system

$$x(k+1) = \Phi x(k) + v(k)$$
$$y(k) = Cx(k) \tag{10.1}$$

It is assumed that the output y is a scalar and that the system is completely observable. The input v is assumed to be zero except at isolated points. If the state of the system is known, it is straightforward to predict the state over any interval where the input is zero. However, when there is a pulse, the state can change in an arbitrary manner, but after a pulse there will always be an interval where the input is zero. Because the system is observable, the process state can then be calculated. Exact predictions can then be given until a new

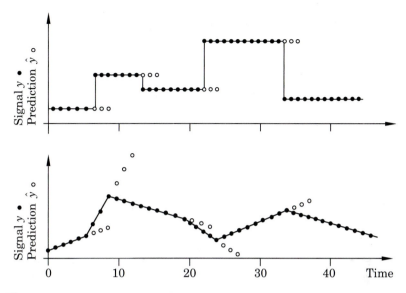

Figure 10.2 Piecewise constant and piecewise linear signals and their m-step predictions when $m = 3$.

pulse occurs. This argument can be converted into mathematics in the following way. From the derivation of the condition for observability in Sec. 3.4, it is found that the state is given by

$$x(k - n + 1) = W_o^{-1} \left(y(k - n + 1) \quad \cdots \quad y(k) \right)^T \tag{10.2}$$

where W_o is the observability matrix given by Eq. (3.22). The following predictor gives the state m steps ahead:

$$\hat{x}(k + m \mid k) = \Phi^{m+n-1} W_o^{-1} \left(y(k - n + 1) \quad \cdots \quad y(k) \right)^T \tag{10.3}$$

The predictor for the signal is thus obtained from a linear combination of n values of the measured signal. The predictor can be expressed as

$$\hat{y}(k + m \mid k) = P^*(q^{-1}) y(k)$$

where P is a polynomial of degree $n - 1$.

The predictor can also be represented by the recursive equation

$$\hat{x}(k \mid k) = \Phi \hat{x}(k - 1 \mid k - 1) + K \left(y(k) - C\Phi \hat{x}(k - 1 \mid k - 1) \right)$$
$$\hat{x}(k + m \mid k) = \Phi^m \hat{x}(k \mid k) \tag{10.4}$$

where the matrix K is chosen so that all eigenvalues of the matrix $(I - KC)\Phi$ are equal to zero.

Simple calculations for an integrator and a double integrator give the same predictors as in Examples 10.1 and 10.2. This is a consequence of the fact that the important characteristics of the disturbances are captured by the dynamics of the systems that generate the disturbances. These dynamics determine the predictors uniquely; it does not matter if the systems are driven by a single pulse or by several pulses. The properties of the predictors are illustrated in Fig. 10.2.

Input-Output Models

Because the predictor for a piecewise deterministic signal becomes a polynomial, it seems natural to obtain it directly by polynomial calculations. For this purpose it is assumed that the signal is generated by the dynamic system

$$y(k) = \frac{C(q)}{A(q)} \, w(k)$$

where it is assumed that $\deg C = \deg A$ and that the input w is a signal that is zero except at isolated points, which are spaced more than $\deg A + m$. Define $F(z)$ and $G(z)$ through the identity

$$z^{m-1}C(z) = A(z)F(z) + G(z)$$

It can be shown that the m-step predictor for y is given by the difference equation

$$C(q)\hat{y}(k + m \mid k) = qG(q)y(k)$$

A reference to the proof of this is given in Sec. 10.9.

Notice that the signals discussed in this section are similar to the classical disturbance signals discussed in Sec. 3.5 in the sense that they are characterized by dynamic systems. The only difference between the signals is that the inputs to the systems are different. This idea is extended in the next section.

10.4 Stochastic Models of Disturbances

It is natural to use *stochastic*, or random, concepts to describe disturbances. By such an approach it is possible to describe a wide class of disturbances, which permits good formulation of prediction problems. The theory of random processes and the prediction theory were in fact developed under close interaction. The general theory of stochastic processes is quite complex. For computer-control theory, it is fortunately often sufficient to work with a special case of the general theory, which requires much less sophistication. This theory is developed in this section. First, some elements of the theory of random processes are given, and then the notion of discrete-time white noise is discussed. Disturbances are then modeled as outputs of dynamic systems with white-noise inputs. The disturbance models are thus similar to the models discussed in the previous sections; the only difference is the character of the input signals to the systems. Tools for analyzing the properties of the models are also given.

Stochastic Processes

The concept of a stochastic process is complex. It took brilliant researchers hundreds of years to find the right ideas. The concept matured in work done by the mathematician Kolmogorov around 1930. A simple presentation of the ideas is given here. Interested readers are strongly urged to consult the references.

A stochastic process (random process, random function) can be regarded as a family of stochastic variables $\{x(t), t \in T\}$. The stochastic variables are indexed with the parameter t, which belongs to the set T, called the *index set*. In stochastic-control theory, the variable t is interpreted as time. The set T is then the real variables. When considering sampled-data systems, as in this book, the set T is the sampling instants, that is, $T = \{\dots, -h, 0, h, \dots\}$ or $T = \{\dots, -1, 0, 1, \dots\}$ when the sampling period is chosen as the time unit. We then have a *stochastic process*.

A random process may be considered as a function $x(t, \omega)$ of two variables. For fixed $\omega = \omega_0$ the function $x(\cdot, \omega_0)$ is an ordinary time function called a *realization*. For fixed $t = t_0$, the function $x(t_0, \cdot)$ is a random variable. A random process can thus be viewed as generated from a random-signal generator. The argument ω is often suppressed.

Completely deterministic stochastic processes.   One possibility of obtaining a random process is to pick the initial conditions of an ordinary differential equation as a random variable and to generate the time functions by solving the differential equations. These types of random processes are, however, not very interesting because they do not exhibit enough randomness. This is clearly seen by considering the stochastic process generated by an integrator with random initial conditions. Because the output of the integrator is constant it follows that

$$x(t, \omega) - x(t - h, \omega) = 0$$

for all t, h, and ω. A stochastic process with this property is called a *completely deterministic stochastic process*, because its future values can be predicted exactly from its past.

In general it will be said that a random process $x(t, \omega)$ is called *completely deterministic* if

$$\ell\Big(x(t, \omega)\Big) = 0 \qquad \text{for almost all } \omega$$

where ℓ is an arbitrary linear operator that is not identically zero. This means that completely deterministic random processes can be predicted exactly with linear predictors for almost all ω. (Almost all ω means all ω except for possibly a set of points with zero measure.)

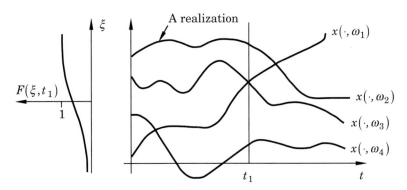

Figure 10.3 A stochastic process and a finite-dimensional distribution function.

The completely deterministic random processes are closely related to the classical disturbance signals discussed in Sec. 3.5. These signals will be completely deterministic random processes if the initial conditions to the dynamic systems are chosen as random processes. The completely deterministic processes are normally excluded because they are too regular to be of interest.

Concepts.   Some important concepts for random processes will now be given. The values of a random process at n distinct times are n-dimensional random variables. The function

$$F(\xi_1, \dots, \xi_n; t_1, \dots, t_n) = P\{x(t_1) \le \xi_1, \dots, x(t_n) \le \xi_n\}$$

where P denotes probabilities, is called the *finite-dimensional distribution function* of the random process. An illustration is given in Fig. 10.3. A random process is called *Gaussian*, or *normal*, if all finite-dimensional distributions are normal. The *mean-value function* of a random process x is defined by

$$m(t) = \mathrm{E}x(t) = \int_{-\infty}^{\infty} \xi \, dF(\xi; t)$$

The mean-value function is an ordinary time function. Higher moments are defined similarly. The *covariance function* of a process is defined by

$$r_{xx}(s, t) = \mathrm{cov}\left(x(s), x(t)\right)$$

$$= \mathrm{E}\left(\left(x(s) - m(s)\right)\left(x(t) - m(t)\right)^T\right)$$

$$= \iint \left(\xi_1 - m(s)\right)\left(\xi_2 - m(t)\right)^T dF(\xi_1, \xi_2; s, t)$$

A Gaussian random process is completely characterized by its mean-value function and its covariance function. The *cross-covariance function*

$$r_{xy}(s, t) = \mathrm{cov}\left(x(s), y(t)\right)$$

of two stochastic processes is defined similarly.

A stochastic process is called *stationary* if the finite-dimensional distribution of $x(t_1), x(t_2), \ldots, x(t_n)$ is identical to the distribution of $x(t_1 + \tau), x(t_2 + \tau), \ldots, x(t_n + \tau)$ for all $\tau, n, t_1, \ldots, t_n$. The process is called *weakly stationary* if the first two moments of the distributions are the same for all τ. The mean-value function of a (weakly) stationary process is constant. The cross-covariance function of weakly stationary processes is a function of the difference $s - t$ of the arguments only. With some abuse of function notation, write

$$r_{xy}(s,t) = r_{xy}(s - t)$$

The cross-covariance function of (weakly) stationary processes is a function of one argument only. Hence

$$r_{xy}(\tau) = \operatorname{cov}\Big(x(t + \tau), y(t)\Big)$$

When x is scalar the function

$$r_x(\tau) = r_{xx}(\tau) = \operatorname{cov}\Big(x(t + \tau), x(t)\Big)$$

called the *autocovariance* function.

The *cross-spectral density* of (weakly) stationary processes is the Fourier transform of its covariance function. Hence,

$$\phi_{xy}(\omega) = \frac{1}{2\pi} \sum_{k=-\infty}^{\infty} r_{xy}(k)e^{-ik\omega} \tag{10.5}$$

and

$$r_{xy}(k) = \int_{-\pi}^{\pi} e^{ik\omega} \phi_{xy}(\omega)\, d\omega \tag{10.6}$$

It is also customary to refer to ϕ_{xx} and ϕ_{xy} as the *autospectral density* and the *cross-spectral density*, respectively. The autospectral density is also called *spectral density* for simplicity.

Interpretation of covariances and spectra.   Stationary Gaussian processes are completely characterized by their mean-value functions and their covariance functions. In applications, it is useful to have a good intuitive understanding of how the properties of a stochastic process are reflected by these functions.

The mean-value function is almost self-explanatory. The value $r_x(0)$ of the covariance function at the origin is the variance of the process. It tells how large the fluctuations of the process are. The standard deviation of the variations is equal to the square root of $r_x(0)$. If the covariance function is normalized by $r_x(0)$, the *correlation function*, which is defined by

$$\rho_x(\tau) = \frac{r_x(\tau)}{r_x(0)}$$

is obtained. It follows from Schwartz's inequality that

$$|r_x(\tau)| \leq r_x(0)$$

The correlation function is therefore less than one in magnitude. The value $\rho_x(\tau)$ gives the correlation between values of the process with a spacing τ. Values close to one mean that there are strong correlations, zero values indicate no correlation, and negative values indicate negative correlation. An investigation of the shape of the correlation function thus indicates the temporal interdependencies of the process.

It is very useful to study realizations of stochastic processes and their covariance functions to develop insight into their relationships. Some examples are shown in Fig. 10.4. All processes have unit variance.

The spectral density has a good physical interpretation. The integral

$$2 \int_{\omega_1}^{\omega_2} \phi(\omega)\, d\omega$$

represents the power of the signal in the frequency band (ω_1, ω_2). The area under the spectral-density curve thus represents the signal power in a certain frequency band. The total area under the curve is proportional to the variance of the signal. In practical work it is useful to develop a good understanding of how signal properties are related to the spectrum (compare with Fig. 10.4).

Notice that the mean-value function, the covariance function, and the spectral density are characterized by the first two moments of the distribution only. Signals whose realizations are very different may thus have the same first moments. The random telegraph wave that switches between the values 0 and 1 thus has the same spectrum as the noise from a simple RC circuit.

Discrete-Time White Noise

A simple and useful random process is now introduced. Let time be the set of integers. Consider a stationary discrete-time stochastic process x such that $x(t)$ and $x(s)$ are independent if $t \neq s$. The stochastic process can thus be considered as a sequence $\{x(t, \omega), \ t = \ldots, -1, 0, 1, \ldots\}$ of independent, equally distributed random variables. The covariance function is given by

$$r(\tau) = \begin{cases} \sigma^2 & \tau = 0 \\ 0 & \tau = \pm 1, \pm 2, \ldots \end{cases}$$

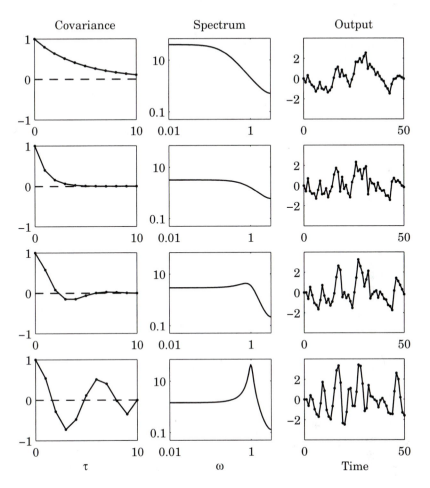

Figure 10.4 Covariance functions, spectral densities, and sample functions for some stationary random processes. All processes have unit variance.

A process with this covariance function is called *discrete-time white noise*. It follows from (10.5) that the spectral density is given by

$$\phi(\omega) = \frac{\sigma^2}{2\pi}$$

The spectral density is thus constant for all frequencies. The analogy with the spectral properties of white light explains the name given to the process. White noise plays an important role in stochastic control theory. All stochastic processes that are needed will be generated simply by filtering white noise. This also implies that only a white-noise generator is needed when simulating stochastic processes. White noise is thus the equivalent of pulses for deterministic systems.

ARMA Processes

Large classes of stochastic processes can be generated by driving linear systems with white noise. Let $\{e(k), k = \ldots, -1, 0, 1, \ldots\}$ be discrete-time white noise. The process generated by

$$y(k) = e(k) + b_1 e(k-1) + \cdots + b_n e(k-n)$$

is called a *moving average*, or an MA process. The process generated by

$$y(k) + a_1 y(k-1) + \cdots + a_n y(k-n) = e(k)$$

is called an *autoregression*, or an AR process. The process

$$y(k) + a_1 y(k-1) + \cdots + a_n y(k-n) = e(k) + b_1 e(k-1) + \cdots + b_n e(k-n)$$

is called an ARMA process. The process

$$y(k) + a_1 y(k-1) + \cdots + a_n y(k-n) = b_0 u(k-d) + \cdots$$
$$+ b_m u(k-d-m) + e(k) + c_1 e(k-1) + \cdots + c_n e(k-n)$$

is called an ARMAX process, that is, an ARMA process with an exogenous signal.

State-Space Models

The concept of state has its roots in cause-and-effect relationships in classical mechanics. The motion of a system of particles is uniquely determined for all future times by the present positions and moments of the particles and the future forces. How the present positions and moments were achieved is not important. The state is an abstraction of this property; it is the minimal information about the history of a system required to predict its future motion.

For stochastic systems, it cannot be required that the future motion be determined exactly. A natural extension of the notion of state for stochastic systems is to require that the probability distribution of future states be uniquely given by the current state. Stochastic processes with this property are called *Markov processes*. Markov processes are thus the stochastic equivalents of state-space models. They are formally defined as follows.

DEFINITION 10.1   MARKOV PROCESS   Let t_i and t be elements of the index set T such that $t_1 < t_2 < \ldots < t_n < t$. A stochastic process $\{x(t), t \in T\}$ is called a Markov process if

$$P\{x(t) \leq \xi \mid x(t_1), \ldots, x(t_n)\} = P\{x(t) \leq \xi \mid x(t_n)\}$$

where $P\{\cdot \mid x(t_1), \ldots, x(t_n)\}$ denotes the conditional probability given $x(t_1), \ldots, x(t_n)$. ∎

A Markov process is completely determined by the *initial probability distribution*

$$F(\xi; t_0) = P\{x(t_0) \le \xi\}$$

and the *transition probability distribution*

$$F(\xi_t; t \mid \xi_s; s) = P\{x(t) \le \xi_t \mid x(s) = \xi_s\}$$

All finite-dimensional distributions can then be generated from these distributions using the multiplication rule for conditional probabilities.

The Markov process is the natural concept to use when extending the notion of state model to the stochastic case.

Linear stochastic-difference equations.   Consider a discrete-time system where the sampling period is chosen as the time unit. Let the state at time k be given by $x(k)$. The probability distribution of the state at time $k + 1$ is then a function of $x(k)$. If the mean value is linear in $x(k)$ and the distribution around the mean is independent of $x(k)$, then $x(k + 1)$ can be represented as

$$x(k + 1) = \Phi x(k) + v(k) \tag{10.7}$$

where $v(k)$ is a random variable with zero mean and covariance R_1 that is independent of $x(k)$ and independent of all past values of x. This implies that $v(k)$ also is independent of all past v's. The sequence $\{v(k), k = \ldots, -1, 0, 1, \ldots\}$ is a sequence of independent equally distributed random variables. The stochastic process $\{v(k)\}$ is thus discrete-time white noise. Equation (10.7) is called a *linear stochastic-difference equation*. To define the random process $\{x(k)\}$ completely, it is necessary to specify the initial conditions. It is assumed that initial state has the mean m_0, and the covariance matrix R_0.

Properties of linear stochastic-difference equations.   The character of the random process defined by the linear stochastic-difference equation of (10.7) will now be investigated and the first and second moments of the process will be calculated. To obtain the mean-value function

$$m(k) = \mathrm{E}x(k)$$

simply take the mean values of both sides of (10.7). Because v has zero mean, the following difference equation is obtained:

$$m(k + 1) = \Phi m(k) \tag{10.8}$$

The initial condition is

$$m(0) = m_0$$

The mean value will thus propagate in the same way as the unperturbed system.
To calculate the covariance function, we introduce

$$P(k) = \text{cov}\Big(x(k), x(k)\Big) = \text{E}\tilde{x}(k)\tilde{x}^T(k)$$

where

$$\tilde{x} = x - m$$

It follows from Eqs. (10.7) and (10.8) that $\tilde{x}$ satisfies Eq. (10.7) with the mean of the initial condition equal zero. The mean value can thus be treated separately. To calculate the covariance, form the expression

$$\tilde{x}(k+1)\tilde{x}^T(k+1) = \Big(\Phi\tilde{x}(k) + v(k)\Big)\Big(\Phi\tilde{x}(k) + v(k)\Big)^T$$
$$= \Phi\tilde{x}(k)\tilde{x}^T(k)\Phi^T + \Phi\tilde{x}(k)v^T(k) + v(k)\tilde{x}^T(k)\Phi^T + v(k)v^T(k)$$

Taking mean values gives

$$P(k+1) = \Phi P(k)\Phi^T + R_1$$

because $v(k)$ and $\tilde{x}(k)$ are independent. The initial conditions are

$$P(0) = R_0$$

The recursive equation for P tells how the covariance propagates.
To calculate the covariance function of the state, observe that

$$\tilde{x}(k+1)\tilde{x}^T(k) = \Big(\Phi\tilde{x}(k) + v(k)\Big)\tilde{x}^T(k)$$

Because $v(k)$ and $\tilde{x}(k)$ are independent and $v(k)$ has zero mean,

$$r_{xx}(k+1, k) = \text{cov}\Big(x(k+1), x(k)\Big) = \Phi P(k)$$

Repeating this discussion,

$$r_{xx}(k+\tau, k) = \Phi^\tau P(k) \qquad \tau \geq 0$$

The covariance function is thus obtained by propagating the variance function through a system with the dynamics given by Φ. The results obtained are so important that they deserve to be summarized.

THEOREM 10.1 FILTERED DISCRETE-TIME WHITE NOISE   Consider a random process defined by the linear stochastic-difference equation (10.7), where $\{v(k)\}$ is a white-noise process with zero mean and covariance R_1. Let the initial state have mean m_0 and covariance R_0. The mean-value function of the process is then given by

$$m(k+1) = \Phi m(k) \qquad m(0) = m_0 \tag{10.9}$$

and the covariance function by

$$r(k + \tau, k) = \Phi^{\tau} P(k) \qquad \tau \geq 0 \tag{10.10}$$

where $P(k) = \text{cov}\Big(x(k), x(k)\Big)$ is given by

$$P(k + 1) = \Phi P(k) \Phi^T + R_1 \qquad P(0) = R_0 \tag{10.11}$$

∎

Remark 1. If the random variables are Gaussian, then the stochastic process is uniquely characterized by its mean-value function m and its covariance function r.

Remark 2. If the system has an output $y = Cx$, then the mean-value function of y is given by

$$m_y = Cm$$

and its covariance is given by

$$r_{yy} = C r_{xx} C^T$$

The cross-covariance between y and x is given by

$$r_{yx} = C r_{xx}$$

Remark 3. Notice that the difference equation in (10.11) for the matrix P is the same as Eq. (3.9), which was used to calculate Lyapunov functions in Chapter 3.

Remark 4. The different terms of (10.11) have good physical interpretations. The covariance P may represent the uncertainty in the state, the term $\Phi P(k) \Phi^T$ tells how the uncertainty at time k propagates due to the system dynamics, and the term R_1 describes the increase of uncertainty due to the disturbance v.

Example 10.3   A first order system

Consider the first-order system

$$x(k + 1) = ax(k) + v(k)$$

where v is a sequence of uncorrelated random variables with zero mean values and covariances r_1. Let the state at time k_0 have the mean m_0 and the covariance r_0. It follows from (10.9) that the mean value

$$m(k) = \text{E}x(k)$$

is given by

$$m(k + 1) = am(k) \qquad m(k_0) = m_0$$

Hence

$$m(k) = a^{k-k_0} m_0$$

Equation (10.11) gives

$$P(k + 1) = a^2 P(k) + r_1 \qquad P(k_0) = r_0$$

Solving this difference equation we get

$$P(k) = a^{2(k-k_0)} r_0 + \frac{1 - a^{2(k-k_0)}}{1 - a^2} r_1$$

Furthermore,

$$r_x(l, k) = a^{l-k} P(k) \qquad l \geq k$$

and

$$r_x(l, k) = a^{k-l} P(l) \qquad l < k$$

If $|a| < 1$ and $k_0 \to -\infty$, it follows that

$$m(k) \to 0$$

$$P(k) \to \frac{r_1}{1 - a^2}$$

$$r_x(k + \tau, k) \to \frac{r_1 a^{|\tau|}}{1 - a^2}$$

The process then becomes stationary because m is constant and the covariance function is a function of τ only. If an output

$$y(k) = x(k) + e(k)$$

is introduced, where e is a sequence of uncorrelated random variables with zero mean and covariance r_2, it follows that the covariance function of y becomes

$$r_y(\tau) = \begin{cases} r_2 + \dfrac{r_1}{1 - a^2} & \tau = 0 \\[2mm] \dfrac{r_1 a^{|\tau|}}{1 - a^2} & \tau \neq 0 \end{cases}$$

The spectral density is obtained from (10.5). Hence

$$\phi_y(\omega) = \frac{1}{2\pi} \left(r_2 + \frac{r_1}{(e^{i\omega} - a)(e^{-i\omega} - a)} \right)$$

$$= \frac{1}{2\pi} \left(r_2 + \frac{r_1}{1 + a^2 - 2a \cos \omega} \right)$$

∎

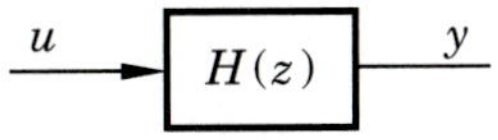

Figure 10.5 Generation of disturbances by driving dynamic systems with white noise.

Input-Output Models

For additional insight an input-output description of signals generated by linear difference equations is given. Notice that the signal x given by (10.7) can be described as the output of a linear dynamic system driven by white noise. From this viewpoint it is then natural to investigate how the properties of stochastic processes change when they are filtered by dynamic systems.

Analysis.   Consider the system shown in Fig. 10.5. For simplicity it is assumed that the sampling period is chosen as the time unit. Assume that the input u is a stochastic process with a given mean-value function m_u and a given covariance function r_u. Let the pulse response of the system be $\{h(k), k = 0, 1, \dots\}$. Notice that h has also been used to denote the sampling period. It is, however, clear from the context what h should be. The input-output relationship is

$$y(k) = \sum_{l=-\infty}^{k} h(k-l)u(l) = \sum_{n=0}^{\infty} h(n)u(k-n) \tag{10.12}$$

Taking mean values

$$m_y(k) = \mathrm{E}y(k) = \mathrm{E}\sum_{n=0}^{\infty} h(n)u(k-n)$$

$$= \sum_{n=0}^{\infty} h(n)\mathrm{E}u(k-n) = \sum_{n=0}^{\infty} h(n)m_u(k-n) \tag{10.13}$$

The mean value of the output is thus obtained by sending the mean value of the input through the system.

To determine the covariance, first observe that a subtraction of (10.13) from (10.12) gives

$$y(k) - m_y(k) = \sum_{n=0}^{\infty} h(n)\Big(u(k-n) - m_u(k-n)\Big)$$

The difference between the input signal and its mean value thus propagates through the system in the same way as the input signal itself. When calculating the covariance, it can be assumed that the mean values are zero. This simplifies the writing. The definition of the covariance function gives

$$r_y(\tau) = \mathrm{E}y(k+\tau)y^T(k)$$

$$= \mathrm{E}\sum_{n=0}^{\infty}h(n)u(k+\tau-n)\left(\sum_{l=0}^{\infty}h(l)u(k-l)\right)^T$$

$$= \sum_{n=0}^{\infty}\sum_{l=0}^{\infty}h(n)\mathrm{E}\Big(u(k+\tau-n)u^T(k-l)\Big)h^T(l) \qquad (10.14)$$

$$= \sum_{n=0}^{\infty}\sum_{l=0}^{\infty}h(n)r_u(\tau+l-n)h^T(l)$$

A similar calculation gives the following formula for the cross-covariance of the input and the output:

$$r_{yu}(\tau) = \mathrm{E}y(k+\tau)u^T(k) = \mathrm{E}\sum_{n=0}^{\infty}h(n)u(k+\tau-n)u^T(k)$$

$$= \sum_{n=0}^{\infty}h(n)\mathrm{E}\Big(u(k+\tau-n)u^T(k)\Big) = \sum_{n=0}^{\infty}h(n)r_u(\tau-n) \qquad (10.15)$$

Notice that it has been assumed that all infinite sums exist and that the operations of infinite summation and mathematical expectation have been freely exchanged in these calculations. This must of course be justified; it is easy to do in the sense of mean-square convergence, if it is assumed that the fourth moment of the input signal is finite.

The relations expressed by Eqs. (10.14) and (10.15) can be expressed in a simpler form if spectral densities are introduced. The definition of spectral density in (10.5) gives

$$\phi_y(\omega) = \phi_{yy}(\omega) = \frac{1}{2\pi}\sum_{n=-\infty}^{\infty}e^{-in\omega}r_y(n)$$

Introducing r_y from (10.14) gives

$$\phi_y(\omega) = \frac{1}{2\pi}\sum_{n=-\infty}^{\infty}e^{-in\omega}\sum_{k=0}^{\infty}\sum_{l=0}^{\infty}h(k)r_u(n+l-k)h^T(l)$$

$$= \frac{1}{2\pi}\sum_{k=0}^{\infty}\sum_{n=-\infty}^{\infty}\sum_{l=0}^{\infty}e^{-ik\omega}h(k)e^{-i(n+l-k)\omega}r_u(n+l-k)e^{il\omega}h^T(l)$$

$$= \frac{1}{2\pi}\sum_{k=0}^{\infty}e^{-ik\omega}h(k)\sum_{n=-\infty}^{\infty}e^{-in\omega}r_u(n)\sum_{l=0}^{\infty}e^{il\omega}h^T(l)$$

Introduce the pulse-transfer function $H(z)$ of the system. This is related to the impulse response $h(k)$ by

$$H(z) = \sum_{k=0}^{\infty} z^{-k} h(k)$$

The equation for the spectral density can then be written as

$$\phi_y(\omega) = H(e^{i\omega}) \phi_u(\omega) H^T(e^{-i\omega})$$

Similarly,

$$\phi_{yu}(\omega) = \frac{1}{2\pi} \sum_{n=-\infty}^{\infty} e^{-in\omega} r_{yu}(n) = \frac{1}{2\pi} \sum_{n=-\infty}^{\infty} e^{-in\omega} \sum_{k=0}^{\infty} h(k) r_u(n-k)$$

$$= \frac{1}{2\pi} \sum_{k=0}^{\infty} e^{-ik\omega} h(k) \sum_{n=-\infty}^{\infty} e^{-in\omega} r_u(n) = H(e^{i\omega}) \phi_u(\omega)$$

Main result.   To obtain the general result, the propagation of the mean value through the system must also be investigated.

THEOREM 10.2   FILTERING OF STATIONARY PROCESSES   Consider a stationary discrete-time dynamic system with sampling period 1 and the pulse-transfer function H. Let the input signal be a stationary stochastic process with mean m_u and spectral density ϕ_u. If the system is stable, then the output is also a stationary process with the mean

$$m_y = H(1) m_u \tag{10.16}$$

and the spectral density

$$\phi_y(\omega) = H(e^{i\omega}) \phi_u(\omega) H^T(e^{-i\omega}) \tag{10.17}$$

The cross-spectral density between the input and the output is given by

$$\phi_{yu} = H(e^{i\omega}) \phi_u(\omega) \tag{10.18}$$
∎

Remark 1.   The result has a simple physical interpretation. The number $|H(e^{i\omega})|$ is the steady-state amplitude of the response of the system to a sine wave with frequency ω. The value of the spectral density of the output is then the product of the power gain $|H(e^{i\omega})|^2$ and the spectral density of the input $\phi_u(\omega)$.

Remark 2.   It follows from Eq. (10.18) that the cross-spectral density is equal to the transfer function of the system if the input is white noise with unit spectral density. This fact can be used to determine the pulse-transfer function of a system.

The result is illustrated by an example.

Example 10.4 Spectral density of a first order system

Consider the process $x(k)$ in Example 10.3. From the input-output point of view, the process can be thought of as generated by sending white noise through a filter with the pulse-transfer function

$$H(z) = \frac{1}{z - a}$$

Because the spectral density of $v(k)$ is

$$\phi_v(\omega) = \frac{r_1}{2\pi}$$

it follows from (10.17) that the spectral density of $x(k)$ is

$$\phi_x(\omega) = H(e^{i\omega})H(e^{-i\omega})\frac{r_1}{2\pi}$$

$$= \frac{r_1}{2\pi} \cdot \frac{1}{(e^{i\omega} - a)(e^{-i\omega} - a)} = \frac{r_1}{2\pi(1 + a^2 - 2a\cos\omega)}$$

Because $x(k)$ and $e(k)$ are independent the process

$$y(k) = x(k) + e(k)$$

has the spectral density

$$\phi_y(\omega) = \frac{1}{2\pi}\left(r_2 + \frac{r_1}{1 + a^2 - 2a\cos\omega}\right)$$

(Compare with the calculation in Example 10.3.)   ∎

Spectral Factorization

Theorem 10.2 gives the spectral density of a stochastic process obtained by filtering another stochastic process. The spectral density of a signal obtained by filtering white noise is obtained as a special case. The inverse problem is discussed next. A linear system that gives an output with a given spectral density when driven by white noise will be determined. This problem is important because it shows how a signal with a given spectral density can be generated by filtering white noise. The solution to the problem will also tell how general the model in (10.7) is. It follows from Theorem 10.2 that the random process generated from a linear system with a white-noise input has the spectral density given by (10.17). If the system is finite-dimensional, H is then a rational function in $\exp(i\omega)$ and the spectral density ϕ will also be rational in $\exp(i\omega)$ or equivalently in $\cos\omega$. With a slight abuse of language, such a spectral density is called *rational*. Introducing

$$z = e^{i\omega}$$

the right-hand side of (10.17) can be written as

$$F(z) = \frac{1}{2\pi} H(z) H^T(z^{-1})$$

If z_i is a zero of $H(z)$, then z_i^{-1} is a zero of $H(z^{-1})$. The zeros of the function F are thus symmetric with respect to the real axis and mirrored in the unit circle. If the coefficients of the rational function H are real, the zeros of the function F will also be symmetric with respect to the real axis. The same argument holds for the poles of H. The poles and zeros of F will thus have the pattern illustrated in Fig. 10.7.

It is now straightforward to find a function H that corresponds to a given rational spectral density as follows: First, determine the poles p_i and the zeros z_i of the function F associated with the spectral density. It follows from the symmetry of the poles and zeros, which has just been established, that the poles and zeros always appear in pairs such that

$$z_i z_j = 1$$
$$p_i p_j = 1$$

In each pair choose the pole or the zero that is less than or equal to one in magnitude; then form the desired transfer function from the chosen poles and zeros as

$$H(z) = K \frac{\prod(z - z_i)}{\prod(z - p_i)} = \frac{B(z)}{A(z)}$$

Because the stochastic process is stationary, the chosen poles p_i will all be strictly less than one in magnitude. There may, however, be zeros that have unit magnitude. The result is summarized as follows.

THEOREM 10.3   SPECTRAL FACTORIZATION THEOREM   Given a spectral density $\phi(\omega)$, which is rational in $\cos \omega$, there exists a linear system with the pulse-transfer function

$$H(z) = \frac{B(z)}{A(z)} \tag{10.19}$$

such that the output obtained when the system is driven by white noise is a stationary random process with spectral density ϕ. The polynomial $A(z)$ has all its zeros inside the unit disc. The polynomial $B(z)$ has all its zeros inside the unit disc or on the unit circle.   ∎

Remark 1.   The spectral factorization theorem is very important. It implies that all stationary random processes can be thought of as being generated by stable linear systems driven by white noise, that is, an ARMA process of a special type. This means a considerable simplification both in theory and practice. It is sufficient to understand how systems behave when excited by white noise. It is only necessary to be able to simulate white noise. All other stationary processes with rational spectral density can then be formed by filtering.

Remark 2.   Because a continuous function can be approximated uniformly arbitrarily well on a compact interval with a rational function, it follows that the models in (10.7) and (10.12) can give signals whose spectra are arbitrarily close to any continuous function. Notice, however, that there are models with nonrational spectral densities. In turbulence theory, for instance, there are spectral densities that decay as fractional powers of ω for large ω.

An important consequence of the spectral factorization theorem is that for systems with one output, it is always possible to represent the net effect of all disturbances with one equivalent disturbance. This disturbance is obtained by calculating the total spectral density of the output signal and applying the spectral factorization theorem.

Remark 3.   It is often assumed that the polynomial $B(z)$ has all its zeros inside the unit disc. This means that the inverse of the system H is stable. The results are illustrated by two examples.

Example 10.5   Spectral factorization

Consider the process $y(k)$ of Examples 10.3 and 10.4. This process has the spectral density

$$\phi_y(\omega) = \frac{1}{2\pi} \left(r_2 + \frac{r_1}{(z-a)(z^{-1}-a)} \right)_{z=e^{i\omega}}$$

$$= \frac{1}{2\pi} \left(\frac{r_1 + r_2(1+a^2) - r_2 a(z+z^{-1})}{(z-a)(z^{-1}-a)} \right)_{z=e^{i\omega}}$$

The denominator is already in factored form. To factor the numerator, we observe that it can be written as

$$\lambda^2(z-b)(z^{-1}-b) \equiv r_1 + r_2(1+a^2) - r_2 a(z+z^{-1})$$

Identification of coefficients of equal powers of z gives

$$z^0: \quad \lambda^2(1+b^2) = r_1 + r_2(1+a^2)$$
$$z^1: \quad \lambda^2 b = r_2 a$$

Elimination of λ gives a second-order algebraic equation for b. This equation has the solution

$$b = \frac{r_1 + r_2(1+a^2) - \sqrt{\left(r_1 + r_2(1+a)^2\right)\left(r_1 + r_2(1-a)^2\right)}}{2ar_2}$$

The other root is discarded because it is outside the unit disc. Furthermore, the variable λ is given by

$$\lambda^2 = \frac{1}{2}\left(r_1 + r_2(1+a^2) + \sqrt{\left(r_1 + r_2(1+a)^2\right)\left(r_1 + r_2(1-a)^2\right)} \right)$$

■

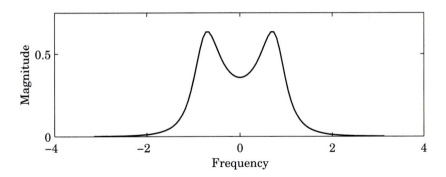

Figure 10.6 The spectral density (10.20) as function of ω, that is, when $z = e^{i\omega}$.

Example 10.6   Generation of a stochastic signal

Assume that we for simulation purposes want to generate a stochastic signal with the spectral density

$$F(z) = \frac{1}{2\pi} \cdot \frac{0.3125 + 0.125(z + z^{-1})}{2.25 - 1.5(z + z^{-1}) + 0.5(z^2 + z^{-2})} \tag{10.20}$$

The spectrum is shown in Fig. 10.6. Factorization of $F(z)$ gives the pole/zero pattern in Fig. 10.7 and the desired noise properties are obtained by filtering white noise through the filter

$$H(z) = \frac{0.5z + 0.25}{z^2 - 1 + 0.5}$$

■

Innovation's Representations

Theorem 10.3 has some conceptually important consequences. It follows from the theorem that a process with rational spectral density can be represented as

$$y(k) = \sum_{n=-\infty}^{k} h(k - n)e(n) \tag{10.21}$$

where e is discrete-time white noise and h is the impulse response that corresponds to the pulse-transfer function (10.19). The system has a stable inverse if the polynomial $B(z)$ has all its zeros inside the unit disc. This means that

$$e(k) = \sum_{n=-\infty}^{k} g(k - n)y(n)$$

where g is the impulse response, which corresponds to the stable pulse-transfer function $A(z)/B(z)$. It thus follows that the sequences $y(k), y(k - 1), \ldots$ and

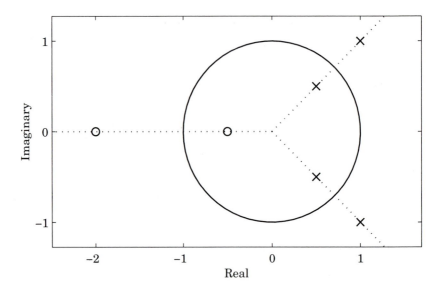

Figure 10.7 Symmetry of the poles and zeros of the spectral-density function (10.20).

$e(k), e(k-1), \ldots$ are equivalent in the sense that one sequence can be calculated
from the other.

Now consider

$$y(k+1) = \sum_{n=-\infty}^{k+1} h(k+1-n)e(n) = \sum_{n=-\infty}^{k} h(k+1-n)e(n) + h(0)e(k+1)$$

$$= \sum_{n=-\infty}^{k} h(k+1-n) \sum_{l=-\infty}^{k} g(n-l)y(l) + h(0)e(k+1)$$

The variable $y(k+1)$ can be written as the sum of two terms: One term is a linear function of $y(k), y(k-1), \ldots$, and the other term is $h(0)e(k+1)$. Thus $e(k+1)$
can be interpreted as the part of $y(k+1)$ that contains new information that
is not available in the past values $y(k), y(k-1), \ldots$. The stochastic process $e(k)$
is therefore called the *innovations* of the process $y(k)$ and the representation in
(10.21) is called the innovation's representation of the process. This representation is important in connection with filtering and prediction problems. The
term

$$\sum_{n=-\infty}^{k} h(k+1-n) \sum_{l=-\infty}^{n} g(n-l)y(l)$$

is in fact the best mean-square prediction of $y(k+1)$ based on $y(k), y(k-1), \ldots$.
This will be discussed in detail in Chapters 11 and 12.

Example 10.7   Innovation's representation

Consider the process $y(k)$ of Example 10.3. The process has the spectral density

$$\phi_y(\omega) = \frac{1}{2\pi} \left(r_2 + \frac{r_1}{1 + a^2 - 2a \cos \omega} \right)$$

It follows from Example 10.5 that the spectral density can be factored as

$$\phi_y(\omega) = \frac{\lambda^2}{2\pi} \cdot \frac{(z - b)(z^{-1} - b)}{(z - a)(z^{-1} - a)}$$

The process y can thus be generated by sending white noise through a system with the pulse-transfer function

$$H(z) = \frac{z - b}{z - a}$$

The input-output relation of such a system can be written as

$$y(k + 1) = ay(k) + e(k + 1) - be(k)$$

where $e(k)$ is white noise with variance λ^2.   ∎

Calculation of Variances

The variance of a signal obtained by filtering white noise can be calculated from the recursive equation of (10.11), if the model is given in state-space form. For a system described by transfer functions it is possible to use the same equations, if the model is first transformed to state-space form. It is naturally convenient to have similar formulas when the system is given in input-output form. Such formulas will now be given.

Consider a signal generated by

$$y(k) = \frac{B(q)}{A(q)} e(k) \tag{10.22}$$

where e is white noise with unit variance. It follows from Theorem 10.2 that the spectral density of the signal y is given by

$$\phi(\omega) = \frac{1}{2\pi} \cdot \frac{B(z)B(z^{-1})}{A(z)A(z^{-1})}$$

where $z = \exp(i\omega)$. It also follows from Theorem 10.2 that the variance of the signal y is given by the complex integral

$$\mathrm{E}y^2 = \int_{-\pi}^{\pi} \phi(\omega) \, d\omega = \frac{1}{i} \int_{-\pi}^{\pi} \phi(\omega) e^{-i\omega} \, d(e^{i\omega})$$

$$= \frac{1}{2\pi i} \oint \frac{B(z)B(z^{-1})}{A(z)A(z^{-1})} \cdot \frac{dz}{z} \tag{10.23}$$

The evaluation of integrals of this form is closely related to Jury's stability test (compare with Sec. 3.2). To evaluate the integral, the following table is formed:

a_0	a_1	$\cdots$	a_{n-1}	a_n			b_0	b_1	$\cdots$	b_{n-1}	b_n		
a_n	a_{n-1}	$\cdots$	a_1	a_0	α_n		a_n	a_{n-1}	$\cdots$	a_1	a_0	β_n	
a_0^{n-1}	a_1^{n-1}	$\cdots$	a_{n-1}^{n-1}				b_0^{n-1}	b_1^{n-1}	$\cdots$	b_{n-1}^{n-1}			
a_{n-1}^{n-1}	a_{n-2}^{n-1}	$\cdots$	a_0^{n-1}		α_{n-1}		a_{n-1}^{n-1}	a_{n-2}^{n-1}	$\cdots$	a_0^{n-1}		β_{n-1}	
$\vdots$													
a_0^1	a_1^1						b_0^1	b_1^1					
a_1^1	a_0^1				α_1		a_1^1	a_0^1				β_1	
a_0^0					1		b_0^0					β_0	

where

$$\alpha_n = a_n/a_0 \qquad \beta_n = b_n/a_0$$
$$\alpha_k = a_k^k/a_0^k \qquad \beta_k = b_k^k/a_0^k$$

and

$$a_i^{k-1} = a_i^k - \alpha_k a_{k-i}^k$$
$$b_i^{k-1} = b_i^k - \beta_k a_{k-i}^k$$

The left half of the table is the same as Jury's stability test. The right half is built up in the same way with the exception that the even rows are taken from the left half of the table. The following theorem results.

THEOREM 10.4 VARIANCE CALCULATION The integral (10.23) is given by

$$I_n = \frac{1}{a_0} \sum_{i=0}^{n} b_i^i \beta_i$$

∎

Application of the theorem gives the following values of the integral for $n = 1$ and 2.

$$I_1 = \frac{(b_0^2 + b_1^2)a_0 - 2b_0 b_1 a_1}{a_0(a_0^2 - a_1^2)}$$

$$I_2 = \frac{B_0 a_0 e_1 - B_1 a_0 a_1 + B_2(a_1^2 - a_2 e_1)}{a_0\left((a_0^2 - a_2^2)e_1 - (a_0 - a_2)a_1^2\right)}$$

where

$$B_0 = b_0^2 + b_1^2 + b_2^2$$
$$B_1 = 2(b_0 b_1 + b_1 b_2)$$
$$B_2 = 2b_0 b_2$$
$$e_1 = a_0 + a_2$$

10.5 Continuous-Time Stochastic Processes

It may be useful to formulate models and specifications in continuous time even if a computer is used to implement the control law. A brief account of continuous-time stochastic processes is therefore given.

Definitions

Continuous-time stochastic processes can be defined in the same way as discrete-time processes. The only difference is that the index set T is the set of real variables instead of a discrete set. Covariance functions and stationary processes are defined as for discrete-time processes using the finite-dimensional distribution functions. A spectral density can also be introduced as the Fourier transform of the covariance function. Equation (10.5) is then replaced by

$$\phi_{xy}(\omega) = \frac{1}{2\pi} \int_{-\infty}^{\infty} e^{-i\omega t} r_{xy}(t) \, dt \tag{10.24}$$

The inverse transform is given by

$$r_{xy}(t) = \int_{-\infty}^{\infty} e^{i\omega t} \phi_{xy}(\omega) \, d\omega \tag{10.25}$$

which replaces (10.6). The spectral density has the same interpretation as for discrete-time systems.

White Noise

White noise is defined as a stationary process with constant spectral density. If

$$\phi(\omega) = \frac{r_0}{2\pi}$$

it follows formally from (10.25) that the corresponding covariance is a delta function, that is,

$$r(t) = r_0 \delta(t)$$

Continuous-time white noise thus has the property that values of the signal at different times are uncorrelated as for discrete-time white noise. Continuous-time white noise has, however, infinite variance. This will cause some mathematical difficulties. Intuitively, continuous-time white noise is analogous to delta functions in the theory of linear systems.

Some of the difficulties with continuous-time white noise can be avoided by introducing a stochastic process that formally is the time integral

$$w(t) = \int_{0}^{t} e(s) \, ds$$

of white noise e. The stochastic process w has zero mean value. Its increments over disjoint intervals are uncorrelated. If the covariance function of e is

$$\text{cov}\Big(e(t), e(s)\Big) = r_0 \delta(t - s)$$

then the variances of the increments of w are given by

$$\text{E}\Big(w(t) - w(s)\Big)^2 = |t - s| \cdot r_0$$

The stochastic process $\{w(t), t \in T\}$ is called a *Wiener process* if it also is Gaussian. The Wiener process is a model for random walk. The infinitesimal increment

$$dw = w(t + dt) - w(t)$$

has the variance

$$\text{E}(dw)^2 = r_0 \, dt$$

The increment dw thus has the magnitude $\sqrt{r_0 \, dt}$ in the mean-square sense. The number $r_0 \, dt$ is called the *incremental covariance* of the Wiener process.

State-Space Models

State models for continuous-time processes can be obtained by a formal generalization of (10.7) to

$$\frac{dx}{dt} = Ax + \dot{v}$$

where $\dot{v}$ is a vector whose elements are white-noise stochastic processes. Because $\dot{v}$ has infinite variance, it is customary to write the equation in terms of differentials as

$$dx = Ax \, dt + dv \qquad (10.26)$$

where v is the integral of $\dot{v}$. The signal v is thus assumed to have zero mean, uncorrelated increments, and the variance

$$\text{cov}\Big(v(t), v(t)\Big) = R_1 t \qquad (10.27)$$

It is also assumed that dv is uncorrelated with x. A precise meaning can be given to (10.26) without any reference to white noise. This form is therefore common in mathematically oriented texts. The form is also useful as a reminder that dv has a magnitude proportional to $\sqrt{dt}$.

Equation (10.26) is called a *stochastic differential equation*. To specify it fully, it is also necessary to give the initial probability distribution of x at the starting time. The following continuous-time analog of Theorem 10.2 is then obtained.

THEOREM 10.5 FILTERING OF CONTINUOUS-TIME PROCESSES   Consider a stochastic process defined by the linear stochastic differential equation (10.26) where the process v has zero mean and incremental covariance $R_1\,dt$. Let the initial state have mean m_0 and covariance R_0. The mean-value function of the process x is then given by

$$\frac{dm(t)}{dt} = Am(t) \qquad m(0) = m_0 \tag{10.28}$$

and the covariance function is given by

$$\text{cov}\Big(x(s), x(t)\Big) = e^{A(s-t)}P(t) \qquad s \geq t \tag{10.29}$$

where $P(t) = \text{cov}\Big(x(t), x(t)\Big)$ is given by

$$\frac{dP(t)}{dt} = AP(t) + P(t)A^T + R_1 \qquad P(0) = R_0 \tag{10.30}$$

Proof.   The formula (10.28) for the mean value is obtained simply by taking the mean value of (10.26). Notice that dv has zero mean.

To obtain the differential equation in (10.30), notice that

$$d(xx^T) = (x + dx)(x + dx)^T - xx^T = x\,dx^T + dx\,x^T + dx\,dx^T$$

Equation (10.26) then gives

$$d(xx^T) = x(Ax\,dt + dv)^T + (Ax\,dt + dv)x^T + (Ax\,dt + dv)(Ax\,dt + dv)^T$$

Taking mean values gives

$$d(\text{E}xx^T) = (\text{E}xx^T)A^T\,dt + A(\text{E}xx^T)\,dt + \text{E}\,dv\,dv^T + A(\text{E}xx^T)A^T(dt)^2$$

because dv is uncorrelated with x. Furthermore, it follows from (10.27) that

$$\text{E}\,dv\,dv^T = R_1\,dt$$

Hence

$$dP = PA^T\,dt + AP\,dt + R_1\,dt + APA^T(dt)^2$$

Dividing by dt and taking the limit as dt goes to zero gives the differential equation in (10.30). To obtain Eq. (10.29), let $s \geq t$ and integrate (10.26). Hence

$$x(s) = e^{A(s-t)}x(t) + \int_t^s e^{A(s-s')}\,dv(s')$$

Multiplying by $x^T(t)$ from the right and taking mathematical expectation give (10.29). Notice that $dv(s')$ is uncorrelated with $x(t)$ if $s' \geq t$. ∎

Example 10.8 First-order continuous-time system

Consider the scalar stochastic differential equation

$$dx = -ax\, dt + dv$$

$$x(t_0) = m_0 \qquad \mathrm{var}\Big(x(t_0)\Big) = r_0$$

where the process $\{v(t), t \in T\}$ has incremental covariance $r_1\, dt$. It follows from
(10.28) that the mean-value function is given by

$$\frac{dm}{dt} = -am \qquad m(t_0) = m_0$$

This equation has the solution

$$m(t) = m_0 e^{-a(t-t_0)}$$

The covariance function is given by

$$r(s,t) = \mathrm{cov}\Big(x(s), x(t)\Big) = e^{-a(s-t)} P(t) \qquad s \geq t$$

and

$$r(s,t) = e^{-a(t-s)} P(s) \qquad s \geq t$$

Equation (10.30) gives the following differential equation for P.

$$\frac{dP}{dt} = -2aP + r_1 \qquad P(t_0) = r_0$$

This differential equation has the solution

$$P(t) = e^{-2a(t-t_0)} r_0 + \int_{t_0}^{t} e^{-2a(t-s)} r_1\, ds$$

$$= e^{-2a(t-t_0)} r_0 + \frac{r_1}{2a}\Big(1 - e^{-2a(t-t_0)}\Big)$$

As $t_0 \to \infty$, the mean-value function goes to zero and the covariance function goes
to

$$r(s,t) = \frac{r_1}{2a}\, e^{-a|t-s|}$$

because the limiting covariance function depends only on the argument difference
$s - t$, the limiting process is (weakly) stationary and its covariance function can be
written as

$$r(\tau) = \frac{r_1}{2a}\, e^{-a|\tau|}$$

Equation (10.24) gives the corresponding spectral density

$$\phi(\omega) = \frac{r_1}{2\pi} \cdot \frac{1}{\omega^2 + a^2}$$

∎

Filtering Continuous-Time Processes

The analysis of linear systems with continuous-time stochastic processes as inputs is analogous to the corresponding analysis for discrete-time systems. Consider a time-invariant stable system with impulse response g. The input-output relationship is

$$y(t) = \int_{-\infty}^{t} g(t-s)u(s)\,ds = \int_{0}^{t} g(s)u(t-s)\,ds \tag{10.31}$$

[compare with Eq. (10.12)]. Let the input signal u be a stochastic process with mean-value function m_u and covariance function r_u.

The following result is analogous to Theorem 10.2 for discrete-time systems.

THEOREM 10.6   FILTERING STATIONARY PROCESSES   Consider a stationary linear system with the transfer function G. Let the input signal be a stationary continuous-time stochastic process with mean value m_u and spectral density ϕ_u. If the system is stable, then the output is also a stationary process with the mean value

$$m_y = G(0)m_u$$

and the spectral density

$$\phi_y(\omega) = G(i\omega)\phi_u(\omega)G^T(-i\omega) \tag{10.32}$$

The cross-spectral density between the input and the output is given by

$$\phi_{yu}(\omega) = G(i\omega)\phi_u(\omega)$$

The result may be interpreted in the same way as the corresponding result for discrete-time systems. Compare with Remarks 1 and 2 of Theorem 10.2.   ∎

Example 10.9   Spectral density of a continuous-time process

Consider the system in Example 10.8. The process x can be considered as the result of filtering white noise with the variance $r_1/2\pi$ through a system having the transfer function

$$G(s) = \frac{1}{s+a}$$

It follows from (10.32) that the spectral density is given by

$$\phi(\omega) = \frac{r_1}{2\pi} \cdot \frac{1}{i\omega+a} \cdot \frac{1}{-i\omega+a} = \frac{r_1}{2\pi} \cdot \frac{1}{\omega^2+a^2}$$

∎

Spectral Factorization

It follows from (10.32) that if the input is white noise with $\phi_u = 1$, then the spectral density of the output is given by

$$\phi_y(\omega) = G(i\omega)G^T(-i\omega) \tag{10.33}$$

This means that any disturbance whose spectral density can be written in this form may be generated by sending continuous-time white noise through a filter with the transfer function G.

Because linear finite-dimensional systems have rational transfer functions, it follows that signals with arbitrary rational spectral densities can be generated from linear finite-dimensional systems. The covariance function is nonnegative and symmetric. It then follows from (10.24) that ϕ is also symmetric. If ϕ is rational it then follows that its poles and zeros are symmetric with respect to the real and imaginary axes. The transfer function G in (10.33) can then be chosen so that all its poles are in the left half-plane and all its zeros in the left half-plane or on the imaginary axis. The following analog of Theorem 10.3 is thus obtained.

THEOREM 10.7   SPECTRAL FACTORIZATION   Given a rational spectral density $\phi(\omega)$, there exists a linear finite-dimensional system with the rational transfer function

$$G(s) = \frac{B(s)}{A(s)}$$

such that the output obtained when the system is driven by white noise is a stationary stochastic process with the given spectral density. The polynomial A has all its zeros in the left half-plane. The polynomial B has no zeros in the right half-plane.   ∎

10.6 Sampling a Stochastic Differential Equation

If process models are presented in continuous time as stochastic differential equations, it is useful to sample these equations to obtain a discrete-time model.

Consider a process described by

$$dx = Ax\,dt + dv_c \tag{10.34}$$

where the process v_c has zero mean value and uncorrelated increments. The incremental covariance of v_c is $R_1\,dt$. Let the sampling instants be $\{t_k;\ k = 0, 1, \ldots\}$. Integration of (10.34) over one sampling period gives

$$x(t_{k+1}) = e^{A(t_{k+1}-t_k)}x(t_k) + \int_{t_k}^{t_{k+1}} e^{A(t_{k+1}-s)}\,dv_c(s)$$

Consider the random variable

$$v(t_k) = \int_{t_k}^{t_{k+1}} e^{A(t_{k+1}-s)} \, dv_c(s)$$

This variable has zero mean because v_c has zero mean. The random variables $v(t_k)$ and $v(t_l)$ are also uncorrelated for $k \neq l$ because the increments of v over disjoint intervals are uncorrelated. The covariance of $v(t_k)$ is given by

$$\begin{aligned} \mathrm{E}\Big(v(t_k), v^T(t_k)\Big) &= \mathrm{E} \iint_{t_k}^{t_{k+1}} e^{A(t_{k+1}-s)} \, dv_c(s) \, dv_c^T(t) e^{A^T(t_{k+1}-t)} \\ &= \iint_{t_k}^{t_{k+1}} e^{A(t_{k+1}-s)} R_1 e^{A^T(t_{k+1}-s)} \, ds \end{aligned} \tag{10.35}$$

It is thus found that the random sequence $\{x(t_k), k = 0, 1, \dots\}$ obtained by sampling the process $\{x(t)\}$ is described by the difference equation

$$x(t_{k+1}) = e^{A(t_{k+1}-t_k)} x(t_k) + v(t_k)$$

where $\{v(t_k)\}$ is a sequence of uncorrelated random variables with zero mean and covariance (10.35).

10.7 Conclusions

The main purpose of this chapter is to develop mathematical models for disturbances. The result is a uniform approach to models for a wide variety of signals. The signals are viewed as being generated from dynamic systems driven by a pulse, a sequence of pulses, or white noise. Equivalently, the signals may be considered as being generated by dynamic systems with initial conditions.

Simple disturbances like step, ramp, and sinusoid can be generated as outputs of linear systems driven by a pulse. More complicated disturbances may be viewed as pulse responses of more complicated systems. Section 10.3 shows that the class of disturbances could be widened by driving the systems with signals composed of several pulses. This leads to the piecewise deterministic signals. A further extension is given in Sec. 10.4, where the input signal to the disturbance-generating system was chosen as white noise.

A unified way of modeling different types of disturbances is obtained. The disturbances are characterized by a dynamic system

$$y(k) = \frac{B(q)}{A(q)} \, \varepsilon(k) \tag{10.36}$$

where the input ε is a pulse, several pulses, or white noise. The system is called the *disturbance generator*. The dynamic system can, of course, also be represented in state-space form.

The problem of prediction is important when controlling systems with disturbances that cannot be measured. The problem of predicting a signal given by (10.36) is in essence to compute ε from y. This is the same as inverting the dynamic system (10.36). To obtain a stable inverse, the polynomial $B(q)$ must then have all its zeros inside the unit disc. This will, in general, not be the case for deterministic disturbances. A consequence is that the performance of the prediction will deteriorate. It takes longer to obtain the prediction. For a stochastic system it follows from the spectral factorization theorem that the polynomial $B(q)$ has all its zeros inside the unit disc or on the unit circle.

The essential point of the discussion is that the predictors for the signals are uniquely given by the pulse-transfer function $H = B/A$. The predictors are thus the same for inputs that are pulses, pulse trains, or white noise. This means that predictors that are designed for deterministic disturbances can work very well also for stochastic disturbances if the disturbance generators for the signals are the same. The unified approach to model disturbances also leads to a substantial simplification of the theory because it is sufficient to work with a few prototypes for disturbances only.

10.8 Problems

10.1 List situations in which it is possible to reduce the influence of disturbances by (a) reduction at the source, (b) local feedback, and (c) prediction.

10.2 Show that the predictor (10.4) is equivalent to the predictor (10.3).

10.3 Determine the m-step predictor for the disturbance model

$$y(k) = \frac{C(q)}{A(q)} w(k)$$

where $w(k)$ is zero except at isolated points that are spaced more than $\deg A$. Use the result to determine the signal and the prediction when $A(q) = q - 0.5$, $C(q) = q$, and $m = 3$ and when $w(k)$ is zero except for $k = 0$ and 5. The initial conditions are assumed to be equal to zero.

10.4 Use Theorem 10.1 to compute the stationary covariance function of the process

$$x(k + 1) = \begin{pmatrix} 0.4 & 0 \\ -0.6 & 0.2 \end{pmatrix} x(k) + v(k)$$

where v is a white-noise process with zero mean and the variance

$$R_1 = \begin{pmatrix} 1 & 0 \\ 0 & 2 \end{pmatrix}$$

10.5 Consider a stationary stochastic process generated by

$$x(k + 1) = \Phi x(k) + v(k)$$
$$y(k) = C x(k)$$

where $v(k)$ is a sequence of equally distributed, independent, zero-mean, stochastic variables. Let Φ have the characteristic equation

$$z^n + a_1 z^{n-1} + \cdots + a_n = 0$$

Show that the autocovariance function of the output $r_y(\tau)$ satisfies

$$r_y(\tau) + a_1 r_y(\tau - 1) + \cdots + a_n r_y(\tau - n) = 0$$

for $\tau \geq n + 1$. (This equation is called the *Yule-Walker equation*.)

10.6 Consider the process

$$x(k + 1) = \begin{pmatrix} -a & 0 \\ 0 & -b \end{pmatrix} x(k) + v(k)$$

$$y(k) = \begin{pmatrix} 1 & 1 \end{pmatrix} x(k)$$

where $v(k)$ is white noise with zero mean and the covariance matrix

$$R_1 = \begin{pmatrix} \sigma_1^2 & 0 \\ 0 & \sigma_2^2 \end{pmatrix}$$

Show that $y(k)$ can be represented in the form

$$y(k) = \lambda \frac{q + c}{(q + a)(q + b)} e(k)$$

where $e(k)$ is white noise with zero mean and unit variance. Find the relationship from which λ and c can be determined.

10.7 A stochastic process $y(k)$ is described by

$$x(k + 1) = ax(k) + v(k)$$
$$y(k) = x(k) + e(k)$$

where v and e are normally distributed white-noise processes with the properties

$$Ev = Ee = 0$$
$$\operatorname{var} v = 1$$
$$\operatorname{var} e = r_2$$
$$Ev(k)e(j) = r_{12} \quad \text{when } k = j \text{ and } 0 \text{ otherwise}$$

Show that $y(k)$ can be represented as the output of a linear filter

$$y(k) = \lambda \frac{q - c}{q - a} \varepsilon(k) \qquad |c| \leq 1$$

where $\varepsilon(k)$ is white noise with zero mean and unit variance. Determine λ and c.

10.8 Determine the covariance function, $r_y(\tau)$, and the spectrum, $\phi_y(\omega)$, of the process $y(k)$ when

$$y(k) - 0.7y(k - 1) = e(k) - 0.5e(k - 1)$$

where $e(k)$ is white noise with unit variance.

10.9 Determine the variance of the stochastic process $y(k)$ defined by

$$y(k) - 1.5y(k - 1) + 0.7y(k - 2) = e(k) + 0.2e(k - 1)$$

where e is white noise with unit variance.

10.10 Calculate the stationary covariance function $r_y(\tau), \tau = 0, 1, 2, \ldots$, for the system

$$y(k) = e(k) - 2e(k - 1) + 3e(k - 2) - 4e(k - 3)$$

when e is zero-mean white noise with unit variance.

10.11 Assume that we want to generate a signal $y(k)$ with the spectral density

$$\phi_y(\omega) = \frac{1}{1.36 + 1.2 \cos \omega}$$

(a) Determine a stable filter $H(q)$ that gives the desired signal $y(k) = H(q)e(t)$, where e is white noise such that $e \in N(0, 1)$.

(b) What is the variance of y?

10.12 Consider the discrete-time system

$$x(k + 1) = \begin{pmatrix} 0.3 & 0.2 \\ 0 & 0.5 \end{pmatrix} x(k) + \begin{pmatrix} 0 \\ 1 \end{pmatrix} u(k) + v(k)$$

$$y(k) = \begin{pmatrix} 1 & 0 \end{pmatrix} x(k) + e(k)$$

Assume that v and e are white-noise processes that are uncorrelated and with the covariances

$$R_1 = \begin{pmatrix} 1 & 0 \\ 0 & 0.5 \end{pmatrix} \quad \text{and} \quad R_2 = 0.2$$

respectively. Assume that the initial value has the zero-mean value and the covariances

$$\text{cov}\left(x(0), x(0)^T\right) = I$$

Compute the stationary value of the covariance of the state vector.

10.13 Assume that a white-noise generator is available that gives a zero mean output with unit variance. Determine a filter that can be used to generate a stochastic signal with the spectral density

$$\phi(\omega) = \frac{3}{2\pi\left(5.43 - 5.40 \cos(\omega)\right)}$$

10.9 Notes and References

The principles for reducing disturbances by feedback and feedforward and the classic disturbance models are a key element of classic feedback theory. See Brown and Campbell (1948), Chestnut and Mayer (1959), and Gille, Pelegrin, and Decaulne (1959).

The notion of piecewise deterministic signals was introduced in Åström (1980) where the formulas for prediction are proven and more details are given. Tables for the integrals for computing the variance for low values of n are given in Jury (1982).

The ideas of representing disturbances as stochastic processes was also part of classic feedback theory. See James, Nichols, and Philips (1947), Tsien (1955), Laning and Battin (1956), and Newton, Gould, and Kaiser (1957).

A reasonably complete treatment of stochastic processes requires a full book. The following books give a good background: Parzen (1962), Papoulis (1965), Karlin (1966), Chung (1974), Kumar and Varaiya (1986), and Caines (1988). There are also shorter summaries in the books on stochastic control listed in what follows.

Prediction theory originated in Kolmogorov (1941), Wiener (1949), Kalman (1960b), and Kalman and Bucy (1961). The papers by Wiener and Kolmogorov are not easy to read. A readable account of Kolmogorov's work is found in Whittle (1963). Wiener's results were originally published as an MIT report in 1942. It became known as the "yellow peril" because of its yellow cover and its style of writing. Kalman (1960b), which deals with discrete-time processes, is easy to read. There are also full books devoted to prediction, filtering theory, and stochastic control: Åström (1970), McGarty (1974), Box and Jenkins (1976), and Anderson and Moore (1979).

11

Optimal Design Methods: A State-Space Approach

11.1 Introduction

In Chapters 4 and 5 the synthesis problem is solved using pole-placement techniques. The main design parameters have been the locations of the closed-loop poles, and the presentations have been limited to single-input–single-output systems. In this chapter a more general control problem is discussed. The process is still assumed to be linear, but it may be time-varying and have several inputs and outputs. Further, process and measurement noise are introduced in the models. The synthesis problem is formulated to minimize a criterion, which is a quadratic function of the states and the control signals. The resulting optimal controller is linear. The problem, which is stated formally in what follows, is called the *Linear Quadratic (LQ) control problem*, or the *Linear Quadratic Gaussian (LQG) control problem* if Gaussian stochastic disturbances are allowed in the process models. The stationary solution to the LQ-problem for time-invariant systems leads to a control law of the same structure as the state-feedback controller in Chapter 4. The LQ-controller can also be interpreted as a pole-placement controller. The degrees of freedom of the multivariable version of the controller in Chapter 4 is resolved by the minimization of a loss function instead of specifying only the closed-loop poles.

LQ-control is a large topic treated in many books. In this chapter, only a brief review of the main ideas and results is given. The problem is stated and some useful results are given in this section. The solution of the LQ-control problem, if all the states are available, is given in Sec. 11.2, where the properties of LQ-controllers are also discussed. If all the states are not measurable, they can be estimated using a dynamic system, as in Sec. 4.4. For the case with Gaussian disturbances, it is possible to determine the optimal estimator, which

408

minimizes the variance of the estimation error. This is called the *Kalman filter*. The estimator has the same structure as in (4.28). However, the gain matrix, K, is determined differently and is in general time-varying. Kalman filters are discussed in Sec. 11.3. The LQG-problem is solved in Sec. 11.4, where the states are estimated using a Kalman filter. The solution is based on the separation theorem or the *certainty equivalence principle*. This implies that the optimal control strategy can be separated into two parts: one state estimator, which gives the best estimates of the states from the observed outputs, and one linear-feedback law from the estimated states. The linear controller used is the same as the one used if there are no disturbances acting on the system. Some practical aspects are discussed in Sec. 11.5.

Problem Formulation

The design problem is specified by giving the process, the criterion, and the admissible control laws.

The process.   It is assumed that the process to be controlled is described by the continuous-time model

$$dx = Ax\,dt + Bu\,dt + dv_c \tag{11.1}$$

where A and B may be time-varying matrices. The process v_c has mean value of zero and uncorrelated increments. The incremental covariance of v_c is $R_{1c}\,dt$ (compare with Sec. 10.5). The model in (11.1) can be sampled as in Sec. 10.6. Some modifications must be made because the system is allowed to be time-varying. The input $u(t)$ is constant over the sampling period; for the noise-free case the solution of (11.1) can be written as

$$x(t) = \Phi(t, kh)x(kh) + \Gamma(t, kh)u(kh) \tag{11.2}$$

where $\Phi(t, kh)$ is the fundamental matrix of (11.2) satisfying

$$\frac{d}{dt}\Phi(t, kh) = A(t)\Phi(t, kh) \qquad \Phi(kh, kh) = I$$

and

$$\Gamma(t, kh) = \int_{kh}^{t} \Phi(t, s)B(s)\,ds$$

Omitting the time arguments of the matrices, the sampled model can be written as

$$x(kh + h) = \Phi x(kh) + \Gamma u(kh) + v(kh)$$
$$y(kh) = Cx(kh) + e(kh) \tag{11.3}$$

where v and e are discrete-time Gaussian white-noise processes with zero-mean value and

$$\mathrm{E}\,v(kh)v^T(kh) = R_1 = \int_0^h e^{A\tau} R_{1c} e^{A^T\tau}\, d\tau$$

$$\mathrm{E}\,v(kh)e^T(kh) = R_{12}$$

$$\mathrm{E}\,e(kh)e^T(kh) = R_2$$

The expression for the covariance matrix R_1 was given in Sec. 10.6. Further, it is assumed that the initial state $x(0)$ is Gaussian distributed with

$$\mathrm{E}x(0) = m_0 \quad \text{and} \quad \mathrm{cov}\Big(x(0)\Big) = R_0$$

The matrices R_0, R_1, and R_2 are positive semidefinite. The covariance matrices may be time-varying. It is assumed that the model (11.3) is reachable and observable.

As discussed in Chapter 4, it is possible to include other types of disturbances and effects from the environment by augmenting the state vector of the process.

The criterion.   The design criteria we will use is a way of weighting the magnitude of the states and control signals. One way can be to look at the power of the state, that is,

$$J = \int_0^{Nh} |x(t)|^2\, dt = \int_0^{Nh} x(t)^T x(t)\, dt$$

The components of the state may have different dimensions and we can instead use a more general weighting

$$J = \int_0^{Nh} x(t)^T Q_{1c} x(t)\, dt$$

where Q_{1c} is a symmetric positive semidefinite matrix. The control signal and the state at the end time can be penalized in a similar way. This leads to a control problem where we want to minimize the loss function

$$J = \mathrm{E}\left(\int_0^{Nh} \Big(x^T(t) Q_{1c} x(t) + 2x^T(t) Q_{12c} u(t) \right.$$

$$\left. + u^T(t) Q_{2c} u(t) \Big)\, dt + x^T(Nh) Q_{0c} x(Nh) \right) \qquad (11.4)$$

$$= \mathrm{E}\left(\int_0^{Nh} \begin{pmatrix} x^T(t) & u^T(t) \end{pmatrix} Q_c \begin{pmatrix} x(t) \\ u(t) \end{pmatrix} dt + x^T(Nh) Q_{0c} x(Nh) \right)$$

with

$$Q_c = \begin{pmatrix} Q_{1c} & Q_{12c} \\ Q_{12c}^T & Q_{2c} \end{pmatrix}$$

and where the matrices Q_{0c}, Q_{1c}, and Q_{2c} are symmetric and at least positive semidefinite. The matrices in the loss function may depend on time.

Admissible control laws.   It is important to specify the data available for determining the control signal. The first assumption is that periodic sampling is used and that the control signal is constant over the sampling periods. The control problem can then easily be translated into a discrete time problem.

If C equals the unit matrix and if $e(kh) = 0$ in (11.3), then the full-state vector is available. The control signal is then allowed to be a function of the state up to and including time kh. This is called *complete state information*. In many cases only the outputs can be measured. This implies that only noise-corrupted measurements are available for the controller. This is called *incomplete state information*. In this case the control signal at time kh is allowed to be a function of the outputs and inputs up to and including either time $kh - h$ or time kh.

The problem.   The optimal control problem is now defined to be finding the admissible control signal that minimizes the loss function of (11.4) when the process is described by the model of (11.1) or the equivalent model of (11.3). The design parameters are the matrices in the loss function and the sampling period.

Sampling the Loss Function

The loss function in (11.4) is expressed in continuous time. It is first transformed into a discrete-time loss function. Integrating (11.4) over intervals of lengths h gives

$$J = \mathrm{E}\left(\sum_{k=0}^{N-1} J(k) + x^T(Nh)Q_{0c}x(Nh) \right)$$

where

$$J(k) = \int_{kh}^{kh+h} \left(x^T(t)Q_{1c}x(t) + 2x^T(t)Q_{12c}u(t) + u^T(t)Q_{2c}u(t) \right) dt \qquad (11.5)$$

Using (11.2) in (11.5) and the fact that $u(t)$ is constant over the sampling period gives

$$J(k) = x^T(kh)Q_1x(kh) + 2x^T(kh)Q_{12}u(kh) + u^T(kh)Q_2u(kh)$$

where

$$Q_1 = \int_{kh}^{kh+h} \Phi^T(s,kh) Q_{1c} \Phi(s,kh)\, ds \tag{11.6}$$

$$Q_{12} = \int_{kh}^{kh+h} \Phi^T(s,kh) \big(Q_{1c} \Gamma(s,kh) + Q_{12c} \big)\, ds \tag{11.7}$$

$$Q_2 = \int_{kh}^{kh+h} \big(\Gamma^T(s,kh) Q_{1c} \Gamma(s,kh) + 2\Gamma^T(s,kh) Q_{12c} + Q_{2c} \big)\, ds \tag{11.8}$$

Minimizing the loss function of (11.4) when $u(t)$ is constant over the sampling period is thus the same as minimizing the discrete-time loss function

$$J = \mathrm{E} \left(\sum_{k=0}^{N-1} \Big(x^T(kh) Q_1 x(kh) + 2 x^T(kh) Q_{12} u(kh) \right.$$

$$\left. + u^T(kh) Q_2 u(kh) \Big) + x^T(Nh) Q_0 x(Nh) \right) \tag{11.9}$$

$$= \mathrm{E} \left(\sum_{k=0}^{N-1} \big(x^T(kh) \quad u^T(kh) \big) Q \left(\begin{array}{c} x(kh) \\ u(kh) \end{array} \right) + x^T(Nh) Q_{0c} x(Nh) \right)$$

where

$$Q = \left(\begin{array}{cc} Q_1 & Q_{12} \\ Q_{12}^T & Q_2 \end{array} \right) \tag{11.10}$$

The matrices Q_1, Q_{12}, and Q_2 are given by (11.6) to (11.8), respectively, and $Q_0 = Q_{0c}$. In the following it is assumed that Q_1 and Q_0 are positive semidefinite and that Q_2 is positive definite. The condition on Q_2 will be relaxed in what follows. Notice that the sampled loss function (11.9) will have a cross-coupling term Q_{12} even if $Q_{12c} = 0$.

When the stochastic case is considered, one additional term depending on the noise is obtained in (11.9). However, this term is independent of the control signal and can thus be disregarded when performing the minimization.

The optimal-control problem has now been transformed into the discrete-time problem of minimizing the loss function (11.9) when the process is described by (11.3). To facilitate the writing in the sequel, it is assumed that the sampling period is used as time unit, that is, $h = 1$.

Completing the Squares

Quadratic functions will be minimized several times in the sequel. The loss functions will have the form

$$J(x,u) = \big(x^T \quad u^T \big) \left(\begin{array}{cc} Q_x & Q_{xu} \\ Q_{xu}^T & Q_u \end{array} \right) \left(\begin{array}{c} x \\ u \end{array} \right) \tag{11.11}$$

and we want to find the minimum with respect to u. Then there exists an L satisfying

$$Q_u L = Q_{xu}^T \tag{11.12}$$

such that the loss function (11.11) can be written as

$$J(x, u) = x^T (Q_x - L^T Q_u L)x + (u + Lx)^T Q_u (u + Lx) \tag{11.13}$$

This is easily shown by inserting (11.12) into (11.13). Rewriting (11.11) as in (11.13) is called *completing the squares*. Because (11.13) is quadratic in u and both terms are greater or equal zero, it is easily seen that (11.11) is minimized for

$$u = -Lx \tag{11.14}$$

and that L is unique if Q_u is positive definite. The minimum is

$$J_{\min} = x^T (Q_x - L^T Q_u L)x \tag{11.15}$$

11.2 Linear Quadratic Control

The LQ-control problem will now be solved for the case of complete state information.

The Deterministic Case

The deterministic case, where $v(k) = 0$ and $e(k) = 0$ in (11.3), is first considered. The system is thus described by

$$x(k + 1) = \Phi x(k) + \Gamma u(k) \tag{11.16}$$

where $x(0)$ is given. The problem is now to determine the control sequence $u(0)$, $u(1), \ldots, u(N - 1)$ such that the loss function in (11.9) is minimized.

The idea behind the derivation of the control law is to use the *principle of optimality* and *dynamic programming*. The principle of optimality states that an optimal policy has the property that whatever the initial state and initial decision are the remaining decisions must be optimal with respect to the state resulting from the first decision. By using this idea and starting from the end time N and going backwards in time, it is possible to determine the best control law for the last step independent of how the state at time $N - 1$ was reached. The remaining loss-to-go will now depend on the state at time $N - 1$. Iterating backwards to the initial time $k = 0$ determines the optimal-control policy. The procedure is called dynamic programming and was introduced by Bellman. The solution is given by the following theorem.

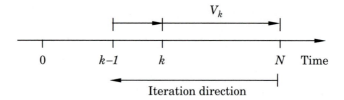

Figure 11.1   Illustration of the iteration procedure using dynamic programming.

THEOREM 11.1   LQ-CONTROL OF A DETERMINISTIC SYSTEM   Consider the system of (11.16). Allow $u(k)$ to be a function of $x(k)$, $x(k-1)$, ... We introduce

$$S(k) = \Phi^T S(k+1)\Phi + Q_1 - \left(\Phi^T S(k+1)\Gamma + Q_{12}\right)$$
$$\times \left(\Gamma^T S(k+1)\Gamma + Q_2\right)^{-1}\left(\Gamma^T S(k+1)\Phi + Q_{12}^T\right)$$

(11.17)

with end condition $S(N) = Q_0$. Assume that Q_0 is positive semidefinite and that $Q_2 + \Gamma^T S(k)\Gamma$ is positive definite. Then there exists a unique, admissible, control strategy

$$u(k) = -L(k)x(k)$$

(11.18)

where

$$L(k) = \left(Q_2 + \Gamma^T S(k+1)\Gamma\right)^{-1}\left(\Gamma^T S(k+1)\Phi + Q_{12}^T\right)$$

(11.19)

that minimizes the loss (11.9). The minimal value of the loss is

$$\min J = V_0 = x^T(0)S(0)x(0)$$

Further $S(k)$ is positive semidefinite.

Proof.   To prove the theorem, dynamic programming will be used. We start from the end point and iterate backwards in time. See Fig. 11.1. Introduce

$$V_k = \min_{u(k),\ldots,u(N-1)}\left(\sum_{i=k}^{N-1}\left(x^T(i)Q_1x(i) + u^T(i)Q_2u(i) + 2x^T(i)Q_{12}u(i)\right)\right.$$
$$\left. + x^T(N)Q_0x(N)\right)$$

V_k can be interpreted as the loss from k to N (loss-to-go) and is a function of the state $x(k)$ at time k. For $k = N$ we have

$$V_N = x^T(N)S(N)x(N)$$

where

$$S(N) = Q_0$$

We will now show that V_k will be quadratic in $x(k)$ for all k. For $k = N - 1$,

$$
\begin{aligned}
V_{N-1} = \min_{u(N-1)} \Big(x^T(N-1)Q_1 x(N-1) + u^T(N-1)Q_2 u(N-1) \\
+ 2x^T(N-1)Q_{12}u(N-1) + V_N \Big)
\end{aligned}
\tag{11.20}
$$

Using (11.16) for $k = N - 1$ gives

$$
\begin{aligned}
V_{N-1} &= \min_{u(N-1)} \Big(x^T(N-1)Q_1 x(N-1) + u^T(N-1)Q_2 u(N-1) \\
&\quad + 2x^T(N-1)Q_{12}u(N-1) \\
&\quad + \Big(\Phi x(N-1) + \Gamma u(N-1) \Big)^T S(N)\Big(\Phi x(N-1) + \Gamma u(N-1) \Big) \Big) \\
&= \min_{u(N-1)} \Big(x^T(N-1)\Big(Q_1 + \Phi^T S(N)\Phi\Big) x(N-1) \\
&\quad + x^T(N-1)\Big(\Phi^T S(N)\Gamma + Q_{12}\Big) u(N-1) \\
&\quad + u^T(N-1)\Big(\Gamma^T S(N)\Phi + Q_{12}^T\Big) x(N-1) \\
&\quad + u^T(N-1)\Big(\Gamma^T S(N)\Gamma + Q_2\Big) u(N-1) \Big) \\
&= \min_{u(N-1)} \begin{pmatrix} x^T(N-1) & u^T(N-1) \end{pmatrix} \\
&\quad \times \begin{pmatrix} Q_1 + \Phi^T S(N)\Phi & \Gamma^T S(N)\Phi + Q_{12}^T \\ \Phi^T S(N)\Gamma + Q_{12} & \Gamma^T S(N)\Gamma + Q_2 \end{pmatrix} \begin{pmatrix} x(N-1) \\ u(N-1) \end{pmatrix}
\end{aligned}
$$

This is a function that is quadratic in $u(N-1)$. By using (11.14) and (11.15), the control law

$$u(N-1) = -L(N-1)x(N-1)$$

gives the minimum loss

$$V_{N-1} = x^T(N-1)S(N-1)x(N-1)$$

which is quadratic in $x(N-1)$ and where

$$S(N-1) = \Phi^T S(N)\Phi + Q_1 - L^T(N-1)\Big(Q_2 + \Gamma^T S(N)\Gamma\Big)L(N-1)$$

and

$$L(N-1) = \left(Q_2 + \Gamma^T S(N)\Gamma\right)^{-1}\left(\Gamma^T S(N)\Phi + Q_{12}^T\right)$$

Because V_{N-1} is positive semidefinite, so is its minimum, that is, $S(N-1)$ is positive semidefinite. Dynamic programming now gives

$$V_{N-2} = \min_{u(N-2),u(N-1)}\left(\sum_{i=N-2}^{N-1}\left(x^T(i)Q_1x(i) + u^T(i)Q_2u(i)\right.\right.$$

$$\left.\left. + 2x^T(i)Q_{12}u(i)\right) + x^T(N)Q_0x(N)\right)$$

$$= \min_{u(N-2)}\left(x^T(N-2)Q_1x(N-2) + u^T(N-2)Q_2u(N-2)\right.$$

$$\left. + 2x^T(N-2)Q_{12}u(N-2) + V_{N-1}\right)$$

This is the same as (11.20), but with the time arguments shifted one step. The procedure can now be repeated, and $V_0 = x^T(0)S(0)x(0)$, which is the minimum of J, is obtained by iterating backward in time. This proves (11.17) to (11.19).
It also follows that (11.17) can be written as

$$S(k) = \left(\Phi - \Gamma L(k)\right)^T S(k+1)\left(\Phi - \Gamma L(k)\right) + \left(I \quad -L(k)^T\right)Q\left(\begin{matrix} I \\ -L(k) \end{matrix}\right)$$

(11.21)

This implies that $S(k)$ is positive semidefinite if $S(N) = Q_0$ is positive semidefinite.   ∎

Remark 1.   Notice that it is not assumed that Q_2 be positive definite, only that $Q_2 + \Gamma^T S(k)\Gamma$ is positive definite.

Remark 2.   The calculations needed to determine the LQ-controller can be made by hand only for very simple examples. In practice it is necessary to have access to interactive programs, which can compute the control law and simulate the systems.

The Riccati Equation

Equation (11.17) is called the *discrete-time Riccati equation*. It is possible to use the Riccati equation to rewrite the loss function of (11.9), which gives the following theorem.

THEOREM 11.2   DISCRETE-TIME RICCATI EQUATION   Assume that the Riccati equation of (11.17) has a solution that is nonnegative definite in the interval

$0 \le k \le N$; then

$$x^T(N)Q_0x(N) + \sum_{k=0}^{N-1}\left(x^T(k)Q_1x(k) + u^T(k)Q_2u(k) + 2x^T(k)Q_{12}u(k)\right)$$

$$= x^T(0)S(0)x(0) + \sum_{k=0}^{N-1}\left(u(k) + L(k)x(k)\right)^T$$

$$\times \left(\Gamma^TS(k+1)\Gamma + Q_2\right)\left(u(k) + L(k)x(k)\right)$$

$$+ \sum_{k=0}^{N-1}\left(v^T(k)S(k+1)\left(\Phi x(k) + \Gamma u(k)\right) + \left(\Phi x(k) + \Gamma u(k)\right)^T S(k+1)v(k)\right)$$

$$+ \sum_{k=0}^{N-1}v^T(k)S(k+1)v(k) \tag{11.22}$$

$$= x^T(0)S(0)x(0) + \sum_{k=0}^{N-1}\left(u(k) + L(k)x(k) + L_v(k)v(k)\right)^T$$

$$\times \left(\Gamma^TS(k+1)\Gamma + Q_2\right)\left(u(k) + L(k)x(k) + L_v(k)v(k)\right)$$

$$+ \sum_{k=0}^{N-1}v^T(k)\left(S(k+1) - L_v^T(k)\left(\Gamma^TS(k+1)\Gamma + Q_2\right)L_v(k)\right)v(k)$$

$$+ \sum_{k=0}^{N-1}v^T(k)S(k+1)\left(\Phi - \Gamma L(k)\right)x(k)$$

$$+ \sum_{k=0}^{N-1}x^T(k)\left(\Phi - \Gamma L(k)\right)^T S(k+1)v(k) \tag{11.23}$$

where $L(k)$ is defined by (11.19) and

$$L_v(k) = \left(\Gamma^TS(k+1)\Gamma + Q_2\right)^{-1}\Gamma^TS(k+1) \tag{11.24}$$

and $x(k+1)$ is given by (11.3).

Proof.   We have the identity

$$x^T(N)Q_0x(N) = x^T(N)S(N)x(N)$$

$$= x^T(0)S(0)x(0) + \sum_{k=0}^{N-1}\left(x^T(k+1)S(k+1)x(k+1) - x^T(k)S(k)x(k)\right)$$

$$\tag{11.25}$$

Consider the different terms in the sum and use (11.3) and (11.17). Then

$$x^T(k+1)S(k+1)x(k+1)$$
$$= \Big(\Phi x(k) + \Gamma u(k) + v(k)\Big)^T S(k+1)\Big(\Phi x(k) + \Gamma u(k) + v(k)\Big) \quad (11.26)$$

and

$$x^T(k)S(k)x(k) = x^T(k)\Big(\Phi^T S(k+1)\Phi + Q_1$$
$$- L^T(k)(\Gamma^T S(k+1)\Gamma + Q_2)L(k)\Big)x(k) \quad (11.27)$$

Introducing (11.26) and (11.27) in (11.25) gives

$$x^T(N)Q_0 x(N) = x^T(0)S(0)x(0) + \sum_{k=0}^{N-1}\Big[\Big(\Phi x(k) + \Gamma u(k)\Big)^T S(k+1)v(k)$$
$$+ v^T(k)S(k+1)\Big(\Phi x(k) + \Gamma u(k)\Big) + v^T(k)S(k+1)v(k)\Big]$$
$$+ \sum_{k=0}^{N-1}\Big[u^T(k)\Big(\Gamma^T S(k+1)\Gamma + Q_2\Big)u(k)$$
$$+ u^T(k)\Big(\Gamma^T S(k+1)\Phi + Q_{12}^T\Big)x(k)$$
$$+ x^T(k)\Big(\Phi^T S(k+1)\Gamma + Q_{12}\Big)u(k)$$
$$+ x^T(k)L^T(k)\Big(\Gamma^T S(k+1)\Gamma + Q_2\Big)L(k)x(k) - x^T(k)Q_1 x(k)$$
$$- u^T(k)Q_2 u(k) - u(k)^T Q_{12}^T x(k) - x^T(k)Q_{12}u(k)\Big]$$

where the terms $u^T Q_2 u$, $u(k)^T Q_{12}x(k)$, and $x^T(k)Q_{12}u(k)$ have been added and subtracted in the last sum. Rearrangement of the terms using (11.19) gives (11.22). To show the second equality use (11.24) and insert that in (11.22). Rearrangement of the terms gives (11.23) and completes the proof. ∎

Mean Value of a Quadratic Form

In the following, expressions of the form

$$Ex^T Sx$$

will be evaluated, where x is a Gaussian random variable with mean m and covariance matrix R. We have

$$Ex^T Sx = E(x-m)^T S(x-m) + Em^T Sx + Ex^T Sm - Em^T Sm$$
$$= E(x-m)^T S(x-m) + m^T Sm$$

Further,

$$\mathrm{E}(x - m)^T S(x - m) = \mathrm{E} \operatorname{tr} (x - m)^T S(x - m) = \mathrm{E} \operatorname{tr} S(x - m)(x - m)^T$$
$$= \operatorname{tr} S\mathrm{E}(x - m)(x - m)^T = \operatorname{tr} SR$$

Thus

$$\mathrm{E} x^T S x = m^T S m + \operatorname{tr} SR \tag{11.28}$$

Complete State Information

Assume that $v(k) \equiv 0$ in (11.3) but that the initial state is uncertain. Theorem 11.2 gives

$$J = \mathrm{E} \left(\sum_{k=0}^{N-1} \left(x^T(k) Q_1 x(k) + u^T(k) Q_2 u(k) + 2x^T(k) Q_{12} u(k) \right) + x^T(N) Q_0 x(N) \right)$$

$$= \mathrm{E} \left(x^T(0) S(0) x(0) \right)$$

$$+ \mathrm{E} \left(\sum_{k=0}^{N-1} \left(u(k) + L(k) x(k) \right)^T \left(\Gamma^T S(k+1) \Gamma + Q_2 \right) \left(u(k) + L(k) x(k) \right) \right)$$

Because $S(k)$ is positive semidefinite, the second term is nonnegative. Further, $S(k)$ is independent of $u(k)$, and it follows that

$$J_{\text{complete}} \geq \mathrm{E} x^T(0) S(0) x(0) = m_0^T S(0) m_0 + \operatorname{tr} S(0) R_0 \tag{11.29}$$

where (11.28) has been used. Equality is obtained for the control law of (11.18). Theorem 11.2 and (11.29) give an alternative way to prove Theorem 11.1.

Now assume that there are stochastic disturbances acting on the system and that the full state is still measurable. Using Theorem 11.2, (11.22), and that $v(k)$ is independent of $u(k)$ and $x(k)$ gives

$$J = \mathrm{E} \left(x^T(0) S(0) x(0) + \sum_{k=0}^{N-1} v^T(k) S(k+1) v(k) \right.$$

$$\left. + \sum_{k=0}^{N-1} \left(u(k) + L(k) x(k) \right)^T \left(\Gamma^T S(k+1) \Gamma + Q_2 \right) \left(u(k) + L(k) x(k) \right) \right)$$
$$\tag{11.30}$$

Using (11.28) gives the relationship

$$J_{\text{noise}} \geq m_0^T S(0) m_0 + \operatorname{tr} S(0) R_0 + \sum_{k=0}^{N-1} \operatorname{tr} S(k+1) R_1 \tag{11.31}$$

Equality is obtained for the control law of (11.18), which is an admissible control law. The difference in the optimal costs of (11.29) and (11.31) is due to the disturbance $v(k)$. The control law of (11.18) thus minimizes the loss for the complete state information case.

Assume on the other hand that $v(k)$ is known when determining $u(k)$. From (11.23) it follows that the loss function is minimized for

$$u(k) = -L(k)x(k) - L_v(k)v(k) \tag{11.32}$$

where L_v is given by (11.24) and the minimum loss is

$$
J = m_0^T S(0)m_0 + \text{tr } S(0)R_0 + \sum_{k=0}^{N-1} \text{tr } S(k+1)R_1
$$
$$
- \sum_{k=0}^{N-1} \text{tr } L_v(k)R_1 L_v^T(k)\left(\Gamma^T S(k+1)\Gamma + Q_2\right)
$$

This loss is less than (11.31) and shows the improved performance if $v(k)$ could be used.

The solution to the LQ-problem gives a time-varying controller. The feedback matrix does not depend on x and can be precomputed from $k = N$ to $k = 0$ and stored in the computer. For time-invariant processes and loss functions, usually only the stationary controller—the constant controller obtained when the Riccati equation is iterated until a constant S is obtained—is used. $S(k)$ will—under quite general assumptions—converge to a constant matrix as the time horizon increases. In general, there exist several solutions resulting from different Q_0.

The stationary solution can be obtained by iterating (11.17) or by solving the algebraic Riccati equation

$$\bar{S} = \Phi^T \bar{S}\Phi + Q_1 - \left(\Phi^T \bar{S}\Gamma + Q_{12}\right)\left(\Gamma^T \bar{S}\Gamma + Q_2\right)^{-1}\left(\Gamma^T \bar{S}\Phi + Q_{12}^T\right) \tag{11.33}$$

Because Q in (11.10) is symmetric and positive semidefinite we can write

$$Q = \begin{pmatrix} C_l & D_l \end{pmatrix}^T \begin{pmatrix} C_l & D_l \end{pmatrix}$$

If the system of (11.16) is reachable and if

$$\begin{pmatrix} -zI + \Phi & \Gamma \\ C_l & D_l \end{pmatrix}$$

has full column rank for $|z| \geq 1$, that is, there are no unstable zeros to the system defined by Φ, Γ, C_l, and D_l, then there exists only one symmetric nonnegative definite solution to the algebraic Riccati equation (11.33).

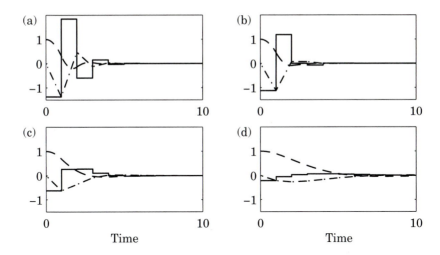

Figure 11.2 Linear quadratic control of the double-integrator plant for different weightings, ρ, on the control signal. The initial value of the state is $x(0) = [1 \quad 0]$. The position x_1 (dashed), velocity x_2 (dashed-dotted), and the control signal u (solid) are shown. (a) $\rho = 0.01563$, (b) $\rho = 0.05$, (c) $\rho = 0.5$, and (d) $\rho = 10$.

Example 11.1   LQ-control of the double integrator

Consider the double integrator (see Example A.1) and use the sampling period $h = 1$. Let the weighting matrices in (11.9) be

$$Q_1 = \begin{pmatrix} 1 & 0 \\ 0 & 0 \end{pmatrix} \quad \text{and} \quad Q_2 = \begin{pmatrix} \rho \end{pmatrix}$$

The influence of the weighting can now be investigated. The stationary feedback vector has been calculated for different values of ρ. Figure 11.2 shows the states and the control signal for some values. When $\rho = 0$, which means there is a penalty only on the output, then the resulting controller is the same as the deadbeat controller in Sec. 4.3. When ρ is increased, then the magnitude of the control signal is decreased.

Figure 11.3 shows the stationary L vector as a function of the control weighting ρ. When ρ increases the gains go to zero and there will be almost no feedback.

■

Example 11.2   Time-varying controller

Consider the integrator process

$$x(k + 1) = x(k) + u(k)$$

Let the loss function be

$$\sum_{k=0}^{4} \left(x^2(k) + 10u^2(k) \right) + q_0 x^2(5)$$

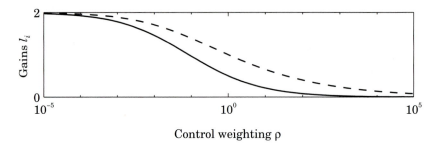

Figure 11.3 Linear quadratic controller for the double integrator. The stationary gains l_1 (solid) and l_2 (dashed) of the feedback vector $L = [l_1, \; l_2]$ for different values of ρ.

That is, the time horizon is only five steps. The Riccati equation and the controller gain become

$$s(k) = s(k + 1) + 1 - \frac{s^2(k + 1)}{s(k + 1) + 10} \qquad\qquad s(5) = q_0$$

$$l(k) = \frac{s(k + 1)}{s(k + 1) + 10}$$

Figure 11.4 shows $s(k)$, $l(k)$, and the trajectory of the state when $x(0) = 1$ for different values of q_0. The value $q_0 = 3.70$ corresponds to the stationary solution of the Riccati equation. When q_0 is increasing $x(5)$ approaches zero.   ∎

Properties of the LQ-Controller

The pole-placement controller in Sec. 4.3 and the stationary LQ-controller have the same structure. However, they are obtained differently, so there are some differences in their properties.

The linear state-feedback controller of (11.18) has n parameters in the single-input case. It is, in general, difficult to tune the parameters directly such that a good performance of the closed-loop system is obtained. Instead, the tuning procedure can be to choose the n eigenvalues of the closed-loop system and use the design procedure in Sec. 4.3. This procedure is well suited for single-input–single-output systems. It is, however, difficult to compromise between the speed of the system and the magnitude of the control signal.

The LQ-controller has several good properties. It is applicable to multivariable and time-varying systems. Also, changing the relative magnitude between the elements in the weighting matrices means a compromise between the speed of the recovery and the magnitudes of the control signals. The following two theorems give properties of the closed-loop system when using the LQ-controller.

THEOREM 11.3 STABILITY OF THE CLOSED-LOOP SYSTEM   Let the system of (11.16) be time-invariant and let the loss function of (11.9) be such that Q in

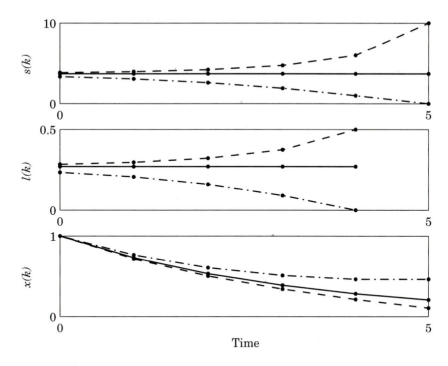

Figure 11.4 Simulation of the process in Example 11.2 for different values of the weighting at the end point; $q_0 = 10$ (dashed), $q_0 = 3.70$ (solid), and $q_0 = 0$ (dashed-dotted).

(11.10) is positive definite. Assume that a positive-definite steady-state solution, $\bar{S}$, to (11.33) exists. Then the steady-state optimal-control strategy

$$u(k) = -Lx(k) = -\left(Q_2 + \Gamma^T \bar{S} \Gamma\right)^{-1} \left(\Gamma^T \bar{S} \Phi + Q_{12}^T\right)x(k)$$

gives an asymptotically stable closed-loop system

$$x(k+1) = (\Phi - \Gamma L)x(k)$$

Proof.   Theorem 3.4 can be used to show that the closed-loop system is asymptotically stable. It is to be shown that the function

$$V\big(x(k)\big) = x^T(k)\bar{S}x(k)$$

is a Lyapunov function. V is positive definite and

$$\begin{aligned}
\Delta V\big(x(k)\big) &= x^T(k+1)\bar{S}x(k+1) - x^T(k)\bar{S}x(k) \\
&= x^T(k)(\Phi - \Gamma L)^T \bar{S}(\Phi - \Gamma L)x(k) - x^T(k)\bar{S}x(k) \\
&= -x^T(k)\big(Q_1 + L^T Q_2 L - L^T Q_{12}^T - Q_{12}L\big)x(k) \\
&= -x^T(k)\begin{pmatrix} I & -L^T \end{pmatrix} Q \begin{pmatrix} I \\ -L \end{pmatrix} x(k)
\end{aligned}$$

Example 11.4   Ship steering

The linearized dynamics that describe the steering of ships can be described by the equation

$$\frac{d}{dt}\begin{pmatrix} v \\ r \\ \Psi \end{pmatrix} = \begin{pmatrix} a_{11} & a_{12} & 0 \\ a_{21} & a_{22} & 0 \\ 0 & 1 & 0 \end{pmatrix} \begin{pmatrix} v \\ r \\ \Psi \end{pmatrix} + \begin{pmatrix} b_1 \\ b_2 \\ 0 \end{pmatrix} \delta \qquad (11.41)$$

where δ is rudder angle, Ψ is the heading angle, r is the turning rate, and v the sway velocity. The relative increase in the drag due to steering may be approximated by the expression

$$\frac{\Delta R}{R} = \frac{\alpha}{T} \int_0^T (vr + \rho \delta^2)\, dt \qquad (11.42)$$

The first term represents the Coriolis force due to coupling of sway velocity and turning rate. The second term represents the drag induced by the rudder deflections.   ■

In many cases it is difficult to find natural quadratic loss functions. LQ-control theory has found considerable use even when this cannot be done. In such cases the control designer chooses a loss function. The feedback law is obtained directly by solving the Riccati equation. The closed-loop system obtained is then analyzed with respect to transient response, frequency response, robustness, and so on. The elements of the loss function are modified until the desired result is obtained. Such a procedure may seem like a strange use of optimization theory.

The fact that other methods, such as direct search over the feedback gain or pole placement, are not used instead might be questioned. It has been found empirically that LQ-theory is quite easy to use in this way. The search will automatically guarantee stable closed-loop systems with reasonable margins.

It is often fairly easy to see how the weighting matrices should be chosen to influence the properties of the closed-loop system. Variables x_i, which correspond to significant physical variables, are chosen first. The loss function is then chosen as a weighted sum of x_i. Large weights correspond to small responses. The responses of the closed-loop system to typical disturbances are then evaluated. A particular difficulty is to find the relative weights between state variables and control variables, which can be done by trial and error.

Sometimes the specifications are given in terms of the maximum allowed deviations in the states and the control signals for a given disturbance. One rule of thumb to decide the weights in (11.4) is to choose the diagonal elements as the inverse value of the square of the allowed deviations. Another way is to consider only penalties on the state variables and constraints on the control deviations. If the constraints are quadratic, a method using a Lagrange multiplier gives a criterion such as (11.9).

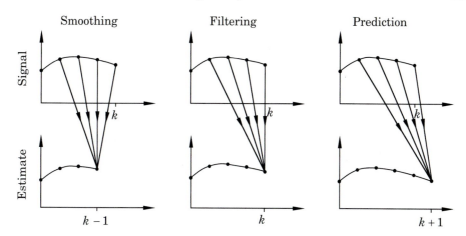

Figure 11.6   Smoothing, filtering, and prediction.

11.3 Prediction and Filtering Theory

When using the LQ-controller, the full-state vector must be measurable. The problem of estimating the states of (11.3) from measurements of the output is discussed in this section. An estimator of the same structure as in Sec. 4.4 is postulated, but the gain vector is now determined differently. The problem is solved as a parametric optimization problem, where the variance of the estimation error is minimized.

Prediction, Filtering, and Smoothing

Different estimators for the states in (11.3) can be derived depending on the available measurements. Assume that the data

$$Y_k = \{ y(i),\, u(i) \mid i \le k \}$$

is known. Using Y_k we want to estimate $x(k + m)$. We have three cases:

- Smoothing $(m < 0)$
- Filtering $(m = 0)$
- Prediction $(m > 0)$

Figure 11.6 illustrates the different cases. In this section the prediction and filtering problems are discussed. The resulting dynamic system is called a filter regardless of which of the problems is solved.

The Kalman Filter

Let the process be described by (11.3) with $h = 1$. Postulate an one-step-ahead estimator of the form

$$\hat{x}(k + 1 \mid k) = \Phi \hat{x}(k \mid k - 1) + \Gamma u(k) + K(k)\Big(y(k) - C\hat{x}(k \mid k - 1)\Big) \qquad (11.43)$$

The reconstruction error $\tilde{x} = x - \hat{x}$ is governed by

$$\tilde{x}(k+1) = \Phi\tilde{x}(k) + v(k) - K(k)\big(y(k) - C\hat{x}(k \mid k-1)\big)$$

$$= \Big(\Phi - K(k)C\Big)\tilde{x}(k) + v(k) - K(k)e(k) \qquad (11.44)$$

$$= \Big(I \quad -K(k)\Big)\left(\begin{bmatrix} \Phi \\ C \end{bmatrix}\tilde{x}(k) + \begin{pmatrix} v(k) \\ e(k) \end{pmatrix}\right)$$

In Sec. 4.4 K is used to give the system of (11.44) desired eigenvalues. The problem is approached differently here: The properties of the noise are taken into account and the criterion is to minimize the variance of the estimation error, which is denoted by $P(k)$.

$$P(k) = \mathrm{E}\,\big(\tilde{x}(k) - \mathrm{E}\tilde{x}(k)\big)\big(\tilde{x}(k) - \mathrm{E}\tilde{x}(k)\big)^{T}$$

The mean value of $\tilde{x}$ is obtained from (11.44)

$$\mathrm{E}\,\tilde{x}(k+1) = \Big(\Phi - K(k)C\Big)\mathrm{E}\,\tilde{x}(k)$$

Because $\mathrm{E}\,x(0) = m_0$, the mean value of the reconstruction error is zero for all times $k \geq 0$ independent of K if $\mathrm{E}\,\hat{x}(0) = m_0$. Because $\tilde{x}(k)$ is independent of $v(k)$ and $e(k)$ Eq. (11.44) now gives

$$P(k+1) = \mathrm{E}\tilde{x}(k+1)\tilde{x}(k+1)^{T}$$

$$= \Big(I \quad -K(k)\Big)\left(\begin{bmatrix} \Phi \\ C \end{bmatrix} P(k) \begin{bmatrix} \Phi \\ C \end{bmatrix}^{T} + \begin{bmatrix} R_1 & R_{12} \\ R_{12}^{T} & R_2 \end{bmatrix}\right)\begin{pmatrix} I \\ -K^{T}(k) \end{pmatrix}$$

$$= \Big(I \quad -K(k)\Big)\begin{bmatrix} \Phi P(k)\Phi^{T} & \Phi P(k)C^{T} + R_{12} \\ CP(k)\Phi^{T} + R_{12}^{T} & CP(k)C^{T} + R_2 \end{bmatrix}\begin{pmatrix} I \\ -K^{T}(k) \end{pmatrix}$$
$$(11.45)$$

Further, $P(0) = R_0$. From (11.45) it follows that if $P(k)$ is positive semidefinite, then $P(k+1)$ is also positive semidefinite. Equation (11.45) has the same form as (11.11) and should be minimized with respect to $K(k)$. By using the idea of completion of squares, it follows that $\alpha^{T}P(k+1)\alpha$ is minimized by $K(k)$ satisfying

$$K(k)\Big(R_2 + CP(k)C^{T}\Big) = \Phi P(k)C^{T} + R_{12}$$

for any α. If $CP(k)C^{T} + R_2$ is positive definite then

$$K(k) = \Big(\Phi P(k)C^{T} + R_{12}\Big)\Big(R_2 + CP(k)C^{T}\Big)^{-1} \qquad (11.46)$$

This inserted into (11.45) or using (11.15) gives

$$P(k + 1) = \Phi P(k)\Phi^T + R_1$$
$$- \left(\Phi P(k)C^T + R_{12}\right)\left(R_2 + CP(k)C^T\right)^{-1}\left(CP(k)\Phi^T + R_{12}^T\right)$$
$$(11.47)$$

$$P(0) = R_0$$

The reconstruction defined by (11.43), (11.46), and (11.47) is called the *Kalman filter*. This is summarized in the following theorem.

THEOREM 11.5   THE KALMAN FILTER—PREDICTOR CASE   Consider the process of (11.3). The reconstruction of the states using the model in (11.43) is optimal in the sense that the variance of the reconstruction error is minimized if the matrix $R_2 + CP(k)C^T$ is positive definite and if the gain matrix is chosen according to (11.46) and (11.47). The variance of the reconstructing error is given by (11.47). ■

Remark 1.   The reconstruction problem has been solved as a parametric optimization problem by assuming the structure in (11.43) of the estimator. It is in fact true that the structure is optimal for Gaussian disturbances.

Remark 2.   Better than the traditional notation for the variance $P(k)$ is $P(k \mid k - 1)$. The latter notation indicates that measurements up to and including time $k - 1$ are used. The different terms in the variance equation of (11.47) can be interpreted in the following way: The term $\Phi P\Phi^T$ shows how the variance is changed due to the system dynamics, and R_1 represents the increase in the variance due to the noise v [compare with (10.11)]. The last term shows how the variance is decreased due to the information obtained through the measurements. Notice that $P(k)$ does not depend on the observations. Thus the gain can be precomputed in forward time and stored in the computer.

Remark 3.   The Kalman filter can also be interpreted as the conditional mean of the state at time $k + 1$ given Y_k; that is,

$$\hat{x}(k + 1 \mid k) = \mathrm{E}\big(x(k + 1) \mid Y_k\big)$$
$$P(k + 1) = \mathrm{E}\left[\big(x(k + 1) - \hat{x}(k + 1 \mid k)\big)\big(x(k + 1) - \hat{x}(k + 1 \mid k)\big)^T \mid Y_k\right]$$

Example 11.5   Kalman filter for a first order system
   Consider the scalar system

$$x(k + 1) = x(k)$$
$$y(k) = x(k) + e(k)$$

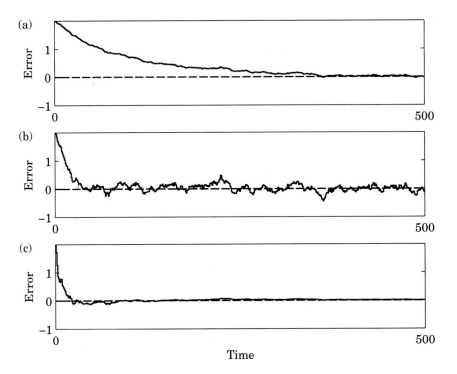

Figure 11.7 Estimation error for the system in Example 11.5 when starting from $x(0) = -2$, and when using $\sigma = 1$ and (a) $K = 0.01$, (b) $K = 0.05$, and (c) the optimal gain of (11.49).

where e has standard deviation σ and $x(0)$ has the variance 0.5. The state is thus constant and has to be reconstructed from noisy measurements. The Kalman filter is given by

$$\hat{x}(k+1 \mid k) = \hat{x}(k \mid k-1) + K(k)\Big(y(k) - \hat{x}(k \mid k-1)\Big) \qquad (11.48)$$

$$K(k) = \frac{P(k)}{\sigma^2 + P(k)} \quad \text{and} \quad P(k+1) = \frac{\sigma^2 P(k)}{\sigma^2 + P(k)} \qquad (11.49)$$

The variance and the gain are decreasing with time. Figure 11.7 shows realizations of the estimation error when the Kalman filter is used and when (11.48) is used with constant gain. A large fixed gain gives a rapid initial decrease in the error, while the steady-state variance is large. A small fixed gain gives a slow decrease in the error, but a better performance in steady state.   ∎

The Filter Problem

The predictor in (11.43) has the property that the state at time k is reconstructed from $y(k-1), y(k-2), \ldots$. It is also possible to derive the filter, which also uses $y(k)$, to estimate $x(k)$. In the filter case $y(k)$ will contain information about $v(k)$, which will be reflected in the equations that follow.

THEOREM 11.6   KALMAN FILTER–FILTER CASE   Consider the process (11.3) and let Y_k be available for the estimation of $x(k)$. If the matrix $R_2 + CP(k \mid k - 1)C^T$ is positive definite then the optimal filter is given by the following equations:

$$\hat{x}(k \mid k) = \hat{x}(k \mid k - 1) + K_f(k)\Big(y(k) - C\hat{x}(k \mid k - 1)\Big)$$

$$\hat{v}(k \mid k) = K_v(k)\Big(y(k) - C\hat{x}(k \mid k - 1)\Big)$$

$$\hat{x}(k + 1 \mid k) = \Phi\hat{x}(k \mid k) + \Gamma u(k) + \hat{v}(k \mid k) \qquad (11.50)$$

$$= \Phi\hat{x}(k \mid k - 1) + \Gamma u(k) + K(k)\Big(y(k) - C\hat{x}(k \mid k - 1)\Big)$$

where

$$K_f(k) = P(k \mid k - 1)C^T\Big(CP(k \mid k - 1)C^T + R_2\Big)^{-1} \qquad (11.51)$$

$$K_v(k) = R_{12}\Big(CP(k \mid k - 1)C^T + R_2\Big)^{-1} \qquad (11.52)$$

$$K(k) = \Phi K_f(k) + K_v(k)$$

$$= \Big(\Phi P(k \mid k - 1)C^T + R_{12}\Big)\Big(CP(k \mid k - 1)C^T + R_2\Big)^{-1} \qquad (11.53)$$

The variance is given by the Riccati equation

$$P(k + 1 \mid k) = \Phi P(k \mid k - 1)\Phi^T + R_1$$

$$- K(k)\Big(CP(k \mid k - 1)C^T + R_2\Big)K^T(k)$$

$$P(k \mid k) = P(k \mid k - 1)$$

$$- P(k \mid k - 1)C^T\Big(CP(k \mid k - 1)C^T + R_2\Big)^{-1}CP(k \mid k - 1)$$

$$P(0 \mid -1) = R_0$$

$$(11.54)$$

Proof.   The proof is based on expressions analogous to (11.45).   ∎

Remark 1.   The notation $P(k \mid k - 1)$ is used here instead of $P(k)$ to specify the available data; $P(k \mid k)$ is the variance of the estimation error at time k given Y_k.

Remark 2.   Notice that the expression for $\hat{x}(k + 1 \mid k)$ in (11.50) is the same as in (11.43).

Remark 3.   Notice that $\hat{v}(k + 1 \mid k) = 0$ because $y(k)$ does not contain any information about $v(k + 1)$.

11.8 Notes and References

LQG-control and optimal filters are the subjects of many textbooks, for instance, Athans and Falb (1966), Bryson and Ho (1969), Åström (1970), Andersson and Moore (1971, 1979, 1990), Kwakernaak and Sivan (1972), and Kučera (1991). The principle of optimality and dynamic programming are discussed, for instance, in Bellman (1957, 1961).

Kalman and Bucy made the main contributions to the development of the recursive optimal filters discussed in Sec. 11.3. See Bucy (1959), Kalman (1960b), and Kalman and Bucy (1961).

A good source for properties of the discrete-time as well as continuous-time Riccati equations is Bittanti, Laub, and Willems (1991). Numerical algorithms for solving the Riccati equation are also discussed, for instance, in Kleinman (1968), Biermann (1977), Pappas, Laub, and Sandell (1980), Van Dooren (1981), Arnold III and Laub (1984), and Benner, Laub, and Mehrmann (1995). The Euler equation is discussed, for instance, in Emami-Naeini and Franklin (1980), Arnold III and Laub (1984), and Hagander and Hansson (1996).

Choice of the sampling interval for LQ-controllers is discussed in Åström (1963), Melzer and Kuo (1971), and Lennartson (1987).

The separation theorem Theorem 11.7 appeared first in economic literature: Simon (1956). Discrete-time versions of the separation theorem can be found in Gunkel and Franklin (1963).

Gain margin for discrete-time LQ-controllers is discussed in Willems and Van De Voorde (1978) and Safonov (1980). Robustness of LQG controllers is discussed in Doyle and Stein (1981).

Many of the modifications for the third edition of the book are based on Gustafsson and Hagander (1991). The cross terms in both the loss function ($Q_{12} \neq 0$) and the Kalman filter ($R_{12} \neq 0$) are also discussed in Kwong (1991).

12

Optimal Design Methods:
A Polynomial Approach

12.1 Introduction

Optimal design methods based on input-output models are considered in this chapter. Design of regulators based on linear models and quadratic criteria is discussed. This is one class of problems that admits closed-form solutions. The problems are solved by other methods in Chapter 11. The input-output approach gives additional insight and different numerical algorithms are also obtained.

The problem formulation is given in Sec. 12.2. This includes discussion of models for dynamics, disturbances, and criteria, as well as specification of admissible controls. The model is given in terms of three polynomials. A very simple example is also solved using first principles. This example shows clearly that optimal control and optimal filtering problems are closely connected. The prediction problem is then solved in Sec. 12.3. The solution is easily obtained by polynomial division. A simple explicit formula for the transfer function of the optimal predictor is given.

The minimum-variance control law is derived in Sec. 12.4. For systems with stable inverses, the control law is obtained in terms of the polynomials that characterize the optimal predictor. For systems with unstable inverses, the solution is obtained by solving a Diophantine equation in polynomials of the type discussed in Chapter 5. The minimum-variance control problem may thus be interpreted as a pole-placement problem. This gives insight into suitable choices of closed-loop poles and observer poles for the pole-placement problem. The LQG-control problem is solved in Sec. 12.5. It is shown that the solution may be expressed in terms of spectral factorization and solution of a Diophantine equation. Practical aspects, such as selection of the sampling period, are given in Sec. 12.6.

12.2 Problem Formulation

It is assumed that the process to be controlled is linear and time-invariant and that it has one input u and one output y. The dynamics of the process are characterized by a combination of a time-delay and a rational-transfer function. It is also assumed that the disturbances may be described as filtered white noise. A steady-state regulation problem is considered. The criterion is based on the mean-square deviations of the control signal and the output signal. In the formal problem statement given next, it is assumed that the model and the criterion are sampled [compare with Sec. 2.3 and (11.1)].

Process Dynamics

Assume that the process dynamics are characterized by

$$x(k) = \frac{B_1(q)}{A_1(q)} u(k) \tag{12.1}$$

where $A_1(q)$ and $B_1(q)$ are polynomials in the forward-shift operator.

Disturbances

Assume that the influence of the environment on the process can be characterized by disturbances that are stochastic processes. Because the system is linear, the principle of superposition can be used to reduce all disturbances to an equivalent disturbance v at the system output. The output of the system is thus given by

$$y(k) = x(k) + v(k) \tag{12.2}$$

Further assume that the disturbance v may be represented as the output of a linear system driven by white noise—that is,

$$v(k) = \frac{C_1(q)}{A_2(q)} e(k) \tag{12.3}$$

where $C_1(q)$ and $A_2(q)$ are polynomials in the forward-shift operator, and $e(k)$ is a sequence of independent or uncorrelated random variables with zero mean and standard deviation σ. The disturbance v may be a stationary random process. It may, however, also be drifting, because the polynomial $A_2(q)$ may be unstable. The model of the process and its environment can be reduced to a standard form. Eliminate v and x among (12.1), (12.2), and (12.3), and introduce

$$\begin{aligned} A &= A_1 A_2 \\ B &= B_1 A_2 \\ C &= C_1 A_1 \end{aligned} \tag{12.4}$$

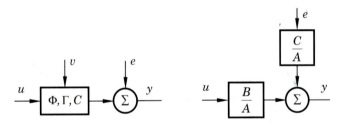

Figure 12.1 Representation of a system with one input and stochastic disturbances using one or two noise sources.

The following model is then obtained.

$$A(q)y(k) = B(q)u(k) + C(q)e(k) \qquad (12.5)$$

This is the canonical model, which will be the basis of the control design. In the special case when there are no disturbances, the model is simply a rational pulse-transfer function (see Sec. 2.6). When there is no control signal, the model is a stochastic process with a rational spectral density or an ARMA process (see Sec. 10.4). The model (12.5) is a convenient canonical representation of a linear system perturbed by noise. In Chapter 11 the process was driven by two noise sources. By using the spectral-factorization theorem (Theorem 10.3) the noise can be reduced to one source. Compare Fig. 12.1.

When the polynomial $C(q)$ has all its zeros inside the unit disc, it is called an *innovation's representation*, because the random variables $e(k)$ represent the innovations of the random process. Notice the symmetry between y and e. If e and u are known up to time k, then $y(k)$ can be computed, and if y and u are known up to time k, the innovation $e(k)$ can also be computed. Notice that the calculations of the residuals are governed by the dynamics of the polynomial $C(q)$. This polynomial can therefore be interpreted as the observer polynomial. Because (12.5) is an innovations model, the solutions to filtering problems become very simple.

Equation (12.5) can be normalized so that the leading coefficients of the polynomials $A(q)$ and $C(q)$ are unity. Such polynomials are called *monic*. The polynomial C may also be multiplied by an arbitrary power of q, as this does not change the correlation structure of $C(q)e(t)$. This may be used to normalize C so that $\deg C = \deg A$. The polynomials $A(q)$ and $B(q)$ may have zeros inside or outside the unit disc. It is assumed that all the zeros of the polynomial $C(q)$ are inside the unit disc. By spectral factorization (Theorem 10.3), the polynomial $C(q)$ may be changed so that all its zeros are inside the unit disc or on the unit circle. An example is used to show this important point.

Example 12.1   Modification of the polynomial C

Consider the polynomial

$$C(z) = z + 2$$

which has the zero $z = -2$ outside the unit disc. Consider the signal

$$n(k) = C(q)e(k)$$

where $e(k)$ is a sequence of uncorrelated random variables with zero mean and unit variance. The spectral density of n is given by

$$\phi(e^{i\omega h}) = \frac{1}{2\pi} C(e^{i\omega h})C(e^{-i\omega h})$$

Because

$$C(z)C(z^{-1}) = (z + 2)(z^{-1} + 2) = (1 + 2z^{-1})(1 + 2z)$$
$$= (2z + 1)(2z^{-1} + 1) = 4(z + 0.5)(z^{-1} + 0.5)$$

the signal n may also be represented as

$$n(k) = C^*(q)e(k)$$

where

$$C^*(z) = 2z + 1$$

is the reciprocal of the polynomial $C(z)$ (see Sec. 2.6).   ∎

If the calculations of (12.4) give a polynomial $C(q)$ that has zeros outside the unit disc, the polynomial C is factored as

$$C = C^+C^-$$

where C^- contains all factors with zeros outside the unit disc. The polynomial C is then replaced by C^+C^{-*}.

Criteria

In steady-state regulation it makes sense to express the criteria in terms of *steady-state variances* of the control variable and the process output. For regulation of systems with one output, the criterion may be to minimize the variance of the output. This is discussed in Sec.6.6 . Also compare with Fig. 6.7. This leads to the criterion

$$J_{mv} = \mathrm{E}y^2(k) \tag{12.6}$$

where it is assumed that the scales are chosen so that $y = 0$ corresponds to the desired set point. A control law that minimizes the criterion (12.6) is called *minimum-variance control*. The criterion may also be expressed as

$$J_\infty = \lim_{N \to \infty} \mathrm{E} \left\{ \frac{1}{N} \sum_{k=1}^{N} y^2(k) \right\}$$

Notice that this criterion is an approximation of the continuous-time loss function

$$J_\infty = \lim_{T \to \infty} \mathrm{E}\left\{ \frac{1}{T} \int_0^T y^2(t)\,dt \right\} \tag{12.7}$$

A more accurate approximation, which takes the behavior of the signals between the sampling instants into account, is given in Sec. 11.1. Some consequences of the approximation are discussed in Sec. 12.6. The properties of the control signal under minimum-variance control depend critically on the sampling period. A short sampling period gives a large variance of the control signal and a long sampling period gives a small variance.

In some cases it is desired to trade variances of control and output signals. This may be done by introducing the loss function

$$J_{lq} = \mathrm{E}\left(y^2(k) + \rho u^2(k) \right) \tag{12.8}$$

The control law that minimizes this criterion is called the *linear quadratic control law*.

Admissible Controls

It is assumed that the control law is such that $u(k)$, that is, the value of the control signal at time k, is a function of $y(k), y(k-1), \ldots$ and $u(k-1), u(k-2), \ldots$. Thus the computational delay is negligible in comparison with the sampling period. It is very easy to modify the results to take delays in the computations into account.

There are two versions of the theory. A linear control law may be postulated. It is then sufficient to assume that the disturbances $e(i)$ and $e(j)$ are uncorrelated for $i \neq j$. If $e(i)$ and $e(j)$ are assumed to be independent, it can be shown that the optimal-control law is linear. The formula for the optimal-control law is the same in both cases.

Minimum-Variance Control: An Example

The optimal-control problem defined by the model of (12.5) and the criterion of (12.6) is solved in a special case. The solution, which is easily obtained from first principles, gives good insight into the assumptions made. It also indicates how the general problem should be solved.

Consider the first-order system

$$y(k+1) + ay(k) = bu(k) + e(k+1) + ce(k) \tag{12.9}$$

where $|c| < 1$ and $e(k)$ is a sequence of independent random variables with unit variance.

Consider the situation at time k. The outputs $y(k), y(k-1), \ldots$ have been observed. The control $u(k)$ should be determined so that the output is as close to zero as possible. It follows from (12.9) that $y(k+1)$ may be changed arbitrarily by a proper choice of $u(k)$. Because $e(k+1)$ is independent of $y(k)$ and of the terms of the right-hand side of (12.9), it follows that

$$\operatorname{var} y(k+1) \geq \operatorname{var} e(k+1) = 1 \tag{12.10}$$

The term $e(k)$ may be computed in terms of the known data $y(k), y(k-1),\ldots$ and $u(k-1), u(k-2),\ldots$. When the variables $y(k)$ and $e(k)$ are known, the control law

$$u(k) = (ay(k) - ce(k))/b \tag{12.11}$$

gives

$$y(k+1) = e(k+1) \tag{12.12}$$

which corresponds to the lower bound in (12.10). If the control law in (12.11) is used in each step, Eq. (12.12) holds for all k. The computation of $e(k)$ from the data available at time k is then trivial and the control law in (12.11) can be written as

$$u(k) = -\frac{c-a}{b} y(k) \tag{12.13}$$

The optimal control is thus a proportional feedback with the gain $(c-a)/b$.

To analyze the properties of the closed-loop system under optimal control, eliminate u between (12.9) and (12.13). This gives

$$y(k+1) + cy(k) = e(k+1) + ce(k)$$

Notice that the closed-loop system has the characteristic polynomial

$$C(z) = z + c$$

This shows the importance of the assumption that the polynomial $C(z)$ is stable. This difference equation has the solution

$$y(k) = e(k) + (-c)^{k-k_0}\left(y(k_0) - e(k_0)\right)$$

Because c is less than one in magnitude, the last term goes to zero as $k - k_0$ increases toward infinity. Thus control law in (12.13) gives the minimum-variance in steady state.

With this result, some observations are possible. The quantity $-ay(k) + bu(k) + ce(k)$ can be interpreted as the best estimate of $y(k+1)$, given the data available at time k. The quantity $e(k+1)$ is the prediction error. The control law in (12.13) implies that the control signal is chosen so that the predicted value is equal to the reference value, which is zero in this case. The control error is then equal to the prediction error. The solution to the minimum-variance control problem is thus closely related to the solution of a prediction problem. Therefore, the prediction problem is solved before the solution of the general minimum-variance control problem is attempted.

12.3 Optimal Prediction

Prediction theory can be stated in many different ways, which differ in the assumptions made on the process, the criterion, and the admissible predictors. One formulation is given in Sec. 11.3. In this section the following assumptions are made:

- The process to be predicted is generated by filtered white Gaussian noise.

- The best predictor is the one that minimizes the mean-square prediction error.

- An admissible m-step predictor for $y(k + m)$ is an arbitrary function of $y(k), y(k - 1), \ldots$.

An intuitive derivation of a predictor is first given. The result is then formalized.

Heuristics

Consider the signal y generated by the model

$$y(k) = \frac{C(q)}{A(q)} e(k) = \frac{C^*(q^{-1})}{A^*(q^{-1})} e(k) \tag{12.14}$$

where A^* and C^* are the reciprocals of A and C, that is, $A^*(q^{-1}) = q^{-n}A(q)$, and q^{-1} is the backward-shift operator. It is convenient to introduce this operator because the discussion is based on causality. It is assumed that A and C are of order n.

Consider the situation at time k. The variables $y(k), y(k - 1), \ldots$ have been observed and it is desired to predict $y(k + m)$. A formal series expansion of C^*/A^* in q^{-1} gives

$$
\begin{aligned}
y(k + m) &= \frac{C^*(q^{-1})}{A^*(q^{-1})} e(k + m) \\
&= \underbrace{e(k + m) + f_1 e(k + m - 1) + \cdots + f_{m-1} e(k + 1)}_{\text{Unknown at time } k} \\
&\quad + \underbrace{f_m e(k) + f_{m+1} e(k - 1) + \cdots}_{\text{Known at time } k}
\end{aligned}
\tag{12.15}
$$

The terms of the right-hand side are all independent because $e(k)$ is a sequence of independent random variables. It follows from the model of (12.14) that if the polynomial C is stable, then $e(i)$ can be computed exactly from $y(i), y(i - 1), \ldots$ using

$$e(k) = \frac{A^*(q^{-1})}{C^*(q^{-1})} y(k)$$

The first terms of (12.15) are independent of the data at time k. The second part is known functions of the data available at time k. Thus it follows that the optimal predictor is given by

$$\hat{y}(k + m \mid k) = f_m e(k) + f_{m+1} e(k-1) + f_{m+2} e(k-2) + \cdots$$

and that the prediction error is

$$\tilde{y}(k + m \mid k) = e(k+m) + f_1 e(k+m-1) + \cdots + f_{m-1} e(k+1)$$

To provide a formal proof it remains to show how the numbers f_i can be computed from A and C and how $e(k)$ can be expressed in terms of past data.

Main Result

The main result can be stated as follows.

THEOREM 12.1 OPTIMAL PREDICTION   Let $y(k)$ be a random process generated by the model in (12.14), where all the zeros of the polynomial $C(z)$ are inside the unit disc, and $e(k)$ is a sequence of independent random variables. The minimum-variance predictor over m steps is given by

$$\hat{y} = \hat{y}(k + m \mid k) = \frac{qG(q)}{C(q)} y(k) = \frac{G^*(q^{-1})}{C^*(q^{-1})} y(k) \qquad (12.16)$$

where the polynomials F and G are the quotient and the remainder when dividing $q^{m-1}C$ by A; that is,

$$q^{m-1}C(q) = A(q)F(q) + G(q) \qquad (12.17)$$

The prediction error is a moving average

$$\tilde{y}(k + m \mid k) = y(k+m) - \hat{y}(k + m \mid k) = F(q)e(k+1) \qquad (12.18)$$

It has zero mean and the variance

$$E\tilde{y}(k + m \mid k)^2 = \left(1 + f_1^2 + \cdots + f_{m-1}^2\right)\sigma^2 \qquad (12.19)$$

Proof.   The polynomial F is monic of degree $m - 1$ and G is of degree less than n. Hence

$$F(q) = q^{m-1} + f_1 q^{m-2} + \cdots + f_{m-1}$$
$$G(q) = g_0 q^{n-1} + g_1 q^{n-2} + \cdots + g_{n-1}$$

We introduce

$$F^*(q^{-1}) = 1 + f_1 q^{-1} + \cdots + f_{m-1} q^{-m+1}$$
$$G^*(q^{-1}) = g_0 + g_1 q^{-1} + \cdots + g_{n-1} q^{-n+1}$$

It follows from (12.17) that

$$C^*(q^{-1}) = A^*(q^{-1})F^*(q^{-1}) + q^{-m}G^*(q^{-1}) \qquad (12.20)$$

Equation (12.15) can then be written as

$$y(k+m) = \frac{C^*(q^{-1})}{A^*(q^{-1})} e(k+m) = F^*(q^{-1})e(k+m) + \frac{G^*(q^{-1})}{A^*(q^{-1})} e(k)$$

By using Equation (12.14) the signal e in the last term can be expressed in terms of the data available at time k. Hence,

$$y(k+m) = F^*(q^{-1})e(k+m) + \frac{G^*(q^{-1})}{C^*(q^{-1})} y(k)$$

The first term of the right-hand side is a linear function of $e(k+1)$, $e(k+2)$, $\dots$, $e(k+m)$, which are all independent of the data $y(k)$, $y(k-1)$, $y(k-2)$, $\dots$ available at time k. The last term is a linear function of the data. Let $\hat{y}$ be an arbitrary function of $y(k)$, $y(k-1)$, $\dots$. Then

$$\mathrm{E}\left(y(k+m) - \hat{y}\right)^2 = \mathrm{E}\left(F^*(q^{-1})e(k+m)\right)^2 + \mathrm{E}\left(\frac{G^*(q^{-1})}{C^*(q^{-1})} y(k) - \hat{y}\right)^2$$

$$+ 2\mathrm{E}\left\{\left(F^*(q^{-1})e(k+m)\right)\left(\frac{G^*(q^{-1})}{C^*(q^{-1})} y(k) - \hat{y}\right)\right\} \qquad (12.21)$$

The last term is zero because $e(k+m)$, $e(k+m-1)$, $\dots$, and $e(k+1)$ have zero mean values and are independent of $y(k)$, $y(k-1)$, $\dots$. The predictor that minimizes the mean-square prediction error is thus given by (12.16) and the prediction error by (12.18). The proof is completed by taking the mean value of the square of the prediction error (12.18). This gives (12.19). ∎

Remark 1. Notice that the best predictor is linear. The linearity does not depend critically on the minimum-variance criterion. If the probability density of $y(k)$ is symmetric, the predictor of (12.16) is optimal for all criteria of the form $\mathrm{E}\, g\left((y(k+m) - \hat{y})^2\right)$ for symmetric g.

Remark 2. The assumption that $e(i)$ and $e(j)$ are independent for $i \neq j$ is essential for the last term in (12.21) to vanish. If the variables are uncorrelated, the term will still vanish if the predictor y is restricted to being linear.

Remark 3. It follows from (12.18) that

$$\tilde{y}(k+1 \mid k) = y(k+1) - \hat{y}(k+1 \mid k) = e(k+1)$$

The random variables $e(k)$ can thus be interpreted as the innovations of the process $y(k)$ (compare with Sec. 10.4).

Remark 4.   Notice that the function

$$J(m) = \sigma^2 \left(1 + f_1^2 + \cdots + f_{m-1}^2\right)$$

is the variance of the prediction error over the time interval mh. The function $J(m)$ approaches the variance of y as $m \to \infty$. A graph of the function J shows how well the process may be predicted over different horizons. See Example 12.3.

Remark 5.   The predictor discussed in this section is equivalent to the steady-state predictor obtained using the Kalman filter in Sec.11.3 (see Example 11.6).

Calculation of the Optimal Predictor

It follows from (12.17) that $F(q)$ is the quotient and $G(q)$ the remainder when dividing $q^{m-1}C(q)$ by $A(q)$. The polynomials F and G can thus be determined by polynomial division. An explicit formula for the coefficients of the polynomials can also be given. Equating the coefficients of equal powers of q in (12.17) gives the following equations:

$$c_1 = a_1 + f_1$$
$$c_2 = a_2 + a_1 f_1 + f_2$$
$$\vdots$$
$$c_{m-1} = a_{m-1} + a_{m-2}f_1 + \cdots + a_1 f_{m-2} + f_{m-1}$$
$$c_m = a_m + a_{m-1}f_1 + \cdots + a_1 f_{m-1} + g_0$$
$$c_{m+1} = a_{m+1} + a_m f_1 + \cdots + a_2 f_{m-1} + g_1$$
$$\vdots$$
$$c_n = a_n + a_{n-1}f_1 + \cdots + a_{n-m+1}f_{m-1} + g_{n-m}$$
$$0 = a_n f_1 + a_{n-1}f_2 + \cdots + a_{n-m+2}f_{m-1} + g_{n-m+1}$$
$$\vdots$$
$$0 = a_n f_{m-1} + g_{n-1}$$

These equations are easy to solve recursively. Compare the solution of the Diophantine equation in Chapter 5.

Example 12.2   Prediction

Consider the system (12.14) defined by the polynomials

$$A(q) = q^2 - 1.5q + 0.7$$
$$C(q) = q^2 - 0.2q + 0.5$$

and where e has unit variance. Determine first the three-step-ahead prediction of the output. The identity (12.17) gives

$$q^2(q^2 - 0.2q + 0.5) = (q^2 - 1.5q + 0.7)(q^2 + f_1 q + f_2) + g_0 q + g_1$$

This gives the triangular linear system of equations

$$
\begin{aligned}
q^3 : \quad -0.2 &= -1.5 + f_1 & f_1 &= 1.3 \\
q^2 : \quad 0.5 &= 0.7 - 1.5f_1 + f_2 & f_2 &= 1.75 \\
q^1 : \quad 0 &= 0.7f_1 - 1.5f_2 + g_0 & g_0 &= 1.715 \\
q^0 : \quad 0 &= 0.7f_2 + g_1 & g_1 &= -1.225
\end{aligned}
$$

The prediction three steps ahead is thus given by

$$
\hat{y}(k+3 \mid k) = \frac{qG(q)}{C(q)} y(k) = \frac{1.715q^2 - 1.225q}{q^2 - 0.2q + 0.5} y(k)
$$

and the variance of the prediction error is

$$
E\tilde{y}^2 = 1 + (1.3)^2 + (1.75)^2 = 5.7525
$$

∎

Example 12.3   Influence of prediction horizon

Consider the process in Example 12.2. From (12.19) it follows that the variance of the prediction error will increase with the prediction horizon. Also (12.17) shows that the F-polynomial is obtained from the division of the C- and A-polynomials. That is, the coefficients f_i are the coefficients of the impulse response of the system. Thus

$$
y(k) = \frac{C(q)}{A(q)} e(k) = \frac{q^2 - 0.2q + 0.5}{q^2 - 1.5q + 0.7} e(k)
$$

$$
= \left(1 + 1.3q^{-1} + 1.75q^{-2} + 1.715q^{-3} + \cdots \right) e(k) = \sum_{j=0}^{\infty} f_j e(k-j)
$$

and the prediction loss is

$$
E\tilde{y}^2(k+m \mid k) = \sigma^2 \sum_{j=0}^{m-1} f_j^2
$$

Figure 12.2 shows the variance of the prediction error for different values of the prediction horizon m. It is seen that the variance of the prediction error is monotonically increasing with m. Figure 12.3 shows the output, the predicted output, and the accumulated prediction loss, $\sum (y(k) - \hat{y}(k \mid k - m))^2$, for different prediction horizons. ∎

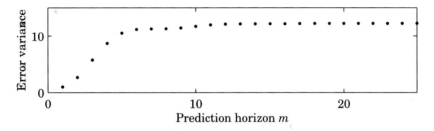

Figure 12.2   The variance of the prediction error as function of the prediction horizon m for the system in Example 12.3.

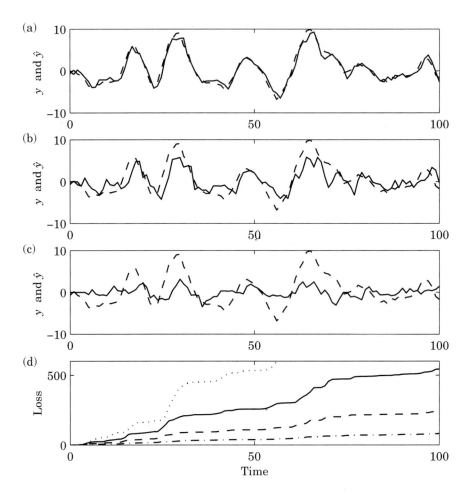

Figure 12.3 The process output (dashed) and the predicted output (solid) for Example 12.3 when (a) $m = 1$, (b) $m = 3$, (c) $m = 5$, and (d) the accumulated prediction loss, $\sum (y(k) - \hat{y}(k \mid k - m))^2$, for $m = 1$ (dashed-dotted), $m = 2$ (dashed), $m = 3$ (solid), and $m = 5$ (dotted).

The Case When C Has Zeros on the Unit Circle

The predictor of (12.16) is a dynamic system with the characteristic polynomial $C(z)$. The assumption that C has all its zeros inside the unit disc thus guarantees that the predictor is stable in steady state. The initial conditions are irrelevant because their influence will decay exponentially.

    It follows from the spectral factorization that C may be chosen to have its zeros inside the unit disc or on the unit circle. The zeros outside the unit disc is mirrored in the unit circle. Compare with Example 12.1. Thus it remains to discuss the case when C has zeros on the unit circle.

Example 12.4   Zeros on the unit circle

Consider the process

$$y(k) = e(k) - e(k - 1) \qquad (12.22)$$

In this case the polynomial $C(z) = z - 1$ has a zero on the unit circle. Applying the previous methods formally gives the one-step predictor

$$\hat{y}(k + 1 \mid k) = -e(k)$$

Attempting to calculate $e(k)$ from $y(k), y(k - 1), \ldots, y(k_0)$ as was done previously gives

$$e(k) = e(k_0 - 1) + \sum_{i=k_0}^{k} y(i) = e(k_0 - 1) + z(k)$$

The presence of the term $e(k_0 - 1)$, which does not go to zero as $k_0 \to -\infty$, shows the consequences of C being unstable. The Kalman filtering theory can, however, be used to determine the optimal predictor. The signal given by (12.22) can be written as

$$x(k + 1) = e(k)$$
$$y(k) = -x(k) + e(k)$$

where $R_1 = R_2 = R_{12} = \sigma^2$ with the notations used in Sec. 11.3. The Kalman filter is

$$\hat{x}(k + 1 \mid k) = K(k)\Big(y(k) + \hat{x}(k \mid k - 1)\Big)$$

$$P(k + 1) = \frac{\sigma^2 P(k)}{P(k) + \sigma^2}$$

$$K(k) = \frac{\sigma^2}{P(k) + \sigma^2}$$

with the initial conditions

$$\hat{x}(k_0 \mid k_0 - 1) = 0$$
$$P(k_0) = \sigma^2$$

The predictor for the output is

$$\hat{y}(k + 1 \mid k) = -\hat{x}(k + 1 \mid k) = -K(k)\Big(y(k) - \hat{y}(k \mid k - 1)\Big)$$

Simple calculations give

$$\hat{y}(k + 1 \mid k) = -\frac{1}{k - k_0 + 2} \sum_{n=0}^{k-k_0} (n + 1) y(k_0 + n)$$

The optimum predictor is thus a time-varying system. Notice that the influence of the initial condition $y(k_0)$ goes to zero at the rate $1/(k - k_0 + 2)$. This is much slower than in the case of stable polynomials C.   ∎

It follows from the example that the optimal predictor is a time-varying system if the polynomial C has zeros on the unit circle. Such models should be avoided if time-invariant predictors are desired. Unfortunately, this fact is not always noticed, as Example 12.5 illustrates.

Example 12.5 How to model offsets

The model

$$A(q)y(k) = C(q)e(k) + b$$

where b is an unknown constant, represents a signal with an offset. The constant b can be eliminated by taking differences. Hence,

$$(q - 1)A(q)y(k) = (q - 1)C(q)e(k)$$

The common factor $q - 1$ can be eliminated by regarding $\Delta y(k) = (q - 1)y(k)$ as the output. The model

$$A(q)\Delta y(k) = (q - 1)C(q)e(k) = \tilde{C}(q)e(k)$$

is then obtained. In this model the polynomial $\tilde{C}$ apparently has a zero on the unit circle. This model is, however, not very desirable because the optimal predictor is a time-varying system. It is much better to model an offset as a Wiener process. This leads to a process model with $A(1) = 0$ that is unstable with a stationary predictor. ∎

Other reasons for avoiding models where the polynomial $C(z)$ has zeros close to the unit circle are given in Sec. 12.6.

12.4 Minimum-Variance Control

To determine the minimum-variance control law, the special case when the polynomial B in (12.5) is stable is discussed first. This means that the process dynamics have a stable inverse. With some abuse of language, this case is also called the minimum-phase case because the pulse-transfer function has all its zeros inside the unit disc. The solution to the control problem is very simple in this special case. The solution also gives insight into the properties of the control problem.

Systems with Stable Inverses

By introducing the backward-shift operator q^{-1}, the model in (12.5) can be written as

$$
\begin{aligned}
y(k) &= \frac{B(q)}{A(q)} u(k) + \frac{C(q)}{A(q)} e(k) \\
&= \frac{B^*(q^{-1})}{A^*(q^{-1})} q^{-d}u(k) + \frac{C^*(q^{-1})}{A^*(q^{-1})} e(k)
\end{aligned}
\tag{12.23}
$$

where

$$d = \deg A - \deg B > 0$$

is the *pole excess* of the system (see Sec. 2.6). Further, $\deg A = \deg C = n$. The reciprocal polynomials are introduced to make the discussion based on causality arguments more transparent.

It follows from (12.23) that

$$
\begin{aligned}
y(k+d) &= \frac{C^*(q^{-1})}{A^*(q^{-1})}\, e(k+d) + \frac{B^*(q^{-1})}{A^*(q^{-1})}\, u(k) \\
&= F^*(q^{-1})e(k+d) + \frac{G^*(q^{-1})}{A^*(q^{-1})}\, e(k) + \frac{B^*(q^{-1})}{A^*(q^{-1})}\, u(k)
\end{aligned}
\tag{12.24}
$$

where Equation (12.20) with $m = d$ has been used to obtain the last equality. The first term of the right-hand side is independent of the data available at time k and thus also of the second and third terms. The second term can be computed exactly in terms of data available at time k. To do this, the variable $e(k)$ is given by (12.23); that is,

$$
e(k) = \frac{A^*}{C^*}\, y(k) - q^{-d}\frac{B^*}{C^*}\, u(k)
$$

where the arguments of the polynomials have been dropped to simplify the writing. Using this expression for e, Eq. (12.24) can be written as

$$
\begin{aligned}
y(k+d) &= F^*e(k+d) + \frac{G^*}{C^*}\, y(k) - q^{-d}\frac{B^*G^*}{A^*C^*}\, u(k) + \frac{B^*}{A^*}\, u(k) \\
&= F^*e(k+d) + \frac{G^*}{C^*}\, y(k) + \frac{B^*F^*}{C^*}\, u(k)
\end{aligned}
\tag{12.25}
$$

Now let $u(k)$ be an arbitrary function of $y(k), y(k-1), \ldots$ and $u(k-1), u(k-2), \ldots$. Then

$$
\mathrm{E}y^2(k+d) = \mathrm{E}\big(F^*e(k+d)\big)^2 + \mathrm{E}\left(\frac{G^*}{C^*}\, y(k) + \frac{B^*F^*}{C^*}\, u(k)\right)^2
\tag{12.26}
$$

The mixed terms vanish because $e(k+d), \ldots, e(k+1)$ are independent of $y(k), y(k-1), \ldots$ and $u(k), u(k-1), \ldots$. Because the last term in (12.26) is nonnegative, it follows that

$$
\mathrm{E}y^2(k+d) \geq \big(1 + f_1^2 + \cdots + f_{d-1}^2\big)\sigma^2
$$

where equality is obtained for

$$
u(k) = -\frac{G^*(q^{-1})}{B^*(q^{-1})F^*(q^{-1})}\, y(k) = -\frac{G(q)}{B(q)F(q)}\, y(k)
\tag{12.27}
$$

which is the desired minimum-variance control law. The result can be summarized as follows.

THEOREM 12.2   MINIMUM-VARIANCE CONTROL—STABLE INVERSE   Consider a process described by (12.5), where $e(k)$ is a sequence of independent random variables with zero mean values and standard deviations σ. Let the polynomials B and C have all their zeros inside the unit disc. The minimum-variance control law is then given by (12.27), where the polynomials F^* and G^* are given by (12.20) with $m = d$. This control law gives the output

$$y(k) = F^*(q^{-1})e(k) = e(k) + f_1 e(k-1) + \cdots + f_{d-1}e(k-d+1)$$

in steady state.   ■

Remark 1.   The theorem still holds when $e(i)$ and $e(j)$ are uncorrelated for $i \neq j$ if a linear control law is postulated.

Remark 2.   The result is closely related to the solution of the prediction problem (Theorem 12.1). Identity (12.17) or (12.20) was used in both cases. The last two terms in (12.25) can be interpreted as the d-step prediction of the output. The minimum-variance strategy is thus obtained by predicting the output d steps ahead and choosing a control that makes the prediction equal to the desired output. The stochastic-control problem can thus be separated into two problems, one stochastic-prediction problem and one deterministic-control problem. Theorem 12.2 can therefore be interpreted as a separation theorem.

Remark 3.   The error under minimum-variance control is a moving average of order $d - 1$. Thus the covariance function of the regulation error will vanish for arguments larger than $d - 1$. This fact can be used for diagnosis to determine if a minimum-variance strategy is used.

Remark 4.   All process zeros are canceled when the control law of (12.27) is used. The consequences of this are discussed later.

It is very easy to calculate the minimum-variance control law for a given model (12.5), as illustrated by the following example.

Example 12.6   Minimum-variance control

Consider a system given by (12.5), where

$$A(q) = q^3 - 1.7q^2 + 0.7q$$
$$B(q) = q + 0.5$$
$$C(q) = q^3 - 0.9q^2$$

The pole excess is $d = 2$. Division of $q^{d-1}C(q)$ by $A(q)$ gives the quotient

$$F(q) = q + 0.8$$

and the remainder

$$G(q) = 0.66q^2 - 0.56q$$

The minimum-variance control law is thus

$$u(k) = -\frac{q(0.66q - 0.56)}{(q + 0.5)(q + 0.8)}\, y(k)$$

The variance of the output when the optimal controller is used is

$$Ey^2 = 1 + (0.8)^2 = 1.64$$

∎

Example 12.7 Influence of the delay

Let the process be described by

$$A^*(q^{-1}) = 1 - 1.5q^{-1} + 0.7q^{-2}$$
$$B^*(q^{-1}) = q^{-d}(1 + 0.5q^{-1})$$
$$C^*(q^{-1}) = 1 - 0.2q^{-1} + 0.5q^{-2}$$

Compute the minimum-variance controller when $d = 1$, 3, or 5. The controller is given by (12.27), where the F-polynomial is given in Example 12.3. Figure 12.4 shows the output and input when the minimum-variance controller is used for different delays in the process. When $d = 1$, $d = 3$, and $d = 5$ the output variance is 1, 5.8, and 10.5, respectively.

∎

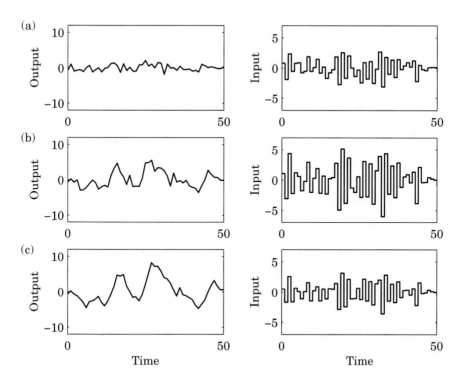

Figure 12.4 Simulation of the system in Example 12.7 with the control law given by Theorem 12.2. The output (left) and the input (right) when (a) $d = 1$, (b) $d = 3$, and (c) $d = 5$.

Interpretation as Pole-Placement Design

The minimum-variance control law can be interpreted in terms of the pole-placement design discussed in Chapter 5. To see the relationships, the closed-loop system obtained when the control law of (12.27) is applied to the system of (12.5) is analyzed. Equations (12.5) and (12.27) can be written as

$$
\begin{pmatrix} A(q) & -B(q) \\ G(q) & F(q)B(q) \end{pmatrix} \begin{pmatrix} y(k) \\ u(k) \end{pmatrix} = \begin{pmatrix} C(q) \\ 0 \end{pmatrix} e(k) \qquad (12.28)
$$

The characteristic polynomial of the closed-loop system is the determinant of the matrix on the left-hand side of (12.28). Hence,

$$
A(q)F(q)B(q) + G(q)B(q) = q^{d-1}B(q)C(q) \qquad (12.29)
$$

where Eq. (12.17), with $m = d$, is used to obtain the first equality. The closed-loop system is of order $2n - 1$. It has $2n - d$ poles at the zeros of B and C and an additional $d - 1$ poles at the origin.

The minimum-variance control strategy can be interpreted as a pole-placement design, where the poles are placed at the zeros given by (12.29). The similarities to pole placement are seen even more clearly if the control law of (12.27) is written as

$$
u(k) = -\frac{G(q)}{B(q)F(q)} y(k) = -\frac{S(q)}{R(q)} y(k)
$$

where $S = G$ and $R = FB$ [compare with Eq. (5.2)]. Multiplication of (12.17) by B gives

$$
q^{d-1}C(q)B(q) = A(q)F(q)B(q) + G(q)B(q) = A(q)R(q) + B(q)S(q)
$$

This equation is a special case of the Diophantine equation in (5.22) when $B^+ = B$ and with $A_c = q^{d-1}B$ and $A_o = C$.

Systems with Unstable Inverses

Remark 4 to Theorem 12.2 mentions that the control law given by (12.27) cancels all process zeros. If there are process zeros outside the unit disc, the closed-loop system will then have unstable modes that are unobservable from the output. The implications of this are discussed first. Other control laws that do not require all zeros of $B(z)$ to be inside the unit disc are then presented.

Solving Eq. (12.28) for y and u gives

$$
y(k) = \frac{F(q)}{q^{d-1}} e(k)
$$

and

$$
u(k) = -\frac{G(q)}{q^{d-1}B(q)} e(k)
$$

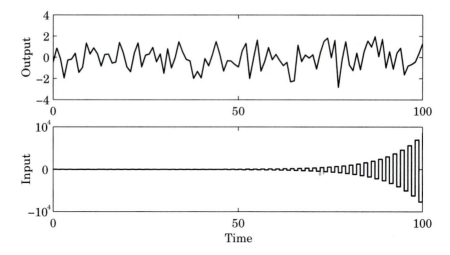

Figure 12.5 Simulation of the system in Example 12.8 with the control law given by Theorem 12.2 that cancels an unstable process zero.

The necessity of the assumption that B is stable is clearly seen from these equations. If the polynomial B is unstable, the system has unstable modes, which are excited by the disturbance. These unstable modes are coupled to the control signal and the control signal grows exponentially. However, the output signal remains bounded because the unstable modes are not coupled to the output. An example illustrates what happens.

Example 12.8   Cancellation of unstable process zero

Consider a system described by the polynomials

$$A(z) = (z - 1)(z - 0.7)$$
$$B(z) = 0.9z + 1$$
$$C(z) = z(z - 0.7)$$

The polynomial $B(z)$ has a zero $z = -10/9$, which is outside the unit disc. A simulation when using (12.27) is shown in Fig. 12.5. The presence of the unstable mode is clearly seen in the control signal, although it is not noticeable in the system output. If the simulation is continued, the control signal will finally be so large that overflow or numerical errors occur. In a practical problem the signal will quickly be so large that the linear approximation is no longer valid. After a short time the unstable mode will then be noticeable in the output.   ∎

The minimum-variance control law is extended to the case when the polynomial B has zeros outside the unit disc in Theorem 12.3.

THEOREM 12.3   MINIMUM-VARIANCE CONTROL—GENERAL CASE   Consider a system described by (12.5). Factor the polynomial $B(z)$ as

$$B(z) = B^+(z)B^-(z) \tag{12.30}$$

where $B^{-*}(z)$ is monic. All zeros of the polynomial $B^+(z)$ are inside the unit disc and all zeros of $B^-(z)$ are outside the unit disc or on the unit circle. Assume that all the zeros of polynomial $C(z)$ are inside the unit disc and that the polynomials $A(z)$ and $B^-(z)$ do not have any common factors. The minimum-variance control law is then given by

$$u(k) = -\frac{G(q)}{B^+(q)F(q)} y(k) \tag{12.31}$$

where $F(q)$ and $G(q)$ are polynomials that satisfy the Diophantine equation

$$q^{d-1}C(q)B^{-*}(q) = A(q)F(q) + B^-(q)G(q) \tag{12.32}$$

in which $\deg F = d + \deg B^- - 1$ and $\deg G < \deg A = n$.

Proof.   The proof is based on a clever trick introduced by Wiener in his original work on prediction. An alternative method is used in the proof of Theorem 12.4. Consider the operator

$$\frac{1}{q+a}$$

where $|a| > 1$. This operator is normally interpreted as a causal unstable (unbounded) operator. Because $|a| > 1$ and the shift operator has the norm $\|q\| = 1$, the series expansion

$$\frac{1}{q+a} = \frac{1}{a}\frac{1}{1+q/a} = \frac{1}{a}\left(1 - \frac{q}{a} + \frac{q^2}{a^2} - \cdots\right)$$

converges. Thus the operator $(q+a)^{-1}$ can be interpreted as a noncausal stable operator; that is,

$$\frac{1}{q+a} y(k) = \frac{1}{a}\left(y(k) - \frac{1}{a}y(k+1) + \frac{1}{a^2}y(k+2) - \cdots\right)$$

With this interpretation, it follows that

$$(q+a)\left(\frac{1}{q+a} y(k)\right) = y(k)$$

The calculations required for the proof are conveniently done using the backward-shift operator. It follows from the process model of (12.5) that

$$y(k+d) = \frac{B^*(q^{-1})}{A^*(q^{-1})} u(k) + \frac{C^*(q^{-1})}{A^*(q^{-1})} e(k+d)$$

We introduce

$$w(k) = \frac{B^-(q^{-1})}{B^{-*}(q^{-1})} y(k)$$

where the operator $1/B^{-*}(q^{-1})$ is interpreted as a noncausal stable operator. The signals y and w have the same steady-state variance because B^- and B^{-*} are reciprocal polynomials and

$$\left| \frac{B^-(e^{-i\omega})}{B^{-*}(e^{-i\omega})} \right| = 1$$

An admissible control law that minimizes the variance of w also minimizes the variance of y. It follows that

$$w(k+d) = \frac{B^{+*}(q^{-1})B^-(q^{-1})}{A^*(q^{-1})} u(k) + \frac{C^*(q^{-1})B^-(q^{-1})}{A^*(q^{-1})B^{-*}(q^{-1})} e(k+d) \qquad (12.33)$$

The assumption that $A(z)$ and $B^-(z)$ are relatively prime guarantees that (12.32) has a solution. Equation (12.32) implies that

$$C^*(q^{-1})B^-(q^{-1}) = A^*(q^{-1})F^*(q^{-1}) + q^{-d}B^{-*}(q^{-1})G^*(q^{-1})$$

Division by A^*B^{-*} gives

$$\frac{C^*(q^{-1})B^-(q^{-1})}{A^*(q^{-1})B^{-*}(q^{-1})} = \frac{F^*(q^{-1})}{B^{-*}(q^{-1})} + q^{-d} \frac{G^*(q^{-1})}{A^*(q^{-1})}$$

By using this equation, (12.33) can be written as

$$w(k+d) = \frac{F^*(q^{-1})}{B^{-*}(q^{-1})} e(k+d) + \frac{B^{+*}(q^{-1})B^-(q^{-1})}{A^*(q^{-1})} u(k) + \frac{G^*(q^{-1})}{A^*(q^{-1})} e(k) \quad (12.34)$$

Because the operator $1/B^{-*}(q^{-1})$ is interpreted as a bounded noncausal operator and because $\deg F^* = d + \deg B^- - 1$, it follows that

$$\frac{F^*(q^{-1})}{B^{-*}(q^{-1})} e(k+d) = \alpha_1 e(k+1) + \alpha_2 e(k+2) + \cdots$$

These terms are all independent of the last two terms in (12.34). Using the arguments given in detail in the proof of Theorem 12.2, we find that the optimal control law is obtained by putting the sum of the last two terms in (12.34) equal to zero. This gives

$$u(k) = -\frac{G^*(q^{-1})}{B^{+*}(q^{-1})B^-(q^{-1})} e(k) \qquad (12.35)$$

and

$$y(k) = \frac{B^{-*}(q^{-1})}{B^-(q^{-1})} w(k) = \frac{F^*(q^{-1})}{B^-(q^{-1})} e(k) = \frac{F(q)}{q^{d-1}B^{-*}(q)} e(k) \qquad (12.36)$$

Elimination of $e(k)$ between (12.35) and (12.36) gives

$$u(k) = -\frac{G^*(q^{-1})}{B^{+*}(q^{-1})F^*(q^{-1})} y(k)$$

The numerator and the denominator have the same degree because $\deg G < n$ and the control law can then be rewritten as (12.31). ∎

Remark 1.   Only the stable process zeros are canceled by the optimal control law.

Remark 2.   It follows from the proofs of Theorems 12.2 and 12.3 that the variance of the output of a system such as (12.5) may have several local minima if the polynomial $B(z)$ has zeros outside the unit disc. There is one absolute minimum given by Theorem 12.2. However, this minimum will give control signals that are infinitely large. The local minimum given by Theorem 12.3 is the largest of the local minima. The control signal is bounded in this case.

Remark 3.   The factorization of (12.30) is arbitrary because B^+ could be multiplied by a number and B^- could be divided by the same number. It is convenient to select the factors so that the polynomial $B^{-*}(q)$ is monic.

Example 12.9   Minimum-variance control with unstable process zero

Consider the system in Example 12.8 where $d = 1$ and

$$B^+(z) = 1$$
$$B^-(z) = B(z)$$
$$B^{-*}(z) = z + 0.9$$

Equation (12.32) becomes

$$z(z - 0.7)(z + 0.9) = (z - 1)(z - 0.7)(z + f_1) + (0.9z + 1)(g_0 z + g_1)$$

Let $z = 0.7$, $z = 1$, and $z = -10/9$. This gives

$$0.7 g_0 + g_1 = 0$$
$$g_0 + g_1 = 0.3$$
$$f_1 = 1$$

The control law thus becomes

$$u(k) = -\frac{G(q)}{B^+(q)F(q)}\, y(k) = -\frac{q - 0.7}{q + 1}\, y(k)$$

The output is

$$y(k) = \frac{F(q)}{B^{-*}(q)}\, e(k - d + 1) = \frac{q + 1}{q + 0.9}\, e(k) = e(k) + \frac{0.1}{q + 0.9}\, e(k)$$

The variance of the output is

$$\mathrm{E}\, y^2 = \left(1 + \frac{0.1^2}{1 - 0.9^2}\right)\sigma^2 = \frac{20}{19}\sigma^2 = 1.05\sigma^2$$

which is about 5% larger than using the controller in Example 12.8. The variance of the control signal is $275\sigma^2/19 = 14.47\sigma^2$. A simulation of the control law is shown in Fig. 12.6. The figure that the controller performs well. Compare also with Fig. 12.5, which shows the effect of canceling the unstable zero. Figure 12.7 shows the accumulated output loss $\sum y^2(k)$ and input loss $\sum u^2(k)$ when the controllers in Example 12.8 and this example are used. The controller (12.27) gives lower output loss, but an exponentially growing input loss, and the controller based on (12.31) gives an accumulated input loss that grows linearly with time.   ∎

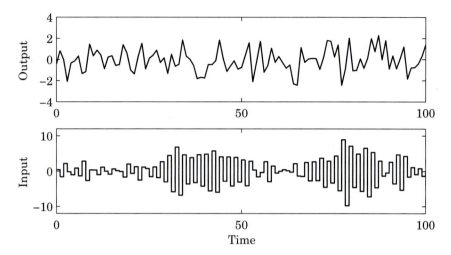

Figure 12.6   Simulation of the system in Example 12.9.

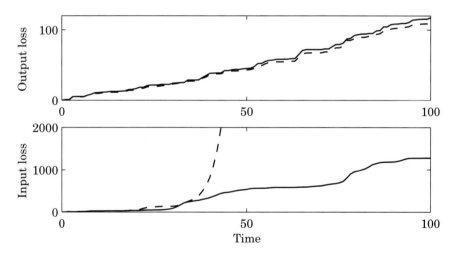

Figure 12.7   The accumulated output loss $\sum y^2(k)$ and input loss $\sum u^2(k)$ when the controllers (12.31) (solid) and (12.27) (dashed) are used.

A Pole-Placement Interpretation

Simple calculations show that the characteristic equation of the closed-loop system obtained from (12.5) and (12.31) is

$$z^{d-1}B^+(z)B^{-*}(z)C(z) = 0$$

Thus the control law of (12.31) can be interpreted as a pole-placement controller, which gives this characteristic equation.

Multiplication of (12.32) by B^+ gives the equation

$$A(z)R(z) + B(z)S(z) = z^{d-1}B^+(z)B^{-*}(z)C(z) \qquad (12.37)$$

where $R(z) = B^+(z)F(z)$ and $S(z) = G(z)$. This equation is the same Diophantine equation that was used in the pole-placement design [compare with Eq. (5.22)]. The closed-loop system has poles corresponding to the observer dynamics, to the stable process zeros, and to the reflections in the unit circle of the unstable process zeros. Notice that the transfer function $B(z)/A(z)$ may be interpreted as having $d = \deg A - \deg B$ zeros at infinity. The reflections of these zeros in the unit circle also appear as closed-loop poles, which are located at the origin.

Equation (12.37) shows that the closed-loop system is of order $2n - 1$ and that $d - 1$ of the poles are in the origin. A complete controller consisting of a full Kalman filter observer and feedback from the observed states gives a closed-loop system of order $2n$. The "missing" pole is due to a cancellation of a pole at the origin in the controller. This is further discussed in Sec. 12.5.

12.5 Linear Quadratic Gaussian (LQG) Control

The optimal control problem for the system of (12.5) with the criterion of (12.8) is now solved. The minimum-variance control law discussed in Sec. 12.4 can be expressed in terms of a solution to a polynomial equation. The solution to the LQG-problem can be obtained in a similar way. Two or three polynomial equations are needed, however. These equations are discussed before the main result is given.

The name *Gaussian* in LQG is actually slightly misleading. The proofs show that the probability distribution is immaterial as long as the random variables $e(k)$ are independent.

Using the state-space solution it is possible to get an interpretation of the properties of the optimal solution. These properties can be expressed in terms of the poles of the closed-loop system. In this way we can establish a connection between LQG design and pole placement.

Properties of the State-Space Solution

The problems discussed in this chapter was solved using state-space methods in Chapter 11. A state-space representation of the model of (12.5) is first given. For this purpose it is assumed that the model is normalized, so that $\deg C(z) = \deg A(z)$. The model of (12.5) can then be represented as

$$x(k + 1) = \Phi x(k) + \Gamma u(k) + Ke(k)$$
$$y(k) = Cx(k) + e(k)$$

where

$$\Phi = \begin{pmatrix} -a_1 & 1 & 0 & \cdots & 0 \\ -a_2 & 0 & 1 & \cdots & 0 \\ \vdots & \vdots & \vdots & \ddots & \vdots \\ -a_{n-1} & 0 & 0 & \cdots & 1 \\ -a_n & 0 & 0 & \cdots & 0 \end{pmatrix} \quad \Gamma = \begin{pmatrix} b_1 \\ b_2 \\ \vdots \\ b_{n-1} \\ b_n \end{pmatrix} \quad K = \begin{pmatrix} c_1 - a_1 \\ c_2 - a_2 \\ \vdots \\ c_{n-1} - a_{n-1} \\ c_n - a_n \end{pmatrix}$$

$$C = \begin{pmatrix} 1 & 0 & \cdots & 0 \end{pmatrix}$$

$$(12.38)$$

Because this is an innovations representation if the matrix $\Phi - KC$ has all its eigenvalues inside the unit disc. The steady-state Kalman filter is then obtained by inspection:

$$\hat{x}(k+1 \mid k) = \Phi \hat{x}(k \mid k-1) + \Gamma u(k) + K\big(y(k) - C\hat{x}(k \mid k-1)\big) \qquad (12.39)$$

The Kalman filter has the characteristic polynomial

$$\det\big(zI - (\Phi - KC)\big) = C(z) \qquad (12.40)$$

This implies that $C(z)$ are some of the closed-loop poles. Assume a computational delay of one sampling period in the control law. The optimal control law is then

$$u(k) = -L\hat{x}(k \mid k-1)$$

and the transfer function of the controller is

$$H_r(z) = -L(zI - \Phi + KC + \Gamma L)^{-1}K = -\frac{S(z)}{R(z)} \qquad (12.41)$$

where $R(z) = \det(zI - \Phi + KC + \Gamma L)$, $\deg R(z) = n$, and $\deg S(z) < n$. It follows from this discussion and Sec. 11.4 that the closed-loop poles are $C(z)$ and

$$P(z) = \det(zI - \Phi + \Gamma L)$$

where $P(z)$ is obtained from the algebraic Riccati equation.

It is more complicated to derive the control law when the admissible control is such that $u(k)$ is a function of $y(k), y(k-1), \ldots$. The loss function (12.8) corresponds to (11.9) with $Q_1 = C^T C$, $Q_{12} = 0$, and $Q_2 = \rho$. From (11.19) and (11.24) it follows that $L = L_v \Phi$. The results from state-space theory (Remark 2 of Theorem 11.7) show that the control law is

$$\begin{aligned} u(k) &= -L\hat{x}(k \mid k) - L_v \hat{v}(k \mid k) \\ &= -L\hat{x}(k \mid k) - L_v K\big(y(k) - C\hat{x}(k \mid k-1)\big) \\ &= -L_v(\Phi - KC)\hat{x}(k \mid k-1) - L_v Ky(k) \end{aligned} \qquad (12.42)$$

where $\hat{x}(k \mid k - 1)$ is given by (12.39). The controller is still of order n. Eliminating $\hat{x}$ between (12.39) and (12.42), we find that the controller can be described by the relation

$$
\begin{aligned}
u(k) &= -L_v(\Phi - KC)(qI - \Phi + KC)^{-1}\Big(\Gamma u(k) + Ky(k)\Big) - L_v Ky(k) \\
&= -L_v(\Phi - KC)(qI - \Phi + KC)^{-1}\Gamma u(k) \\
&\quad - L_v(\Phi - KC + qI - \Phi + KC)(qI - \Phi + KC)^{-1}Ky(k) \qquad (12.43) \\
&= -L_v(\Phi - KC)(qI - \Phi + KC)^{-1}\Gamma u(k) \\
&\quad - L_v q(qI - \Phi + KC)^{-1}Ky(k)
\end{aligned}
$$

Introducing $R_2(q) = \det(qI - \Phi + KC)$ we get

$$
u(k) = -\frac{R_1(q)}{R_2(q)}\, u(k) - \frac{S(q)}{R_2(q)}\, y(k)
$$

where $\deg R_1(z) = n$, $\deg R_2(z) < n$ and $\deg S(z) = n$ with $S(0) = 0$. Hence

$$
u(k) = -\frac{S(q)}{R_1(q) + R_2(q)}\, y(k) = -\frac{S(q)}{R(q)}\, y(k) \qquad (12.44)
$$

We thus find that the controller has the property $\deg R(z) = \deg S(z) = n$. Furthermore the condition $S(0) = 0$ implies that $\deg S^*(z) < n$.

Spectral Factorization

The LQ-problem is solved in Sec. 11.4 using the state-space approach, which led to a steady-state Riccati equation. It follows from the Riccati equation that

$$
rP(z)P(z^{-1}) = \rho A(z)A(z^{-1}) + B(z)B(z^{-1}) \qquad (12.45)
$$

where the monic polynomial $P(z)$ is the characteristic polynomial of the closed-loop system. [see Eq. (11.40)]. The closed-loop characteristic polynomial can be obtained by solving a steady-state Riccati equation. An alternative is to find a polynomial $P(z)$ that satisfies (12.45) directly. A feedback that gives the desired closed-loop poles can then be determined by pole placement. The problem of finding a polynomial $P(z)$ that satisfies (12.45) is called *spectral factorization*.

First, consider a polynomial of the form

$$
F(z) = f_0 z^{2n} + f_1 z^{2n-1} + \cdots + f_{n-1} z^{n+1} + f_n z^n + f_{n-1} z^{n-1} + \cdots + f_1 z + f_0
$$

Such a polynomial is self-reciprocal because

$$
F^*(z) = z^{2n} F(z^{-1}) = F(z)
$$

It then follows that if $z = a$ is a zero of $F(z)$, then $z = 1/a$ is also a zero. Moreover, if the coefficients f_i are real, then $z = \bar{a}$ and $z = 1/\bar{a}$ are also zeros, where $\bar{a}$ is the complex conjugate of a. The following result can now be established.

LEMMA 12.1   Let the real polynomials $A(z)$ and $B(z)$ be relatively prime with $\deg A(z) > \deg B(z)$. Then there exists a unique polynomial $P(z)$ with $\deg P(z) = \deg A(z) = n$ and all its zeros inside the unit disc or on the unit circle such that (12.45) holds. If $\rho > 0$, then $P(z)$ has no zeros on the unit circle.

Proof.   A self-reciprocal polynomial is obtained if the right-hand side of (12.45) is multiplied by z^n. The zeros of the right-hand side are thus mirror images with respect to the unit circle. Because the coefficients are real, the zeros are also symmetric with respect to the real axis. The right-hand side of (12.45) cannot have zeros on the unit circle because if $z = e^{i\omega}$ is such a zero, then

$$\rho A(e^{i\omega})A(e^{-i\omega}) + B(e^{i\omega})B(e^{-i\omega}) = \rho |A(e^{i\omega})|^2 + |B(e^{i\omega})|^2 = 0$$

As $\rho > 0$, this implies that $z = \exp(i\omega)$ is a zero of both $A(z)$ and $B(z)$, which contradicts the assumption that $A(z)$ and $B(z)$ are relatively prime. The condition $\deg P(z) = n$ ensures a unique $P(z)$.   ∎

Remark 1.   By introducing reciprocal polynomials, Eq. (12.45) can be written as

$$rP(z)P^*(z) = \rho A(z)A^*(z) + z^d B(z)B^*(z) \qquad (12.46)$$

where $P^*(z) = z^n P(z^{-1})$, and so on.

Remark 2.   If $P(z)$ satisfies (12.45) so does $z^l P(z)$, where l is an arbitrary integer. To obtain a unique P we can either specify the degree of P or choose P as the polynomial of lowest degree that satisfies (12.45). For a control problem it is natural to interprete $P(z)$ as the closed loop characteristic polynomial under state feedback. With this interpretation it is natural to require that $\deg P(z) = \deg A(z) = n$. Notice that it is possible to find a P of lower degree when $\rho = 0$ or when $A(0) = 0$.

Conceptually the spectral-factorization problem can be solved by finding the zeros of the right-hand side of (12.45) and sorting them. There are also efficient recursive algorithms for solving the problem.

Heuristic Discussion

The LQG-problem will now be related to the pole-placement problem. We will first give the solution heuristically. A formal solution will be given later. First, recall that the pole-placement problem required specifications of the closed-loop characteristic polynomial, which were chosen as $A_c(z)A_o(z)$ when A_o was interpreted as the observer polynomial. In the LQG-problem the observer polynomial is simply $A_o(z) = C(z)$. Compare Theorem 12.1. The polynomial $A_c(z)$ is equal to the polynomial $P(z)$ obtained from the spectral factorization. When the polynomials $A_o(z) = C(z)$ and $A_c(z) = P(z)$ are specified we can now expect that the optimal control law is given by

$$u(k) = -\frac{S(q)}{R(q)} y(k)$$

where $R(z)$ and $S(z)$ are solutions to the Diophantine equation

$$A(z)R(z) + B(z)S(z) = P(z)C(z) \qquad (12.47)$$

The structure of the admissible control laws is determined by the polynomials $R(z)$ and $S(z)$. To describe a control law such that $u(k)$ is a function of $y(k)$, $y(k-1)$, ..., and $u(k-1)$, $u(k-2)$, ..., that is, no delay in the controller, the polynomials $R(z)$ and $S(z)$ should have the same degree. To describe a control law such that $u(k)$ is a function of $y(k-1), y(k-2), \ldots$, and $u(k-1), u(k-2), \ldots$, that is, one sampling period delay in the controller, the pole excess of $S(z)/R(z)$ should be one. The complexity of the control law is determined by the orders of the polynomials $R(z)$ and $S(z)$.

There are many polynomials $R(z)$ and $S(z)$ that satisfy (12.47). Compare the discussion in Sec. 5.3. Among all choices we will determine solutions that minimize the loss function (12.8). Before making a formal solution we will discuss the problem heuristically.

The solution to the LQG problem based on the state space approach gives the additional constraints that have to be imposed on the solution to (12.47). Equation (12.41) gave a polynomial interpretation of the state space solution. The optimal controller was in fact characterized by the following conditions on the controller polynomials: $\deg R(z) = n$ and $\deg S(z) < n$. If A and B are relative prime the optimal LQG-controller is thus the unique solution to (12.47) with $\deg S(z) < \deg A(z)$.

The problem is more complicated when there is no delay in the controller. The transfer function of the optimal controller in this case was given by Eq. (12.44) with $\deg R(z) = \deg S(z) = n$, and and $\deg S^*(z) < n$. These conditions are more conveniently expressed using another version of the Diophantine equation (12.47). Assuming $\deg R(z) = \deg S(z) = n$, writing (12.47) with argument z^{-1} and multiplying it by z^n we find that

$$A^*(z)R^*(z) + z^d B^*(z)S^*(z) = P^*(z)C^*(z) \qquad (12.48)$$

where

$$d = \deg A(z) - \deg B(z)$$

If $\deg A^*(z) = n$ the optimal controller is then the unique solution to (12.48) with $\deg S^*(z) < \deg A^*(z)$. Notice, however, that this does not give the optimal solution when $\deg A^*(z) < n$, i.e. when $A(0) = 0$. This case will be discussed in the next section where we give a direct solution of the LQG problem with polynomial calculations.

Formal Proof

After the informal discussion we will now give a formal proof of the statements. For this purpose we will first prove a preliminary result.

LEMMA 12.2   Let the polynomial $P(z)$ be a solution to the spectral factorization problem (12.46) and let $A(z)$ be monic. Assume that the polynomials $A(z)$ and $B(z)$ do not have common roots outside the unit disc or on the unit circle; then there exists a unique solution to the equations

$$A^*(z)X(z) + rP(z)S^*(z) = B(z)C^*(z)$$
$$z^d B^*(z)X(z) - rP(z)R^*(z) = -\rho A(z)C^*(z)$$

(12.49)

with $\deg X(z) < n$, $\deg R^*(z) \le n$ and $\deg S^*(z) < n$, where $n = \deg A(z)$.

Proof.   First, assume that polynomial $P(z)$ has distinct zeros z_i. Since $P(z)$ is stable we have $|z_i| < 1$. The values $A^*(z_i)$ and $B^*(z_i)$ cannot vanish simultaneously because this would contradict the assumption that $A(z)$ and $B(z)$ do not have common unstable factors. Evaluating (12.49) for $z = z_i$ we get

$$A^*(z_i)X(z_i) = B(z_i)C^*(z_i)$$
$$z_i^d B^*(z_i)X(z_i) = -\rho A(z_i)C^*(z_i)$$

(12.50)

If both $A^*(z_i)$ and $B^*(z_i)$ are different from zero, both equations give the same result, since it follows from (12.46) that

$$\frac{B(z_i)}{A^*(z_i)} = -\frac{\rho A(z_i)}{z_i^d B^*(z_i)}$$

If $A^*(z_i) = 0$ and $B^*(z_i) \ne 0$ it follows from (12.46) that $B(z_i) = 0$. Since $A(z)$ is monic it also follows that $A^*(0) = 1$. This implies that $|z_i| \ne 0$. The equation

$$A^*(z_i)X(z_i) = B(z_i)C^*(z_i)$$

is trivially satisfied and the solution to (12.50) is

$$X(z_i) = -\frac{\rho A(z_i)C^*(z_i)}{z_i^d B^*(z_i)}$$

A similar argument shows that $X(z_i)$ is unique also when $B^*(z_i) = 0$ and $A(z_i) \ne 0$. We can thus determine $\deg P$ values $X(z_i)$. Using Lagrange's interpolation formula the polynomial $X(z)$ of degree $\deg P - 1$ which satisfies (12.50) is thus unique.

It follows from the construction of the polynomial $X(z)$ that the polynomial $A^*(z)X(z) - B(z)C^*(z)$ vanishes for the zeros z_i of $P(z)$. This implies that it is divisible by $P(z)$. The quotient

$$S^*(z) = \frac{A^*(z)X(z) - B(z)C^*(z)}{rP(z)}$$

is thus a polynomial. It has degree

$$\deg S^* \leq \max(\deg A^* + \deg P - 1, \deg B + \deg C^*) - \deg P < n \qquad (12.51)$$

Using the same argument we also find that

$$R^*(z) = \frac{z^d B^*(z) X(z) + \rho A(z) C^*(z)}{r P(z)}$$

is a polynomial of degree

$$\deg R^* \leq \max(d + \deg B^* + \deg P - 1, \deg A + \deg C^*) - \deg P \leq n \qquad (12.52)$$

The solution $X(z)$, $S^*(z)$ and $R^*(z)$ to (12.49) is continuous in the coefficients of polynomials $A(z)$ and $B(z)$. If polynomial $P(z)$ has multiple zeros we can the perturb the coefficients of $A(z)$ and $B(z)$ to obtain a $P(z)$ with distinct zeros and obtain the results by a limiting procedure. The details of this argument are delicate.   ∎

Remark 1.   Notice that if one solution, X_0, R_0^*, S_0^*, to Eq. (12.49) has been obtained all other solutions are given by

$$\begin{aligned}
X(z) &= X_0(z) + Q(z) r P(z) \\
R^*(z) &= R_0^*(z) + Q(z) z^d B^*(z) \\
S^*(z) &= S_0^*(z) - Q(z) A^*(z)
\end{aligned} \qquad (12.53)$$

where $Q(z)$ is an arbitrary polynomial. This is easily verified by direct insertion into the equation.

Remark 2.   The polynomials $R(z)$ and $S(z)$ are given by $R(z) = z^n R^*(z^{-1})$ and $S(z) = z^n S^*(z^{-1})$. The conditions $A^*(0) = P^*(0) = C^*(0) = 1$ together with Eq. (12.53) imply that $R^*(0) = 1$, hence $\deg R(z) = n$ and $\deg S(z) \leq n$ and $\deg S^*(z) < n$.

Remark 3.   Eliminating X by multiplying the first equation by $z^d B^*(z)$ and the second by $A^*(z)$ and subtracting gives

$$r P S^* z^d B^* + r P A^* R^* = B C^* z^d B^* + \rho A^* C^* = r P P^* C^*$$

where the second equality follows from (12.46). Dividing by rP shows that the the polynomials R^* and S^* satisfy the Diophantine equation (12.48).

Remark 4.   In the following we will need another property of the solutions to Eq. (12.49). Adding the first equation multiplied by ρA and the second multiplied by B gives:

$$(\rho A A^* + z^d B B^*) X + \rho r A P S^* - r P B R^* = 0$$

Using the spectral factorization condition (12.46) and dividing by rP now gives:

$$P^*(z) X(z) = B(z) R^*(z) - \rho A(z) S^*(z) \qquad (12.54)$$

After these preliminaries we will now solve the LQG-problem with polynomial calculations.

THEOREM 12.4   LINEAR QUADRATIC GAUSSIAN CONTROL   Consider the system in (12.5) with $\deg A(z) = \deg C(z) = n$. Assume that all the zeros of polynomial $C(z)$ are inside the unit disc, that there are no factors common to all three of the polynomials $A(z)$, $B(z)$, and $C(z)$, and that a possible common factor of $A(z)$ and $B(z)$ has all its zeros inside the unit disc. Let the monic polynomial $P(z)$, which has all its zeros inside the unit disc, be the solution to (12.45) with $\deg P(z) = n$. The admissible control law with no delay that minimizes the criterion of (12.8) is given by

$$u(k) = -\frac{S^*(q^{-1})}{R^*(q^{-1})}\, y(k) = -\frac{S(q)}{R(q)}\, y(k) \tag{12.55}$$

where polynomials $R^*(z)$ and $S^*(z)$ are the unique solution to Equation (12.49) with $\deg X(z) < n$. With the control law of (12.55), the output becomes

$$y(k) = \frac{R(q)}{P(q)}\, e(k) \tag{12.56}$$

and the control signal is

$$u(k) = -\frac{S(q)}{P(q)}\, e(k) \tag{12.57}$$

The minimal value of the loss function is

$$\min \mathrm{E}(y^2 + \rho u^2) = \frac{\sigma^2}{2\pi i} \oint \frac{R(z)R(z^{-1}) + \rho S(z)S(z^{-1})}{P(z)P(z^{-1})}\, \frac{dz}{z} \tag{12.58}$$

Proof.   Introduce

$$u = v - \frac{S}{R}\, y \tag{12.59}$$

where v may be regarded as a transformed control variable, which has to be determined. Equations (12.5), (12.47), and (12.59) give

$$y = \frac{BRv + CRe}{AR + BS} = \frac{BRv + CRe}{PC} = \frac{BR}{PC}\, v + \frac{R}{P}\, e \tag{12.60}$$

It then follows from (12.60) that

$$u = v - \frac{SBv + SCe}{PC} = \frac{PC - BS}{PC}\, v - \frac{S}{P}\, e = \frac{AR}{PC}\, v - \frac{S}{P}\, e \tag{12.61}$$

The loss function of (12.8) can be written as

$$J = \mathrm{E}(y^2 + \rho u^2) = \mathrm{E}\left(\frac{BR}{PC}\, v + \frac{R}{P}\, e\right)^2 + \rho \mathrm{E}\left(\frac{AR}{PC}\, v - \frac{S}{P}\, e\right)^2$$

$$= J_1 + 2J_2 + J_3$$

where

$$
J_1 = \mathrm{E}\left(\left(\frac{BR}{PC}v\right)^2 + \rho\left(\frac{AR}{PC}v\right)^2\right)
$$

$$
J_2 = \mathrm{E}\left(\left(\frac{BR}{PC}v\right)\left(\frac{R}{P}e\right) - \rho\left(\frac{AR}{PC}v\right)\left(\frac{S}{P}e\right)\right)
$$

$$
J_3 = \mathrm{E}\left(\left(\frac{R}{P}e\right)^2 + \rho\left(\frac{S}{P}e\right)^2\right)
$$

It follows from Remark 2 of Theorem 10.2 and (12.45) that

$$
J_1 = \frac{1}{2\pi i}\oint \frac{(B(z)B(z^{-1}) + \rho A(z)A(z^{-1}))R(z)R(z^{-1})}{P(z)P(z^{-1})C(z)C(z^{-1})} V(z)V(z^{-1})\frac{dz}{z}
$$

$$
= \frac{r}{2\pi i}\oint \frac{R(z)R(z^{-1})}{C(z)C(z^{-1})} V(z)V(z^{-1})\frac{dz}{z} = r\mathrm{E}\left(\frac{R(q)}{C(q)}v\right)^2
$$

For causal controllers with no time delay $v(t)$ can be expressed as $v(t) = V(q)e(t)$, where $V(q)$ is a rational function with zero pole excess.

$$
J_2 = \frac{\sigma^2}{2\pi i}\oint \frac{B(z)R(z)R(z^{-1}) - \rho A(z)R(z)S(z^{-1})}{P(z)C(z)P(z^{-1})} V(z)\frac{dz}{z}
$$

It follows from Equation (12.54) that

$$
B(z)R(z^{-1}) - \rho A(z)S(z^{-1}) = P(z^{-1})X(z)
$$

Hence

$$
J_2 = \frac{\sigma^2}{2\pi i}\oint \frac{R(z)X(z)}{P(z)C(z)} V(z)\frac{dz}{z} = \mathrm{E}\left(\left(\frac{R(q)X(q)}{P(q)C(q)}v(k)\right)e(k)\right)
$$

It was assumed that $P(z)$ and $C(z)$ are stable and it follows from Lemma 12.2 that $\deg X(z) < n$. This implies that

$$
\deg R(z)X(z) < \deg P(z)C(z) = 2n
$$

The quantity

$$
\frac{R(q)X(q)}{P(q)C(q)}v(k)
$$

is thus a function of $v(k-1), v(k-2), \ldots$. Because all these terms are independent of $e(k)$, J_2 becomes zero. The loss function can thus be written as

$$
J = r\mathrm{E}\left(\frac{R(q)}{C(q)}v(k)\right)^2 + \mathrm{E}\left(\frac{R(q)}{P(q)}e(k)\right)^2 + \rho\mathrm{E}\left(\frac{S(q)}{P(q)}e(k)\right)^2
$$

where P and C are stable polynomials. It follows that the loss function achieves its minimum (12.58) for $v = 0$, which by (12.59) corresponds to the control law of (12.55). Equations (12.56) and (12.57) follow from (12.60) and (12.61), and Theorem 10.2 and (10.23) give the formula of (12.58). ∎

Remark 1.   The minimum-variance control law is a special case of Theorem 12.4 with $\rho = 0$. It follows from (12.49) that $R^*(z)P(z) = -z^d B^*(z)X(z)$. Because $\deg X(z) < n$, we have $\deg R^*(z) < n$ for $\rho = 0$. Because also $\deg S^*(z) < n$ the polynomials $R(z)$ and $S(z)$ have z as a common factor. Introducing $B(z) = B^+(z)B^-(z)$, where B^+ has all its zeros inside the unit disc and B^- all its zeros outside the unit disc, we get

$$\sqrt{r}\, P(z) = z^d B^+(z)B^{-*}(z)$$

where $\sqrt{r} = B^-(0)$. The Diophantine equation (12.47) then becomes

$$A(z)R(z) + B(z)S(z) = z^d B^+(z)B^{-*}(z)C(z)/\sqrt{r}$$

Cancelling the common factor z in $R(z)$ and $S(z)$ to give $R(z)$ and $S(z)$ we get

$$A(z)\tilde{R}(z) + B(z)\tilde{S}(z) = z^{d-1}B^+(z)B^{-*}(z)C(z)/\sqrt{r}$$

which is identical to (12.32). Theorem 12.3 has thus been proven in a different way. The pole-zero cancellation at the origin of the control law explains that there are $d-1$ instead of d closed-loop poles at the origin. Compare with (12.17).

Remark 2.   If the polynomial $A(z)$ has the form $A(z) = z^l A_1(z)$, where $l \leq d = \deg A(z) - \deg B(z)$, it follows from (12.45) that $P(z) = z^l P_1(z)$. Equation (12.47) then implies that $S(z) = z^l S_1(z)$.

The LQG controller will now be illustrated by an example.

Example 12.10   LQG control with unstable process zero

Consider the same system as in Examples 12.8 and 12.9. Instead of using a minimum-variance control law we will now use an LQG strategy. To do this the parameter ρ in the control strategy must be chosen. To guide this choice we will first calculate the variances of the output and control signals obtained for different values of the loss function. The results are shown in Fig. 12.8. The value $\rho = 0$ corresponds to a minimum-variance strategy. This gives a control signal with large variance. Compare with Example 12.9. The variance of the control signal decreases rapidly with increasing ρ. The variance of the output increases slowly.

By choosing a reasonable value of ρ it is possible to have a control strategy that gives an output variance that is only marginally higher than with minimum-variance control and a variance of the control signal that is substantially lower. A reasonable value is $\rho = 1$. This gives $Ey^2 = 1.39$ and $Eu^2 = 0.22$, which can be compared with minimum-variance control that gives $Ey^2 = 1.05$ and $Eu^2 = 14.47$.

The input- and output signals obtained with $\rho = 1$ are shown in Fig. 12.9. Compare with the corresponding curves for minimum-variance control in Example 12.9. The fluctuations in the output are a little larger, but the fluctuations in the control signal are substantially smaller. This way of applying LQG control where the control weighting is used as a design parameter is very typical.   ∎

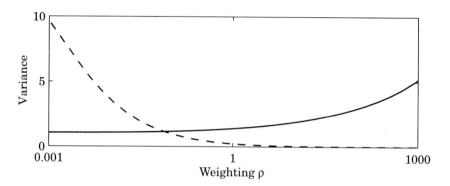

Figure 12.8 Variances of input u (dashed line) and output y (solid line) for LQG controllers having different values of the control weighting ρ for the system in Example 12.10

An Interpretation

Theorem 12.4 establishes the relation between LQG-control and pole-placement control because the polynomial $C(z)$ is the observer polynomial $A_o(z)$ and $P(z)$ is the polynomial $A_c(z)$. The LQG-controller may thus be considered as a pole-placement controller where the observer polynomial $A_o(z)$ is obtained from the noise characteristics and the polynomial $A_c(z)$ from the solution to an optimization problem. The solution to the optimization problem also tells what solution of the Diophantine equation we should choose.

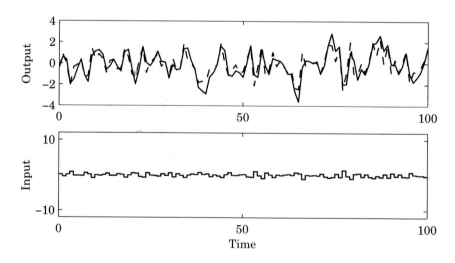

Figure 12.9 Simulation of the for the system in Example 12.10 using the LQG-controller with $\rho = 1$. The output obtained with the minimum-variance controller ($\rho = 0$) is shown in dashed. Also compare with Fig. 12.6.

A Computational Procedure

Theorem 12.4 gives a convenient way to compute the LQG-control law for SISO systems, which can be described as follows.

1. Rewrite the model of the process and the disturbance in the standard form (12.5), where $C(z)$ is a stable polynomial. It may be necessary to use a spectral factorization to obtain this form.

2. Use a spectral factorization to calculate $P(z)$. If the polynomials $A(z)$ and $B(z)$ have a stable common factor $A_2(z)$, the calculations of the control law can be simplified by first factoring $A(z)$ and $B(z)$ as $A(z) = A_1(z)A_2(z)$ and $B(z) = B_1(z)A_2(z)$. It follows from (12.45) that $A_2(z)$ also divides $P(z)$. This polynomial can thus be written as $P(z) = P_1(z)A_2(z)$, where $P_1(z)$ is given by

$$rP_1(z)P_1(z^{-1}) = \rho A_1(z)A_1(z^{-1}) + B_1(z)B_1(z^{-1})$$

The polynomial $P(z)$ is then equal to $P_1(z)A_2(z)$, which is stable, because $A_2(z)$ was assumed stable. Equation (12.47) can also be divided by $A_2(z)$ to give

$$P_1(z)C(z) = A_1(z)R(z) + B_1(z)S(z)$$

where $\deg R(z) = \deg S(z) = \deg C(z) = n$, and $S(0) = 0$.

3a. If there are no common factors between A and B and if $A(0) \neq 0$ then the controller is given by a unique solution to the Diophantine equation (12.47) such that $\deg R(z) = \deg S(z) = n$, and $S(0) = 0$.

3b. If there are stable common factors of A and B or if $A(0) = 0$ the solution is obtained from the Equation (12.49) or Diophantine equation (12.47), and (12.54).

The computational procedure shows that when there are no common factors between A and B and when $A(0) \neq 0$ then it is sufficient to solve only one Diophantine equation with the extra constraint $S(0) = 0$ to obtain a unique solution. In other cases it is necessary to solve the coupled equations (12.49). Theorem 12.4 is illustrated by two examples.

Example 12.11   LQG for first order system

Consider a system characterized by

$$A(z) = z + a \qquad a \neq 0$$
$$B(z) = b$$
$$C(z) = z + c$$

To find the control law that minimizes the criterion of (12.8), the spectral-factorization problem is first solved. Equation (12.45) can be written as

$$r(z + p_1)(z^{-1} + p_1) = \rho(z + a)(z^{-1} + a) + b^2$$

Equating coefficients of equal powers of z gives

$$rp_1 = \rho a$$

$$r(1 + p_1^2) = \rho(1 + a^2) + b^2$$

Elimination of p_1 gives

$$r^2 - r(\rho(1 + a^2) + b^2) + \rho^2 a^2 = 0 \tag{12.62}$$

This equation has the solution

$$r = \frac{1}{2}\left(\rho(1 + a^2) + b^2 + \sqrt{\rho^2(1 - a^2)^2 + 2\rho b^2(1 + a^2) + b^4}\right)$$

where the positive root is chosen to give $|p_1| < 1$. Furthermore

$$p_1 = \frac{\rho a}{r}$$

Because A and B are relative prime and $A(0) \neq 0$, the solution can be found from the Diophantine equation (12.47). With $\deg S = 1$ and $S(0) = 0$, Eq. (12.47) becomes

$$(z + a)(z + r_1) + bs_0 z = (z + p_1)(z + c)$$

Putting $z = -a$ we get

$$s_0 = -\frac{(p_1 - a)(c - a)}{ab}$$

It follows from (12.62) that

$$\rho a p_1^2 - \rho p_1(1 + a^2) - p_1 b^2 + \rho a = 0$$

Hence

$$\rho(a p_1^2 - a^2 p_1 - p_1 + a) = p_1 b^2$$

or

$$\rho\left(a p_1(p_1 - a) - (p_1 - a)\right) = p_1 b^2$$

which gives

$$p_1 - a = -\frac{p_1 b^2}{\rho(1 - a p_1)}$$

We thus get

$$s_0 = \frac{p_1 b^2(c - a)}{\rho ab(1 - a p_1)} = \frac{b(c - a)}{r(1 - a p_1)}$$

Furthermore, equating the constant terms in (12.47) gives

$$r_1 = \frac{p_1 c}{a} = \frac{\rho c}{r}$$

The control law thus becomes

$$u(k) = -\frac{S(q)}{R(q)} y(k) = -\frac{b(c-a)}{r(1-ap_1)} \frac{q}{q + p_1 c/a} y(k)$$

∎

The calculations in Example 12.11 do not work when $a = 0$, because in this case the solution to the LQG-problem is not uniquely determined by the Diophantine equation (12.47) and it is necessary to use (12.49).

Example 12.12   LQG for system with a time-delay

Consider the case

$$A(z) = z$$
$$B(z) = b$$
$$C(z) = z + c$$

The spectral factorization problem (12.45) has the solution

$$P(z) = z \qquad r = \rho + b^2$$

Assuming that it is desired to have a controller with no extra delay we require that $\deg S(z) = \deg R(z) = 1$. The Diophantine equation (12.47) with the constraint $\deg S^*(z) = 0$ becomes

$$z(z + r_1) + bs_0 z = z(z + c)$$

Identification of coefficients of equal power of z gives only one equation

$$r_1 + bs_0 = c$$

to determine two parameters r_1 and s_0. The approach with the Diophantine equation thus does not work in this case. Equation (12.49) gives

$$x_0 + rs_0 z = b(1 + cz)$$
$$bx_0 z - rz(1 + r_1 z) = -\rho z(1 + cz)$$

Identification of coefficients of equal power of z gives linear equations which have the solution

$$x_0 = b$$
$$r_1 = \frac{\rho c}{r} = \frac{\rho c}{\rho + b^2}$$
$$s_0 = \frac{bc}{r}$$

∎

Uncontrollable and Unstable Modes

Models with the property that polynomials $A(z)$ and $B(z)$ have a common factor that is not a factor of $C(z)$ are important in practice. They appear when there are modes that are excited by disturbances and uncontrollable from the input. Compare Sec. 12.2. Because the modes are not controllable, they are not influenced by feedback.

Theorem 12.4 covers the case of stable common factors, but it does not work for unstable common factors. Unstable common factors are important in practice because they give one way of obtaining regulators with integral action.

To see what happens when there are unstable common factors, let A_2 denote the greatest common divisor of A and B and let A_2^- denote the factor of A_2 with zeros outside the unit disc or on the unit circle. Let the feedback be

$$u(k) = -\frac{S(q)}{R(q)} y(k)$$

where $R(z)$ and $S(z)$ are relatively prime. It follows from (12.5) that

$$y(k) = \frac{R(q)C(q)}{A(q)R(q) + B(q)S(q)} e(k) \tag{12.63}$$

$$u(k) = -\frac{S(q)C(q)}{A(q)R(q) + B(q)S(q)} e(k) \tag{12.64}$$

The unstable factor $A_2^-(z)$ divides the denominators of the right-hand sides of (12.63) and (12.64). Both y and u will be unbounded unless $R(z)$ or $S(z)$ are chosen in special ways. The signal y will be bounded if $R(z)$ is divisible by $A_2^-(z)$, and u will be bounded if $A_2^-(z)$ divides $S(z)$. Because $R(z)$ and $S(z)$ are relatively prime, it is not possible to make both y and u bounded. This is natural because infinitely large control actions are necessary to compensate for infinitely large disturbances.

To describe a problem of this type as a meaningful optimization problem, the criterion of (12.8) must be modified. One possibility is to introduce the variable

$$w(k) = q^{-m}A_2^-(q)u(k) \tag{12.65}$$

where $m = \deg A_2^-(z)$, and to introduce the criterion

$$J'_{lq} = E\left(y^2(k) + \rho w^2(k)\right) \tag{12.66}$$

Example 12.13   Integral action

Let the system be described by

$$y(k) = \frac{B_1(q)}{A_1(q)} u(k) + \frac{C_1(q)}{q-1} e(k)$$

which is a special case of Eqs. (12.1) to (12.4) with a drifting disturbance. Hence

$$A(q) = (q - 1)A_1(q)$$
$$B(q) = (q - 1)B_1(q)$$
$$C(q) = A_1(q)C_1(q)$$

Unbounded control signals are necessary to compensate for the unbounded disturbance. This implies that the modified loss function (12.66) becomes

$$J'_{lq} = E\left[y^2(k) + \rho\left(\Delta u(k)\right)^2\right]$$

where

$$\Delta u(k) = u(k) - u(k-1)$$

This means that the difference and not the absolute value of the control signal is penalized. The solution to the LQG-problem gives a controller with integral action. ∎

The following result can then be established.

THEOREM 12.5 LQG-CONTROL WITH UNSTABLE COMMON FACTORS   Consider the system described by (12.5), where $A(z)$ and $C(z)$ are monic polynomials of degree n. Assume that all zeros of $C(z)$ are inside the unit disc and that there is no nontrivial polynomial that divides $A(z)$, $B(z)$, and $C(z)$. Let $A_2(z)$ be the greatest common divisor of $A(z)$ and $B(z)$, let $A_2^+(z)$ of degree l be the factor of $A_2(z)$ with all its zeros inside the unit disc, and let $A_2^-(z)$ of degree m be the factor of $A(z)$ that has zeros on the unit circle or outside the unit disc. The admissible control law that minimizes (12.66) is given by

$$u(k) = -\frac{S(q)}{R(q)}y(k)$$

where $R(z)$ and $S(z)$ are of degree $n + m$

$$R(z) = A_2^-(z)\tilde{R}(z)$$
$$S(z) = z^m\tilde{S}(z) \tag{12.67}$$

and $\tilde{R}(z)$ and $\tilde{S}(z)$ satisfies

$$A_1(z)A_2^-(z)\tilde{R}(z) + z^m B_1(z)\tilde{S}(z) = P_1(z)C(z)$$
$$A^*(z)X(z) + rP(z)\tilde{S}^*(z) = q^m\tilde{B}(z)C^*(z) \tag{12.68}$$

with $\deg \tilde{R}(z) = \deg \tilde{S}(z) = n$, $\deg X(z) < n$ and $\tilde{S}(0) = 0$. Furthermore

$$A(z) = A_1(z)A_2(z)$$
$$B(z) = B_1(z)A_2(z)$$
$$\tilde{B}(z) = B_1(z)A_2^+(z)$$

and $P_1(z)$ is the solution of the spectral-factorization problem

$$rP_1(z)P_1(z^{-1}) = \rho A_1(z)A_2^-(z)A_1(z^{-1})A_2^-(z^{-1}) + B_1(z)B_1(z^{-1}) \qquad (12.69)$$

with $\deg P_1(z) = \deg A_1(z) + \deg A_2^-(z)$.

    Proof.   Introducing the signal (12.65), the model (12.5) can be written as

$$A(q)y(k) = \tilde{B}(q)q^m w(k) + C(q)e(k)$$

The polynomials $A(q)$ and $\tilde{B}(q)$ have the common factor $A_2^+(z)$, which has all its zeros inside the unit disc, but no other common factors with zeros outside the unit disc or on the unit circle. It then follows from Theorem 12.4 that the optimal control law

$$w(k) = -\frac{\tilde{S}(q)}{\tilde{R}(q)}y(k)$$

is obtained from (12.47). Because $A(z)$ and $\tilde{B}(z)$ have the stable common factor $A_2^+(z)$, the polynomial $P(z)$ has the form

$$P(z) = A_2^+(z)P_1(z)$$

where $P_1(z)$ is the solution to the spectral-factorization problem (12.69). From Lemma 12.2 the polynomials $\tilde{R}(z)$ and $\tilde{S}(z)$ satisfy the equations

$$A(z)\tilde{R}(z) + z^m\tilde{B}(z)\tilde{S}(z) = A_2^+(z)P_1(z)C(z)$$
$$A^*(z)X(z) + rP(z)\tilde{S}^*(z) = q^m\tilde{B}(z)C^*(z)$$

with $\deg \tilde{R}(z) = \deg \tilde{S}(z) = n$. Because A_2^+ divides $A(z)$ and $\tilde{B}(z)$ we get (12.68). Using (12.65) to express the control law in terms of the control variable u gives the result.   ■

    Remark.   Notice that using (12.67), Eq. (12.68) can be written as

$$A(z)R(z) + B(z)S(z) = A_2(z)P_1(z)C(z)$$

The LQG-solution can thus be interpreted as a pole-placement controller, where the poles are positioned at the zeros of A_2, P_1, and C. The controller also has the property that A_2^- divides R. This is an example of the internal model principle.

Command Signals

The discussion in this chapter has so far been limited to the regulator problem. To introduce command signals, refer to the discussion in Chapter 5. The key issue is to introduce the command signals in such a way that they do not generate unnecessary reconstruction errors. This is achieved by the control law

$$R(q)u(k) = t_0 A_o(q)u_c(k) - S(q)y(k)$$

where $A_o(q)$ is the observer polynomial and to a constant. For the optimal Kalman filter $A_o(q) = C(q)$, where $C(q)$ is given by (12.40). It then follows from (12.5) that the output of the system is given by

$$y(k) = t_0 \frac{B(q)}{P(q)} u_c(k) + \frac{R(q)}{P(q)} e(k)$$

where $\deg R = n$.

The pulse-transfer function from the command signal is $B(z)/P(z)$. This response may be shaped further by cascading with a precompensator that has an arbitrary stable transfer function $H_f(z)$. The control law becomes

$$u(k) = \frac{A_o(q)}{R(q)} H_f(q)u_c(k) - \frac{S(q)}{P(q)} y(k)$$

which gives

$$y(k) = \frac{B(q)}{P(q)} H_f(q)u_c(k) + \frac{R(q)}{P(q)} e(k)$$

Because the polynomial P is stable, this may be canceled by the precompensator. It thus follows that the response for disturbances and command signals may be shaped differently.

The feedback S/R is first designed to ensure a good response to disturbances. The precompensator H_f is then chosen to obtain the desired response to command signals.

12.6 Practical Aspects

Much of the arbitrariness of design seems to disappear when design problems are formulated as optimization problems. The model and the criteria are stated, and the control law is obtained simply as the solution to an optimization problem. This simplicity is deceptive because the arbitrariness is instead transferred to the modeling and the formulation of criteria. A successful application of optimization theory requires insight into how the properties of the model and the criteria are reflected in the control law. Typical questions are: What should the model look like in order to get a regulator with integral action? What problem statements give regulators with a PID-structure? Some of these issues are discussed in this section, which also gives insight into the properties of the optimal control laws. It turns out that some results can be formulated as design rules.

The polynomial approach, which operates directly with the transfer functions, is well suited to do this.

Other aspects of practical relevance, such as sensitivity and robustness, are also discussed. A brief treatment of the intersample ripple of the loss function is given, together with some aspects of the choice of the sampling period.

Properties of the Optimal Regulator

Some properties of the model influence the optimal-control laws. The basic model used is given by (12.5)—that is,

$$A(q)y(k) = B(q)u(k) + C(q)e(k) \tag{12.70}$$

The ratio B/A represents the pulse-transfer function of the process, and the ratio C/A represents the pulse-transfer function that generates the disturbance of the process output. The polynomials A, B, and C may have common factors that reflect the way the control signal and the disturbance are coupled to the system. There are, however, no factors common to all three polynomials. Compare this with the discussion in Sec. 12.2, where the model is derived. The presence of common factors that will directly influence the properties of the regulators will now be investigated.

The internal-model principle.   Factors that are common to polynomials A and B correspond to disturbance modes that are not controllable from u. Such modes will appear as factors of P. Let

$$A_2 = \gcd(A, B)$$

be the greatest common divisor of polynomials A and B. If A_2 is stable, it follows from Theorem 12.4 that A_2 also divides P. If A_2 has a factor A_2^- with all its zeros outside the unit disc, the corresponding result follows from Theorem 12.5. In this case it also follows from Theorem 12.5 that A_2 divides R. This observation is called the *internal-model principle*; it says that to regulate a system with unstable disturbances, the disturbance dynamics must also appear in the dynamics of the regulator. A few examples illustrate this idea.

Example 12.14   Integral action

A regulator has integral action if $z - 1$ divides $R(z)$. It follows from Theorem 12.5, and the internal-model principle, that this will occur if $z - 1$ divides both A and B, which means that the model is of the form

$$A_1(q)(q - 1)y(k) = B_1(q)(q - 1)u(k) + C(q)e(k)$$

This means that there is a drifting disturbance.   ∎

Example 12.15 Elimination of a sinusoidal disturbance

A narrow-band sinusoidal disturbance with frequency centered at ω may be represented as white noise driving a system with the denominator

$$D(q) = q^2 - 2q \cos \omega h + 1$$

If the poles of the system dynamics do not correspond to D, the model becomes

$$A_1(q)D(q)y(k) = B_1(q)D(q)u(k) + C(q)e(k)$$

The optimal regulator is then such that $D(z)$ divides $R(z)$.   ∎

Cancellation of process poles.   A common factor of A and C corresponds to controllable modes that are not excited by the disturbances. Let A_2 be the greatest common divisor of A and C. The polynomial A_2 is stable because C is stable, and it does not divide B because there is no factor that divides all of A, B, and C. It follows from (12.47) that A_2 also divides the polynomial S, which is the numerator of the regulator transfer function. Thus *stable process poles that are not excited by the disturbances may be canceled*.

Cancellation of process zeros.   Common factors of B and C correspond to process zeros that block transmission both for the control signal u and for the disturbance e. Let B_2 be the greatest common divisor of B and C. The polynomial B_2 is stable and it does not divide A. It then follows from (12.47) that B_2 divides R. This means that the zeros corresponding to $B_2 = 0$ are canceled by the regulator. Therefore, *process zeros that are also transmission zeros for the disturbance C are canceled by the regulator*.

For the minimum-variance control, it follows from (12.46) with $\rho = 0$ that

$$\sqrt{r}P = q^d B^+ B^{-*}$$

where $\sqrt{r} = B^-(0)$ and from (12.47) that B^+ divides R. All stable zeros are thus canceled by the minimum-variance control law.

An analysis of the properties of the optimal-control law thus gives partial answers to the classic cancellation problem.

Sensitivity and Robustness

It is important that a control system be insensitive or robust with respect to measurement errors, plant disturbances, and modeling errors. This may be analyzed as in Sec. 5.5 for the pole-placement problem. The robustness properties are conveniently expressed in terms of the loop gain:

$$\mathcal{L} = \frac{BS}{AR}$$

or the return difference

$$H_{rd} = \frac{1}{S} = 1 + \frac{BS}{AR} = \frac{AR + BS}{AR} = \frac{PC}{AR}$$

The loop gain $\mathcal{L}(\exp i\omega h)$ is normally high for low frequencies and small for high frequencies. The crossover frequency ω_c is the lowest frequency, where

$$\left| \mathcal{L}\left(e^{i\omega_c h} \right) \right| = 1$$

The closed-loop system is insensitive to plant disturbances at those frequencies where the loop gain is high. To have low sensitivity to poor modeling of the high-frequency dynamics of the plant, it is desirable that the loop gain decreases rapidly above the crossover frequency. It is possible to make sure that the loop gain is high for certain frequencies by choosing models with special structure, as was done in Examples 12.14 and 12.15. Plots similar to those in Fig. 5.6 are also useful in evaluating the sensitivity. In a properly designed sample-data system, there will be antialiasing filters, which eliminate signal transmission above the Nyquist frequency. The selection of a proper sampling rate is one way to make sure that the loop gain is low over a given frequency. This also means that high-frequency modeling errors have little influence. Notice, however, that plots of the loop gain and the return difference will not give the complete picture because there may be pole-zero cancellations that do not show up in these plots.

An analysis of the characteristic equations is useful in such a case. To perform such an analysis, assume that the system is governed by

$$A^0(q)y(k) = B^0(q)u(k) + C^0(q)e(k) \tag{12.71}$$

but that a regulator is designed based on a different model, as in (12.70). The regulators given by Theorems 12.4 and 12.5 give a closed-loop system with the characteristic polynomial

$$A^0 R + B^0 S = A^0 R - AR + B^0 S - BS + AR + BS$$
$$= PC + (A^0 - A)R + (B^0 - B)S$$

When the model of (12.70) is equal to the system of (12.71) the characteristic polynomial is $PC = P_1 A_2 C$, as expected. By continuity it also follows that small changes in the system give small changes in the closed-loop poles. The system is sensitive to changes in the parameters if polynomial P_1 or C have zeros close to the unit circle.

To guarantee systems with a low sensitivity, it is necessary to impose further constraints. Recall that both C and P were obtained as solutions to a spectral-factorization problem.

Closed-Loop Systems with Guaranteed Exponential Stability

The control laws given by Theorems 12.2, 12.3, 12.4, and 12.5 give closed-loop systems with poles inside the unit disc. It is sometimes desirable to have control laws such that the closed-loop system has its poles inside a circle with radius $\bar{r}$. It is straightforward to formulate optimization problems that give such control laws.

Introduce the criterion

$$J = E\bar{r}^{-2k}\left(y^2(k) + \rho u^2(k)\right) \tag{12.72}$$

If a control law that minimizes this criterion can be found, the variables $y(k)$ and $u(k)$ must converge to zero at least as fast as $\bar{r}^k$ when k increases. To obtain such a result, it must be assumed that the model of (12.5) is such that the covariance of $e(k)$ also goes to zero as $\bar{r}^k$.

Introduce the scaled variables η, μ, and ε defined by

$$y(k) = \bar{r}^k\eta(k)$$
$$u(k) = \bar{r}^k\mu(k)$$
$$e(k) = \bar{r}^k\varepsilon(k)$$

Because

$$q^l y(k) = q^l\left(\bar{r}^k\eta(k)\right) = \bar{r}^{k+l}\eta(k+l) = \bar{r}^k(\bar{r}q)^l\eta(k)$$

it follows that

$$A(q)y(k) = A(q)\left(\bar{r}^k\eta(k)\right) = \bar{r}^k A(\bar{r}q)\eta(k)$$

Introducing the transformed polynomials

$$\tilde{A}(z) = A(\bar{r}z)$$
$$\tilde{B}(z) = B(\bar{r}z)$$
$$\tilde{C}(z) = C(\bar{r}z)$$

the model of (12.5) can be written as

$$\tilde{A}(q)\eta(k) = \tilde{B}(q)\eta(k) + \tilde{C}(q)\varepsilon(k) \tag{12.73}$$

and the criterion of (12.72) becomes

$$J = E\left(\eta^2(k) + \rho\mu^2(k)\right) \tag{12.74}$$

The control law that minimizes (12.74) for the system of (12.73) is then given by Theorem 12.4. This control law gives a closed-loop system in which all the zeros of the characteristic equation

$$\tilde{P}(z)\tilde{C}(z) = 0$$

are inside the unit disc. Going back to the original variables results in the characteristic equation

$$P(z)C(z) = \tilde{P}\left(\frac{z}{\bar{r}}\right)\tilde{C}\left(\frac{z}{\bar{r}}\right) = 0$$

All the zeros of this equation are inside the circle $|z| = \bar{r}$.

A simple procedure for obtaining feedback laws that give closed-loop systems with all poles inside the circle $|z| = \bar{r}$ has thus been devised.

Disturbance Reduction

The return difference is

$$H_{rd} = 1 + \mathcal{L} = 1 + \frac{BS}{AR} = \frac{AR + BS}{AR}$$

The inverse of the return difference is a measure of how effectively the closed-loop system eliminates disturbances.

Consider the model of (12.70). Without control the output is

$$y_{ol} = \frac{C}{A}e$$

With the LQ-control law, the output becomes

$$y_{lqg} = \frac{R}{P}e$$

Elimination of e between these equations gives

$$y_{lqg} = \frac{AR}{PC}y_{ol} = \frac{1}{\dfrac{PC}{AR}}y_{ol} = \frac{1}{1 + \dfrac{BS}{AR}}y_{ol} = \frac{1}{H_{rd}}y_{ol} = S\,y_{ol}$$

The sensitivity function thus tells how much disturbances of different frequencies are attenuated.

Selection of the Sampling Period

There is a substantial difference between the minimum-variance control law discussed in Sec. 12.4 and the LQG-control law discussed in Sec. 12.5 in terms of the influence of the sampling period. The choice of sampling period is critical for the minimum-variance control. A short sampling period gives a high-bandwidth system, which settles quickly. The control actions will also be large when the sampling period is short. In this respect, the minimum-variance control law is similar to the deadbeat control law discussed in Sec. 4.3. The sampling period is less critical for LQG-control. It follows from the analysis of Sec. 11.5 that the control law approaches continuous-time control as the sampling period h goes to zero. The following discussion therefore concentrates on the minimum-variance control law.

Intersample Variation of the Output Variance

The minimum-variance control law minimizes the variance of the output *at the sampling instants*. However, the main objective may be to minimize the continuous-time loss function of (12.7). This may be achieved by first sampling the continuous-time loss function and to minimize the corresponding discrete-time loss function as was discussed in Section 11.1. This results in a complicated design procedure. The minimum-variance control laws are in many cases a sufficiently good approximation. It is useful to investigate the intersample variation of the loss function. This analysis is similar to the analysis of intersample ripple for deterministic systems of Sec. 3.5. An example is used to illustrate the idea.

Example 12.16   Intersample variation of the loss function

Consider the continuous-time system

$$dx = u \, dt + dv \tag{12.75}$$

where $v(t)$ is a Wiener process with incremental covariance $\sigma_v^2 \, dt$. Assume that the output is observed without antialiasing filters at times $t_k = k \cdot h$, where h is the sampling period. Hence,

$$y(t_k) = x(k) + \varepsilon(t_k)$$

where $\varepsilon(t_k)$ is a sequence of independent random variables with zero mean and covariance σ_ε^2. Sampling of the system gives

$$x(kh + h) = x(kh) + hu(kh) + v(kh + h) - v(kh)$$

$$y(kh) = x(kh) + \varepsilon(kh)$$

Hence,

$$y(kh + h) = y(kh) + hu(kh) + \varepsilon(kh + h) - \varepsilon(kh) + v(kh + h) - v(kh)$$

The disturbance on the right-hand side may be represented as

$$w(kh + h) = e(kh + h) + ce(kh)$$

where $e(kh)$ is a sequence of independent zero-mean random variables with standard deviation σ.

Simple calculations give

$$c = -1 - \frac{h\sigma_v^2}{2\sigma_\varepsilon^2} + \sqrt{\frac{h\sigma_v^2}{\sigma_\varepsilon^2} + \frac{h^2\sigma_v^4}{4\sigma_\varepsilon^4}}$$

$$\sigma^2 = -\frac{\sigma_\varepsilon^2}{c}$$

The minimum-variance control law for the system is

$$u(kh) = -\frac{1 + c}{h} \, y(kh)$$

The standard deviation of the output under minimum-variance control is

$$\mathrm{E}y^2(t) = \sigma^2 \qquad t = h, 2h, \ldots$$

The standard deviation of the state variable x is

$$\mathrm{E}x^2(t) = \sigma^2 - \sigma_\varepsilon^2 \qquad t = h, 2h, \ldots$$

Equation (12.75) is integrated to determine the variance of the state variable between the sampling instants. This gives

$$x(kh + s) = x(kh) + su(kh) + v(kh + s) - v(kh)$$
$$= (1 - \alpha s)x(kh) - \alpha s\varepsilon(kh) + v(kh + s) - v(kh)$$

where

$$\alpha = (1 + c)/h$$

We now introduce

$$P_x(s) = \mathrm{E}x^2(kh + s)$$

It then follows that the output variance is

$$P_y(s) = P_x(s) + \sigma_\varepsilon^2 = (1 - \alpha s)^2(\sigma^2 - \sigma_\varepsilon^2) + (\alpha s)^2\sigma_\varepsilon^2 + s\sigma_v^2 + \sigma_\varepsilon^2$$

The function $P_y(s)$ is shown in Fig. 12.10 when $\sigma_\varepsilon = \sigma_v = 1$. Notice that

$$\max_s\left(P_y(0) - P_y(s)\right) = h^2\sigma_v^2/2$$

The variation in P_y over a sampling interval thus decreases with decreasing h. ∎

The analysis is similar in the general case. The only difference is that Theorem 10.5 must be used to compute the state covariance. In the example the variance is largest at the sampling instants. This is not always the case. Also notice that the correct way of dealing with intersample ripple is to sample the continuous-time system and the continuous-time loss functions, as was discussed in Section 11.1.

Computational Aspects

The LQ-control law can be determined by a combination of spectral factorization and solution of linear Diophantine equations. Recall, however, the fundamental difficulty that arises from poor numerical conditioning of polynomial equations (see Sec. 9.6).

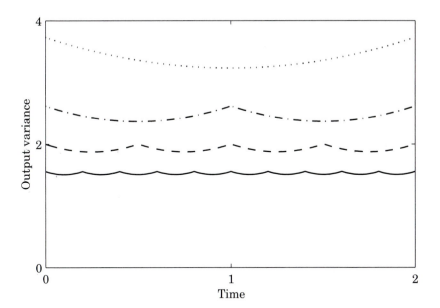

Figure 12.10   Variations of the output variance P_y in Example 12.16 with time for regulators having the sampling periods $h = 0.2$ (solid), $h = 0.5$ (dashed), $h = 1$ (dashed-dotted), and $h = 2$ (dotted).

12.7 Conclusions

In this chapter optimal-control problems are solved for systems described by input-output models. The results given are limited to single-input–single-output systems. A canonical model for the system, Eq. (12.5), is derived first. This model is characterized by three polynomials, A, B, and C. The underlying continuous-time model may be described as a combination of a time delay and a system with rational transfer functions. The disturbances are characterized as filtered white noise. There are many physical systems that can be described by such models.

Optimal-control problems characterized by quadratic loss functions are solved for the system. A special case where the loss function simply is the variance of the output is considered first. The general problem, in which there is also a penalty on the control variable, is then treated. Both these problems are closely related to the prediction problem for a random process with rational spectral density. This problem is also solved. Practical aspects, such as selection of the sampling period, are also discussed.

The solutions to the optimal-control problems give design tools. The solutions also give insight into the character of the optimal solutions. In particular, they tell that the optimal regulator always cancels stable process zeros that are also zeros for the process disturbances. Stable process poles are canceled only if they are not excited by disturbances. The results also give insight into the relationships between the different design methods. For instance, the LQG-solutions can be interpreted as pole-placement regulators, where the process poles and the observer poles are chosen in special ways.

Calculation of the optimal solution is expressed in terms of two polynomial operations, spectral factorization and solution of Diophantine equations.

12.8 Problems

12.1 Consider the process

$$y(k) = 2\frac{q^2 - 1.4q + 0.5}{q^2 - 1.2q + 0.4}e(k)$$

where $e(k)$ is white noise with zero mean and unit variance. Determine the optimal m-step-ahead predictor and the variance of the prediction error when $m = 1, 2,$ and 3.

12.2 Determine the m-step-ahead predictor for the process

$$y(k) + ay(k-1) = e(k) + ce(k-1)$$

Determine also the variance of the prediction error as a function of m.

12.3 A stochastic process is described by

$$y(k) - 0.9y(k-1) = e(k) + 5e(k-1)$$

(a) Determine an equivalent description such that the zero of a corresponding polynomial C is inside the unit circle. How large is the variance of y?

(b) Determine the two-step-ahead predictor for the process and the variance of the prediction error.

12.4 Assume that the demand for a product in an inventory, $z(k)$, can be described as

$$z(k) = 300 + 10k + y(k)$$

where the time unit is months, and $y(k)$ is described by the process

$$y(k) - 0.7y(k-1) - 0.1y(k-2) = 5e(k)$$

where $e(k)$ is white noise with zero mean and unit variance. Make a prediction and determine the expected standard deviation of the prediction error for August through November when the following data are available:

Month	k	$z(k)$
January	1	320
February	2	320
March	3	325
April	4	330
May	5	350
June	6	370
July	7	375

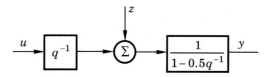

Figure 12.11

12.5 Consider the process

$$y(k) - y(k-1) + 0.5y(k-2) = u(k-2) + 0.5u(k-3)$$
$$+ 0.5\Big(e(k) + 0.8e(k-1) + 0.25e(k-2)\Big)$$

Determine the minimum-variance controller and the minimum achievable variance.

12.6 Determine the minimum-variance controller for the system

$$y(k) - 0.5y(k-1) = u(k-2) + e(k) - 0.7e(k-1)$$

where $e(k)$ is white noise with mean 2 and unit variance.

12.7 Consider the process

$$y(k) + ay(k-1) = u(k-2) + e(k) + ce(k-1)$$

(a) Determine the minimum-variance controller.

(b) Discuss the special case $a = 0$.

12.8 Given the system

$$y(k) - 1.7y(k-1) + 0.7y(k-2) = u(k-d) + 0.5u(k-d-1)$$
$$+ e(k) + 1.5e(k-1) + 0.9e(k-2)$$

(a) Determine the minimum-variance controller and the variance of the output for $d = 1$ and 2.

(b) Simulate the open-loop system and the system controlled with the minimum-variance controller. Compare the output and the control signal for the different cases.

12.9 Consider the process in Fig. 12.11. The disturbance z has the spectral density

$$\phi_z(\omega) = \frac{1}{2\pi} \cdot \frac{1}{1.36 + 1.2 \cos \omega}$$

(a) Determine a pulse-transfer function $H(z)$ that gives an output with spectral density ϕ when driven by zero-mean white noise with unit variance.

(b) What is the steady-state variance of y when

$$u(k) = -Ky(k)$$

for $K = 1$?

(c) What is the minimum achievable variance for a proportional controller and how large is the corresponding value of K?

(d) How large is the variance of y when a minimum-variance controller is used?

12.10 Given the system

$$y(k) - 0.25y(k - 1) + 0.5y(k - 2) = u(k - 1) + e(k) + 0.5e(k - 1)$$

where $e(k)$ is white noise with unit variance. Assume that the process is controlled with the proportional controller

$$u(k) = -Ky(k)$$

(a) Show that the variance of the output is

$$\frac{2.125 - K}{0.5(1.75 - K)(1.25 + K)}$$

and that the lowest variance is obtained for $K = 1$, which gives the variance $4/3$.

(b) The expression above is zero for $K = 2.125$. Explain the paradox.

(c) Compute the minimum-variance controller and the resulting output variance.

12.11 Given the process

$$y(k) - 1.5y(k - 1) + 0.7y(k - 2) = u(k - 2) - 0.5u(k - 3) + v(k)$$

(a) Assume that $v(k) = 0$ and compute the deadbeat controller for the system.

(b) Assume that

$$v(k) = e(k) - 0.2e(k - 1)$$

where $e(k)$ is white noise. Compute the minimum-variance control law.

(c) What is the steady-state variance of y when the deadbeat and the minimum-variance controllers are used on the system when v is as in b)?

(d) Simulate the system using the different controllers. Study the output and the accumulated loss, that is, the sum of the square of the output.

12.12 Consider the dynamic system

$$y(k) = \frac{B(q)}{A(q)} u(k) + \lambda \frac{C(q)}{D(q)} e(k)$$

where $e(k)$ is white noise and B is stable. The polynomials A, C, and D are assumed to be monic. Determine the minimum-variance controller for the system.

12.13 Use the result from Problem 12.12 to determine the minimum-variance controller for the system

$$y(k) = \frac{bq^{-1}}{1 + aq^{-1}} u(k) + (1 + cq^{-1})e(k)$$

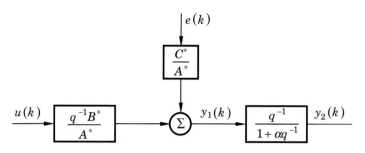

Figure 12.12

12.14 Consider the process in Problem 12.13. Assume that the sampling period is doubled; that is, the control signal can be changed only at every second time unit. Determine the minimum-variance controller and compare with the case when the control period is one time unit.

12.15 Consider the system in Fig. 12.12, where e is white noise with zero mean and unit variance. Further,

$$A(q) = q - 0.7 \qquad B(q) = q$$
$$C(q) = 1 - 0.5q \qquad \alpha = -0.8$$

(a) Determine a controller that minimizes the variance of y_1.

(b) Determine the variances of y_1 and y_2 when the controller in (a) is used.

(c) Determine a controller that minimizes the variance of y_2 if only y_2 is measurable, and compute the variances of y_1 and y_2.

(d) Determine a controller that minimizes the variance of y_2 if both y_1 and y_2 are measurable.

(e) What are the variances of y_1 and y_2 when the controller in (d) is used?

12.16 Given the process

$$A(q)y(k) = B(q)u(k) + C(q)e(k) + D(q)v(k)$$

where $v(k)$ is a known disturbance. Determine the minimum-variance controller for the process when $\deg D = \deg B$.

12.17 Determine the LQG-controller given by Theorem 12.4 for the process

$$(1 - 0.9q^{-1})y(k) = u(k - 1) + (1 - 0.5q^{-1})e(k)$$

when $\rho = 1$. Calculate the variance of the output and the input for different values of ρ.

12.18 Consider a system with stable inverse. Derive the minimum-variance controller, where the control signal $u(k)$ is allowed to be a function of $y(k - 1)$, $y(k - 2)$, ..., $u(k - 1)$, ... Derive the characteristic equation of the closed-loop system.

12.19 Show that the pulse-transfer function from e to y for (12.5) and (12.55) is given by (12.56). Use (12.45) to derive the minimum-variance controller for a system where

$$A(q) = q^2 - 1.5q + 0.7$$
$$B(q) = q + 0.5$$
$$C(q) = q^2 - q + 0.24$$

Compare with the controller obtained through the identity in (12.17).

12.20 Determine for which systems a digital PID-controller has the same structure as the optimal minimum-variance controller.

12.21 Consider a system described by

$$y(k) = \frac{1}{q-a}\left(bu(k) + \varepsilon(k)\right) + \frac{1}{q-1} w(k)$$

where ε and w are white-noise processes with zero mean and standard deviations σ_ε and σ_w, respectively.

(a) Reduce the system to standard form and determine the minimum-variance controller.

(b) Interpret the controller in (a) as a PI-controller and determine how the gain and the reset time depend on the ratio $\sigma_w^2/\sigma_\varepsilon^2$.

12.22 Consider the minimum-variance control law of (12.31) for a system with an unstable inverse. The output of the closed-loop system is given by

$$y(k) = \frac{F(q)}{q^{d-1}B^{-*}(q)} e(k)$$

Show that the function F/B^{-*} has the series expansion

$$\frac{F(q)}{B^{-*}(q)} = q^{d-1} + f_1 q^{d-2} + \cdots + f_{d-1} + \frac{F_2(q)}{B^{-*}(q)}$$

where $\deg F_2(q) < \deg B^{-*}$ and

$$F_1(q) = q^{d-1} + f_1 q^{d-2} + \cdots + f_{d-1}$$

is the quotient of $q^{d-1}C(q)$ and $A(q)$. Give a convenient way of computing F_2. Use the results of the problem to determine the increase of the minimum-variance due to unstable system zeros.

12.23 Determine the intersample ripple of the loss function when the process

$$dx_1 = x_2\, dt$$
$$dx_2 = u\, dt + dv$$
$$y(t_k) = x_1(t_k) + \varepsilon(t_k)$$

is controlled by the minimum-variance regulator. The process $v(t)$ is a Wiener process with incremental covariance $\sigma_v^2\, dt$, and $\varepsilon(t_k)$ is white measurement noise with zero mean and variance σ_ε^2.

12.24 Consider the process in Example 12.16. Determine the control law with sampling period h that minimizes

$$\lim_{T \to \infty} \mathrm{E} \, \frac{1}{T} \int_0^T x^2(s) \, ds$$

and compare it with the minimum-variance control.

12.25 Consider a process subject to a disturbance that is characterized as a Wiener process with incremental covariance dt. Determine the prediction error of the minimum-variance in each case. Use different prediction horizons and sampling periods.

(a) The process has an unstable zero $z = b > 1$.

(b) The process has an unstable pole $z = a > 1$.

12.26 Consider the system in Problem 12.23 with an extra time delay of 1 s. Determine the minimum-variance as a function of the sampling period.

12.27 Consider the system in Problem 12.23. Determine the output variance as a function of the input covariance for different sampling periods.

12.28 Consider the system

$$y(k) = \frac{1}{q - 0.999} u(k) + \frac{q}{q - 0.7} e(k)$$

Determine the minimum-variance control law for the system. Compare it with a proportional feedback that gives a corresponding response rate. Discuss the relative merits of the control laws by calculating their loop gains and return differences. Explain why the minimum-variance control is inferior. (*Hint:* A bad optimization problem gives a bad optimal regulator.)

12.29 Given the system

$$y(k) = 1.4y(k-1) - 0.65y(k-2) + u(k-1) - 0.2u(k-2)$$
$$+ e(k) + 0.4e(k-1)$$

where $e \in N(0, 2)$

(a) Determine the minimum-variance controller.

(b) Determine the deadbeat controller.

(c) Compute the variance of y when the controllers in (a) and (b), respectively, are used.

12.30 Consider the system

$$y(k) + ay(k-1) = u(k-1) + e(k) + ce(k-1)$$

where $e \in N(0, 1)$. We want to determine the minimum-variance controller for the process but the value of c is unknown.

(a) Assume in the design that $c = 0$ and determine the minimum-variance controller for the system

$$y(k) + ay(k-1) = u(k-1) + e(k)$$

How large will the output variance be if this controller is used on the true system?

(b) Assume instead that $c = \hat{c}$ and redo the calculations in (a).

12.31 Consider the stochastic process

$$y(k+2) - 1.1y(k+1) + 0.3y(k) = e(k+2) - 1.25e(k+1)$$

where $e \in N(0,1)$.

(a) Determine the two-step-ahead predictor for $y(k)$.

(b) Calculate the variance of the prediction error.

12.32 Given the system

$$A(q)y(k) = B(q)u(k) + C(q)e(k)$$

where

$$A(q) = q^3 - 1.7q^2 + 0.8q - 0.1$$
$$B(q) = 2(q - 0.9)$$
$$C(q) = q^2(q - 0.1)$$

and $e(k) \in N(0,1)$.

(a) Determine the minimum-variance controller for the system.

(b) Determine the variance of the output when controlling the system with the controller in (a).

(c) Redo the calculations in (a) and (b) when

$$B(q) = 2(0.9q - 1)$$

12.33 Consider the process in Example 12.9. Compute the output variance when the controller does not cancel the zero, that is, when the controller is obtained from the identity

$$zC = AR + BS$$

Compare the variances.

12.34 Consider the process in Example 12.9. Compute the controller that minimizes the loss function (12.7).

12.35 Show that a system with the input-output description

$$A(q)y(k) = B(q)u(k) + C(q)e(k)$$

where

$$A(q) = q^n + a_1 q^{n-1} + \cdots + a_n$$
$$B(q) = b_1 q^{n-1} + \cdots + b_n$$
$$C(q) = q^n + c_1 q^{n-1} + \cdots + c_n$$

has the following state-space description

$$x(k + 1) = \Phi x(k) + \Gamma u(k) + Ke(k + 1)$$
$$y(k) = Cx(k)$$

where the state vector has dimension $n + 1$ and

$$\Phi = \begin{pmatrix} -a_1 & 1 & 0 & \cdots & 0 \\ -a_2 & 0 & 1 & \cdots & 0 \\ \vdots & \vdots & \vdots & \ddots & \vdots \\ -a_n & 0 & 0 & \cdots & 1 \\ 0 & 0 & 0 & \cdots & 0 \end{pmatrix} \quad \Gamma = \begin{pmatrix} b_1 \\ b_2 \\ \vdots \\ b_n \\ 0 \end{pmatrix} \quad K = \begin{pmatrix} 1 \\ c_1 \\ \vdots \\ c_{n-1} \\ c_n \end{pmatrix}$$

$$C = \begin{pmatrix} 1 & 0 & 0 & \cdots & 0 \end{pmatrix}$$

12.36 Consider the system in Problem 12.35. Assume that the polynomial $C(z)$ has all its zeros inside the unit disc. Show that the Kalman filter for the system can be written as

$$\hat{x}(k + 1 \mid k) = \Phi \hat{x}(k \mid k) + \Gamma u(k)$$
$$\hat{x}(k + 1 \mid k + 1) = \hat{x}(k + 1 \mid k) + K\Big(y(k + 1) - C\hat{x}(k + 1 \mid k)\Big)$$

and that the characteristic polynomial of the filter is $zC(z)$.

12.37 Consider the system in Problem 12.35. Assume that minimization of a quadratic loss function gives the feedback law

$$u(k) = -L\hat{x}(k \mid k)$$

Show that the controller has the pulse-transfer function

$$H_c(z) = zL\Big(zI - (I - KC)(\Phi - \Gamma L)\Big)^{-1}\Gamma$$

Show that the results are the same as those given by

$$H_c(z) = L_v(\Phi - KC)\Big(zI - (I - \Gamma L_v)(\Phi - KC)\Big)^{-1}(I - \Gamma L_v)K + L_v K$$

$$= zL_v\Big(zI - (\Phi - KC)(I - \Gamma L_v)\Big)^{-1}K$$

(12.76)

12.38 Consider the system in Problem 12.35. Assume that $b_1 \neq 0$. Determine the minimum-variance strategy using the state-space representations in (12.38) and in Problem 12.37. Compare the results. (*Hint:* The minimum-variance control corresponds to $L = [-a_1 \ 1 \ 0 \ \cdots \ 0]$.)

12.39 Derive the expressions for the transfer function $H_c(z)$ in Eq. (12.76) using the matrix inversion Lemma B.1 in Appendix B.

12.40 Show that the transfer function $H_c(z)$ in Eq. (12.76) can be written as

$$H_c(z) = \frac{S(z)}{R(z)} = \alpha + (L - \alpha C)(zI - \Phi + \Gamma L + KC - \alpha \Gamma C)^{-1}(K - \Gamma \alpha)$$

where $\alpha = L_v K$. Show that this expression is equivalent to

$$\frac{S(z)}{R(z)} = \frac{S_0(z) + \alpha A(z)}{R_0(z) - \alpha B(z)}$$

where $S_0(z)$ and $R_0(z)$ is the solution to the Diophantine equation

$$A(z)R(z) + B(z)S(z) = P(z)C(z)$$

with $\deg R(z) = n$ and $\deg S(z) < n$.

12.9 Notes and References

The treatment of the linear quadratic case is in the spirit of Wiener's work; see Wiener (1949), Newton, Gould, and Kaiser (1957), and Youla, Bongiorno, and Jabr (1976).

    A thorough discussion of prediction and minimum-variance control is found in Åström (1970), which is based on Åström (1965, 1967). A similar approach to the stochastic-control problem is found in Box and Jenkins (1970). The theorem for minimum-variance control of systems with unstable inverses was first published in Peterka (1972). An algebraic approach to the multivariable LQ- and minimum-variance control problems is given in Kučera (1979). Also see Kučera (1984, 1991), and Mosca, Giarre, and Casavola (1990). Choice of sampling interval for stochastic control is discussed in the books mentioned before, and also in MacGregor (1976).

    The intersample variation of the variance is discussed in De Souza and Goodwin (1984) and Lennartson and Söderström (1986).

13

Identification

13.1 Introduction

The notion of a *mathematical model* is fundamental to science and engineering. A model is a very useful and compact way to summarize the knowledge about a process. A model is also a very effective tool for education and communication. The design methods in the previous chapters assume that models for the process and the disturbances are given. The process models can sometimes be obtained from first principles of physics. It is more difficult to get the models of the disturbances, which are equally important. These models often have to be obtained from experiments. The types of models that are needed for the design methods presented here are either state-space models (internal models) or input-output models (external models). The models for the disturbances are for the internal models given as dynamic systems driven by white noise. For external models the disturbances are given in terms of spectral densities and covariance functions. Models for disturbances can, however, only rarely be determined from first principles. Experiments are thus often the only way to get models for the disturbances.

A process cannot be characterized by *one* mathematical model. A process should be represented by a *hierarchy* of models ranging from detailed and complex simulation models to very simple models, which are easy to manipulate analytically. The simple models are used for exploratory purposes and to obtain the gross features of the system behavior. The complicated models are used for a detailed check of the performance of the control system. The complicated models take a long time to develop. Between the two extremes, there may be many different types of models. The trademark of good engineering is to choose the right model for each specific purpose.

Example 13.1 Model hierarchies

To describe a drum boiler power unit, several different models may be needed. For production planning and frequency control, it may be sufficient to characterize the unit using two or three states describing the energy storage in the drum and the superheaters. To construct security systems and control systems, it may be necessary to have a model with 20 to 50 states. Finally, to model temperature and stresses in the turbine unit, several hundred states must be used.   ■

In principle, there are two different ways in which models can be obtained: from prior knowledge—for example, in terms of physical laws or by experimentation on a process. When attempting to obtain a specific model, it is often beneficial to combine both approaches.

Mathematical model building based on physical laws is discussed briefly in Sec. 13.2. In most cases it is not possible to make a complete model only from physical knowledge. Some parameters must be determined from experiments. This approach is called *system identification* and is discussed in Sec. 13.3. There are many methods for analyzing data obtained from experiments. One basic approach is *the principle of least squares* (LS), discussed in Sec. 13.4. Recursive ways to make the computations are given in Sec. 13.5. Examples are given in Sec. 13.6.

13.2 Mathematical Model Building

There are no general methods that always can be used to get a complete model. Each process or problem has its own characteristics. Some general guidelines can be given, but under no circumstances can they replace experience. Model building using physical laws requires knowledge and insight about the process.

The main problem when making a mathematical model is to find the states of the system. The state variables essentially describe storage of energy and mass in the system. Typical variables that are chosen as states are positions and velocities (mechanical systems); voltages and currents (electrical systems); levels and flows (hydraulic systems); and temperatures, pressures, and densities (thermal systems). The relationship between the states is determined using balance equations for force, moment, mass, energy, and constitutive equations.

The advantage of model building from physics is that it gives insight; also, the different parameters and variables have physical interpretations. The drawback is that it may be difficult and time-consuming to build the model from first principles. Mathematical model building often has to be combined with experiments. The references give a more detailed treatment of mathematical model building.

13.3 System Identification

System identification is the experimental approach to process modeling. System identification includes the following:

- Experimental planning
- Selection of model structure

- Criteria

- Parameter estimation

- Model validation

In practice, the procedure of system identification is iterative. When investigating a process where the a priori knowledge is poor, it is reasonable to start with transient or frequency-response analysis to get crude estimates of the dynamics and the disturbances. The results can be used to plan further experiments. The data obtained are then used to estimate the unknown parameters in the model. Based on the results, the model structure can be improved and new experiments may be necessary.

Experimental Planning

It is often difficult and costly to experiment with industrial processes. Therefore, it is desirable to have identification methods that do not require special input signals. Many "classic" methods depend strongly on having the input be of a precise form, for example, sinusoids or impulses. Other techniques can handle virtually any type of input signal, at the expense of increased computations. One requirement of the input signal is that it should excite all the modes of the process sufficiently. A good identification method should thus be insensitive to the characteristics of the input signal.

It is sometimes possible to base system identification on data obtained under closed-loop control of the process. This is useful from the point of view of applications. For instance, adaptive controllers are based mostly on closed-loop identification. The main difficulty with data obtained from a process under feedback is that it may be impossible to determine all the parameters in the desired model; that is, the system is not identifiable, even if the parameters can be determined from an open-loop experiment. Identifiability can be recovered if the feedback is sufficiently complex. It helps to make the feedback nonlinear and time-varying and to change the set points.

Selection of Model Structure

The model structures are derived from prior knowledge of the process and the disturbances. In some cases the only a priori knowledge is that the process can be described as a linear system in a particular operating range. It is then natural to use general representations of linear systems. Such representations are called *black-box models*. A typical example is the difference-equation model

$$A(q)y(k) = B(q)u(k) + C(q)e(k) \qquad (13.1)$$

where u is the input, y is the output, and e is a white-noise disturbance. The parameters, as well as the order of the models, are considered as the unknown parameters.

Sometimes it is possible to apply physical laws to derive models of the process that contain only a few unknown parameters. The model may then be of the form

$$\frac{dx}{dt} = f(x, u, v, \theta)$$

$$y = g(x, u, e, \theta)$$

where θ is a vector of unknown parameters, x is the state of the system, and v and e are disturbances.

Criteria

When formulating an identification problem, a criterion is postulated to give a measure of how well a model fits the experimental data. By making statistical assumptions, it is also possible to derive criteria from probabilistic arguments. The criteria for discrete-time systems are often expressed as

$$J(\theta) = \sum_{k=1}^{N} g\left(\varepsilon(k)\right)$$

where ε is the input error, the output error, or a generalized error. The prediction error is a typical example of a generalized error. The function g is frequently chosen to be quadratic, but it is possible for it to be of many other forms.

The first formulation, solution, and application of an identification problem were given by Gauss in his famous determination of the orbit of the asteroid Ceres. Gauss formulated the identification problem as an optimization problem and introduced the *principle of least squares*, a method based on the minimization of the sum of the squares of the error. Since then, the least-squares criterion has been used extensively.

The least-squares method is very simple and easy to understand. Under some circumstances it gives estimates with the wrong mean values (bias). However, this can be overcome by using various extensions. The least-squares method is restricted to model structures that are linear in the unknown parameters.

When the disturbances of a process are described as stochastic processes, the identification problem can be formulated as a statistical parameter-estimation problem. It is then possible to use the maximum-likelihood method, for example; this method has many attractive statistical properties. It can be interpreted as a least-squares criterion if the quantity to be minimized is taken as the sum of squares of the prediction error. The maximum-likelihood method is a very general technique that can be applied to a wide variety of model structures.

Parameter Estimation

Solving the parameter-estimation problem requires the following:

- Input-output data from the process

- A class of models

- A criterion

Parameter-estimation problem can then be formulated as an optimization problem, where the best model is the one that best fits the data according to the given criterion.

The result of the estimation problem depends, of course, on how the problem is formulated. For instance, the obtained model depends on the amplitude and frequency content of the input signal. There are many possibilities for combining experimental conditions, model classes, and criteria. There are also many different ways to organize the computations. Consequently, there is a large number of different identification methods available. One broad distinction is between *on-line* methods and *off-line* methods. The on-line methods give estimates recursively as the measurement are obtained and are the only alternative if the identification is going to be used in an adaptive controller or if the process is time-varying. In many cases the off-line methods give estimates with higher precision and are more reliable, for instance, in terms of convergence.

The large number of methods is confusing for an industrial engineer who is primarily interested in having a tool to obtain a model. Several attempts to compare different identification methods have been made. The comparisons are largely inconclusive in the sense that there is no method that is universally best. Fortunately, it appears that the choice of method is not crucial. Therefore, it can be recommended that a prospective user learn the classic methods (frequency- and transient-response analysis and correlation and spectral analysis), the least-squares method with extensions, and the maximum-likelihood method.

Model Validation

When a model has been obtained from experimental data, it is necessary to check the model in order to reveal its inadequacies. For model validation, it is useful to determine such factors as step responses, impulse responses, poles and zeros, model errors, and prediction errors. Because the purpose of the model validation is to scrutinize the model with respect to inadequacies, it is useful to look for quantities that are sensitive to model changes.

13.4 The Principle of Least Squares

According to Gauss the principle of least squares is that the unknown parameters of a model should be chosen in such a way that

> the sum of the squares of the differences between the actually observed and computed values multiplied by numbers that measure the degree of precision is a minimum.

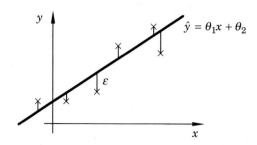

Figure 13.1 Illustration of the variables in the least-squares problem when estimating the parameters of a straight line.

To be able to give an analytic solution, the computed values must be linear functions of the unknown parameters. In the framework of the general formulation of the identification problem given in the previous sections, the class of models is such that the model output is linear in the parameters and the criterion is a quadratic function. The purpose of this section is to formulate the least-squares problem and to give its solution.

The General Problem

In the general least-squares problem, it is assumed that "the computed variable," y, in Gauss' terminology is given by the model

$$\hat{y} = \theta_1 \varphi_1(x) + \theta_2 \varphi_2(x) + \cdots + \theta_n \varphi_n(x) \tag{13.2}$$

where $\varphi_1, \varphi_2, \ldots, \varphi_n$ are known functions, and $\theta_1, \theta_2, \ldots, \theta_n$ are unknown parameters. Pairs of observations $\{(x_i, y_i), i = 1, 2, \ldots, N\}$ are obtained from an experiment. The problem is to determine the parameters in such a way that the variables $\hat{y}$, computed from the model of (13.2) and the experimental values x_i agree as closely as possible with the measured variables y_i. Assuming that all measurements have the same precision, the principle of least squares says that the parameters should be selected in such a way that the loss function

$$J(\theta) = \frac{1}{2} \sum_{i=1}^{N} \varepsilon_i^2$$

is minimal where

$$\varepsilon_i = y_i - \hat{y}_i = y_i - \theta_1 \varphi_1(x_i) - \cdots - \theta_n \varphi_n(x_i) \qquad i = 1, 2, \ldots, N$$

Compare with Fig. 13.1. To simplify the calculations, the following vector notations are introduced:

$$\varphi = \left(\begin{array}{cccc} \varphi_1 & \varphi_2 & \cdots & \varphi_n \end{array} \right)^T$$

$$\theta = \left(\begin{array}{cccc} \theta_1 & \theta_2 & \cdots & \theta_n \end{array} \right)^T$$

$$y = \left(\begin{array}{cccc} y_1 & y_2 & \cdots & y_N \end{array} \right)^T$$

$$\varepsilon = \left(\begin{array}{cccc} \varepsilon_1 & \varepsilon_2 & \cdots & \varepsilon_N \end{array} \right)^T$$

$$\Phi = \left(\begin{array}{c} \varphi^T(x_1) \\ \vdots \\ \varphi^T(x_N) \end{array} \right)$$

The least-squares problem can now be formulated in a compact form. The loss function J can be written as

$$J(\theta) = \frac{1}{2} \varepsilon^T \varepsilon = \frac{1}{2} \|\varepsilon\|^2 \tag{13.3}$$

where

$$\varepsilon = y - \hat{y}$$

and

$$\hat{y} = \Phi\theta$$

Determine the parameter θ in such a way that $\|\varepsilon\|^2$ is minimal. The solution to the least-squares problem is given by the following theorem.

THEOREM 13.1   LEAST-SQUARES SOLUTION   The function of (13.3) is minimal for parameters $\hat{\theta}$ such that

$$\Phi^T \Phi \hat{\theta} = \Phi^T y \tag{13.4}$$

If the matrix $\Phi^T \Phi$ is nonsingular, the minimum is unique and given by

$$\hat{\theta} = (\Phi^T \Phi)^{-1} \Phi^T y = \Phi^\dagger y \tag{13.5}$$

Proof.   The loss function of (13.3) can be written as

$$2J(\theta) = \varepsilon^T \varepsilon = (y - \Phi\theta)^T (y - \Phi\theta)$$
$$= y^T y - y^T \Phi\theta - \theta^T \Phi^T y + \theta^T \Phi^T \Phi\theta$$

Because the matrix $\Phi^T \Phi$ is always nonnegative definite, the function J has a minimum. Assuming that $\Phi^T \Phi$ is nonsingular and using (11.12) the minimum is obtained for

$$\theta = \hat{\theta} = (\Phi^T \Phi)^{-1} \Phi^T y$$

and the theorem is proved.   ∎

Remark 1. Equation (13.4) is called the *normal equation*.

Remark 2. The matrix $\Phi^{\dagger} = (\Phi^T\Phi)^{-1}\Phi^T$ is called the *pseudo-inverse* of Φ if the matrix $\Phi^T\Phi$ is nonsingular.

System Identification

The least-squares method can be used to identify parameters in dynamic systems. Let the system be described by (13.1) with $C(q) = q^n$. Further, assume that A and B are of order n and $n - 1$, respectively. Assume that a sequence of inputs $\{u(1), u(2), \dots, u(N)\}$ has been applied to the system and that the corresponding sequence of outputs $\{y(1), y(2), \dots, y(N)\}$ has been observed. The unknown parameters are then

$$\theta = \begin{pmatrix} a_1 & \dots & a_n & b_1 & \dots & b_n \end{pmatrix}^T \tag{13.6}$$

Further, we introduce

$$\varphi^T(k+1) = \begin{pmatrix} -y(k) & \dots & -y(k-n+1) & u(k) & \dots & u(k-n+1) \end{pmatrix} \tag{13.7}$$

and

$$\Phi = \begin{pmatrix} \varphi^T(n+1) \\ \vdots \\ \varphi^T(N) \end{pmatrix}$$

The least-squares estimate is then given by (13.5) if $\Phi^T\Phi$ is nonsingular. For instance, this is the case if the input signal is, loosely speaking, sufficiently rich.

Example 13.2   Least-squares estimate of first-order systems

Determine the least-squares estimate of the parameters a and b in the model

$$y(k) = -ay(k-1) + bu(k-1)$$

in such a way that the criterion

$$J(a, b) = \frac{1}{2} \sum_{i=2}^{N} \varepsilon(k)^2$$

is minimal, where

$$\varepsilon(k) = y(k) - \hat{y}(k) = y(k) + ay(k-1) - bu(k-1)$$
$$= y(k) - \varphi^T(k)\theta$$

A comparison with the general case gives

$$
y = \begin{pmatrix} y(2) \\ y(3) \\ \vdots \\ y(N) \end{pmatrix} \quad
\Phi = \begin{pmatrix} -y(1) & u(1) \\ -y(2) & u(2) \\ \vdots & \vdots \\ -y(N-1) & u(N-1) \end{pmatrix} \quad
\varepsilon = \begin{pmatrix} \varepsilon(2) \\ \varepsilon(3) \\ \vdots \\ \varepsilon(N) \end{pmatrix}
$$

and

$$
\theta = \begin{pmatrix} a & b \end{pmatrix}^T
$$

Hence

$$
\Phi^T \Phi = \begin{pmatrix} \sum_{k=1}^{N-1} y(k)^2 & -\sum_{k=1}^{N-1} y(k)u(k) \\ -\sum_{k=1}^{N-1} y(k)u(k) & \sum_{k=1}^{N-1} u(k)^2 \end{pmatrix}
$$

$$
\Phi^T y = \begin{pmatrix} -\sum_{k=1}^{N-1} y(k+1)y(k) \\ \sum_{k=1}^{N-1} y(k+1)u(k) \end{pmatrix}
$$

Provided the matrix $\Phi^T \Phi$ is nonsingular, the least-squares estimate of the parameters a and b is now easily obtained. The matrix $\Phi^T \Phi$ will be nonsingular if conditions (*sufficient richness* or *persistent excitation*) are imposed on the input signal.   ∎

Statistical Interpretation

To analyze the properties of the least-squares estimator, it is necessary to make some assumptions. Let the data be generated from the process

$$
y = \Phi \theta_0 + \varepsilon \tag{13.8}
$$

where θ_0 is the vector of "true" parameters, and ε is a vector of noise with zero-mean value. The following theorem is given without proof.

THEOREM 13.2   PROPERTIES OF LEAST-SQUARES ESTIMATE   Consider the estimate (13.5) and assume that the data are generated from (13.8), where ε is white noise with variance σ^2. Then, if n is the number of parameters of $\hat{\theta}$ and θ_0 and N is the number of data, the following conditions hold.

1. $E\hat{\theta} = \theta_0$

2. $\operatorname{var} \hat{\theta} = \sigma^2 (\Phi^T \Phi)^{-1}$

3. $s^2 = 2J(\theta)/(N-n)$ is an unbiased estimate of σ^2 ∎

Theorem 13.2 implies that the parameters in (13.1) can be estimated without bias if $C(q) = q^n$. If $C(q) \neq q^n$, then the estimates will be biased. This is due to the correlation between the noise $C^*(q^{-1})e(k)$ and the data in $\varphi(k)$.

Extensions of the Least-Squares Method

The least-squares method gives unbiased results of the parameters in (13.1) only if $C(q) = q^n$. However, the maximum likelihood method can be used for the general case. It can be shown that maximizing the likelihood function is equivalent to minimizing the loss function of (13.3), where the residuals, ε, are related to the inputs and outputs by

$$C(q)\varepsilon(k) = A(q)y(k) - B(q)u(k)$$

The residuals can be interpreted as the one-step-ahead prediction error. However, the loss function is not linear in the parameters and it has to be minimized numerically. This can be done using a Newton-Raphson gradient routine, which involves computation of the gradient of J with respect to the parameters, as well as the matrix of second partial derivatives. The maximum-likelihood method is thus an off-line method. It is possible to make approximations of the maximum-likelihood method that allow on-line computations of the parameters of the model in (13.1). Some common methods are Extended Least Squares (ELS), Generalized Least Squares (GLS), and Recursive Maximum Likelihood (RML).

13.5 Recursive Computations

In many cases the observations are obtained sequentially. It may then be desirable to compute the least-squares estimate for different values of N. If the least-squares problem has been solved for N observations, it seems to be a waste of computational resources to start from scratch when a new observation is obtained. Hence, it is desirable to arrange the computations in such a way that the results obtained for N observations can be used in order to get the estimates for $N + 1$ observations. An analogous problem occurs when the number of parameters is not known in advance. The least-squares estimate may then be needed for a different number of parameters. The possibility of calculating the least-squares estimate recursively is pursued in this section.

Recursion in the Number of Observations

Recursive equations can be derived for the case when the observations are obtained sequentially. The procedure is often referred to as recursive identification. The solution in (13.5) to the least-squares problem can be rewritten to give recursive equations. Let $\theta(N)$ denote the least-squares estimate based on N measurements. To derive the equations, N is introduced as a formal parameter in the functions, that is,

$$\Phi(N) = \begin{pmatrix} \varphi^T(1) \\ \vdots \\ \varphi^T(N) \end{pmatrix} \qquad y(N) = \begin{pmatrix} y_1 \\ \vdots \\ y_N \end{pmatrix}$$

It is assumed that the matrix $\Phi^T\Phi$ is nonsingular for all N. The least-squares estimate $\hat{\theta}(N)$ is then given by Eq. (13.5):

$$\hat{\theta}(N) = \left(\Phi^T(N)\Phi(N)\right)^{-1}\Phi^T(N)y(N)$$

When an additional measurement is obtained, a row is added to the matrix Φ and an element is added to the vector y. Hence

$$\Phi(N+1) = \begin{pmatrix} \Phi(N) \\ \varphi^T(N+1) \end{pmatrix} \qquad y(N+1) = \begin{pmatrix} y(N) \\ y_{N-1} \end{pmatrix}$$

The estimate $\hat{\theta}(N+1)$ given by (13.5) can then be written as

$$
\begin{aligned}
\hat{\theta}(N+1) &= \left(\Phi^T(N+1)\Phi(N+1)\right)^{-1}\Phi^T(N+1)y(N+1) \\
&= \left(\Phi^T(N)\Phi(N) + \varphi(N+1)\varphi^T(N+1)\right)^{-1} \\
&\quad \times \left(\Phi^T(N)y(N) + \varphi(N+1)y_{N+1}\right)^{-1}
\end{aligned}
\tag{13.9}
$$

The solution is given by the following theorem.

THEOREM 13.3   RECURSIVE LEAST-SQUARES ESTIMATION   Assume that the matrix $\Phi^T(N)\Phi(N)$ is positive definite. The least-squares estimate $\hat{\theta}$ then satisfies the recursive equation

$$\hat{\theta}(N+1) = \hat{\theta}(N) + K(N)\left(y_{N+1} - \varphi^T(N+1)\hat{\theta}(N)\right) \tag{13.10}$$

$$
\begin{aligned}
K(N) &= P(N+1)\varphi(N+1) \\
&= P(N)\varphi(N+1)\left(1 + \varphi^T(N+1)P(N)\varphi(N+1)\right)^{-1}
\end{aligned}
\tag{13.11}
$$

$$P(N+1) = \left(I - K(N)\varphi^T(N+1)\right)P(N) \tag{13.12}$$

Proof.   To simplify the notation in the manipulations that follow, the argument N of $\Phi(N)$ and $y(N)$ and the argument $N+1$ of $\varphi^T(N+1)$ will be suppressed. Equation (13.9) can then be written as

$$
\begin{aligned}
\hat{\theta}(N+1) &= (\Phi^T\Phi + \varphi\varphi^T)^{-1}(\Phi^Ty + \varphi y_{N+1}) \\
&= (\Phi^T\Phi)^{-1}\Phi^Ty + \left((\Phi^T\Phi + \varphi\varphi^T)^{-1} - (\Phi^T\Phi)^{-1}\right)\Phi^Ty \\
&\quad + (\Phi^T\Phi + \varphi\varphi^T)^{-1}\varphi y_{N-1}
\end{aligned}
\tag{13.13}
$$

Observe that

$$\hat{\theta}(N) = (\Phi^T\Phi)^{-1}\Phi^Ty$$

and

$$\left((\Phi^T\Phi + \varphi\varphi^T)^{-1} - (\Phi^T\Phi)^{-1}\right)\Phi^T y$$

$$= (\Phi^T\Phi + \varphi\varphi^T)^{-1}(\Phi^T\Phi - \Phi^T\Phi - \varphi\varphi^T)(\Phi^T\Phi)^{-1}\Phi^T y$$

$$= -(\Phi^T\Phi + \varphi\varphi^T)^{-1}\varphi\varphi^T(\Phi^T\Phi)^{-1}\Phi^T y$$

$$= -(\Phi^T\Phi + \varphi\varphi^T)^{-1}\varphi\varphi^T\hat{\theta}$$

Equation (13.13) can be written as

$$\hat{\theta}(N+1) = \hat{\theta}(N) + K(N)\left(y_{N+1} - \varphi^T(N+1)\hat{\theta}(N)\right)$$

In order to obtain a recursive equation for the weighting factor $K(N)$, it is convenient to introduce the quantity P defined by

$$P(N) = \left(\Phi^T(N)\Phi(N)\right)^{-1}$$

P is proportional to the variance of the estimates (compare with Theorem 13.2). Applying the matrix inversion lemma (Lemma B.1) to the matrix $P(N+1)$ gives

$$P(N+1) = \left(\Phi^T(N+1)\Phi(N+1)\right)^{-1} = (\Phi^T\Phi + \varphi\varphi^T)^{-1}$$

$$= (\Phi^T\Phi)^{-1} - (\Phi^T\Phi)^{-1}\varphi\left(I + \varphi^T(\Phi^T\Phi)^{-1}\varphi\right)^{-1}\varphi^T(\Phi^T\Phi)^{-1}$$

Hence

$$P(N+1) = P(N) - P(N)\varphi(N+1)$$

$$\times \left(I + \varphi^T(N+1)P(N)\varphi(N+1)\right)^{-1}\varphi^T(N+1)P(N)$$

Simple calculations now give

$$K(N) = P(N+1)\varphi(N+1)$$

$$= P(N)\varphi(N+1)\left(I + \varphi^T(N+1)P(N)\varphi(N+1)\right)^{-1}$$

Notice that a matrix inversion is necessary to compute P. However, the matrix to be inverted is of the same dimension as the number of measurements; that is, for a single-output system, it is a scalar.   ■

Remark 1.   Equation (13.10) has a strong intuitive appeal. The estimate $\hat{\theta}(N+1)$ is obtained by adding a correction to the previous estimate $\hat{\theta}(N)$. The correction is proportional to $y_{N+1} - \varphi^T(N+1)\hat{\theta}(N)$, where the last term can be interpreted as the value of y at time $N+1$ predicted by the model (13.2). The correction term is thus proportional to the difference between the measured value of y_{N+1} and the prediction of y_{N+1} based on the previous estimates of the parameters. The components of the vector $K(N)$ are weighting factors that tell how the correction and the previous estimate should be combined. Notice that the ith component of $K(N)$ is proportional to $\varphi_i^T(N+1)$.

Remark 2.   The least-squares estimate can be interpreted as a Kalman filter for the process

$$\theta(k+1) = \theta(k)$$
$$y(k) = \varphi^T(k)\theta(k) + e(k)$$

See Section 11.3.

Notice that the matrix $P(N)$ is defined only when the matrix $\Phi^T(N)\Phi(N)$ is nonsingular. Because

$$\Phi^T(N)\Phi(N) = \sum_{k=1}^{N} \varphi(k)\varphi^T(k)$$

it follows that $\Phi^T\Phi$ is always singular if N is sufficiently small. In order to obtain an initial condition for P, it is necessary to choose an $N = N_0$ such that $\Phi^T(N_0)\Phi(N_0)$ is nonsingular and determine

$$P(N_0) = \left(\Phi^T(N_0)\Phi(N_0)\right)^{-1}$$
$$\hat{\theta} = P(N_0)\Phi^T(N_0)y(N_0)$$

The recursive equations can then be used from $N \geq N_0$. It is, however, often convenient to use the recursive equations in all steps. If the recursive equations are begun with the initial condition

$$P(0) = P_0$$

where P_0 is positive definite, then

$$P(N) = \left(P_0^{-1} + \Phi^T(N)\Phi(N)\right)^{-1}$$

This can be made arbitrarily close to $\left(\Phi^T(N)\Phi(N)\right)^{-1}$ by choosing P_0 sufficiently large.

Using the statistical interpretation of the least-squares method shows that this way of starting the recursion corresponds to the situation when the parameters have a prior covariance proportional to P_0.

Time-Varying Systems

Using the loss function of (13.3), all data points are given the same weight. If the parameters are time-varying, it is necessary to eliminate the influence of old data. This can be done by using a loss function with exponential weighting, that is,

$$J(\theta) = \sum_{k=1}^{N} \lambda^{N-k}\left(y(k) - \varphi^T(k)\theta\right)^2 \tag{13.14}$$

The *forgetting factor*, λ, is less than one and is a measure of how fast old data are forgotten. The least-squares estimate when using the loss function of (13.14) is given by

$$\hat{\theta}(k+1) = \hat{\theta}(k) + K(k)\left(y_{k+1} - \varphi^T(k+1)\hat{\theta}(k)\right)$$

$$K(k) = P(k)\varphi(k+1)\left(\lambda + \varphi^T(k+1)P(k)\varphi(k+1)\right)^{-1} \qquad (13.15)$$

$$P(k+1) = \left(I - K(k)\varphi^T(k+1)\right)P(k)/\lambda$$

It is also possible to model the time-varying parameters by a Markov process,

$$\theta(k+1) = \Phi\theta(k) + v(k)$$

and then use a Kalman filter to estimate θ. See Remark 2 of Theorem 13.3.

Recursion in the Number of Parameters

When extra parameters are introduced, the vector $\hat{\theta}$ will have more components and there will be additional rows in the matrix Φ. The calculations can be arranged so that it is possible to make a recursion in the number of parameters in the model. The recursion involves an inversion of a matrix of the same dimension as the number of added parameters.

U-D Covariance Factorization

Equation (13.15) is one way to mechanize the recursive update of the estimates and the covariance matrix. These equations are not well-conditioned from a numerical point of view, however. A better way of doing the calculation is to update the square-root of P instead of updating P. Another way to do the calculations is to use the U-D algorithm by Bierman and Thorton. This method is based on a factorization of P as

$$P = UDU^T$$

where D is diagonal and U is an upper-triangular matrix. This method is a square-root type as $UD^{1/2}$ is the square root of P. The U-D factorization method does not include square-root calculations and is therefore well suited for small computers and real-time applications. Details about the algorithm can be found in the References.

    A Pascal program for least-squares estimation based on U-D factorization is given in Listing 13.1. The program gives estimates of the parameters of the process

$$y(k) + a_1 y(k-1) + \cdots + a_{na}y(k-na)$$
$$= b_1 u(k-1) + \cdots + b_{nb}u(k-nb) + e(k) \qquad (13.16)$$

The notations used in the program are

Variable	Notation in the program
$u(k)$	u
$y(k)$	y
na	na
$na + nb$	n
$n(n-1)/2$	noff
$\hat{\theta}(k)$ compare (13.6)	theta
$\varphi^T(k)$ compare (13.7)	fi
λ	lambda

Listing 13.1 Pascal program for least-squares estimation of the parameters of the process of (13.16) using U-D factorization.

```
const  npar=10;{maximum number of estimated parameters}
       noff=45;{noff=npar*(npr-1)/2}

type vec1=array[1..npar] of real;
     vec2=array[1..noff] of real;
     estpartyp = record
       n,na:integer;
       theta:vec1;
       fi:vec1;
       diag:vec1;
       offdiag:vec2;
       end;
var y,u,lambda:real;
      eststate:estpartyp;

Procedure LS(u,y,lambda:real;var eststate:estpartyp);
{Computes the least-squares estimate using the U-D method
after Bierman and Thornton}

var kf,ku,i,j:integer;
    perr,fj,vj,alphaj,ajlast,pj,w:real;
    k:vec1;

begin
  with eststate do {Calculate prediction error}
  begin
    perr = y;
    for i:=1 to n do perr:=perr-theta[i]*fi[i];
```

Listing 13.1   (Continued)

```
{Calculate gain and covariance using the U-D method}
fj:=fi[1];
vj:=diag[1]*fj;
k[1]:=vj;
alphaj:=1.0+vj*fj;
diag[1]:=diag[1]/alphaj/lambda;
if n>1 then
begin
  kf:=0;
  ku:=0;
  for j:=2 to n do
  begin
    fj:fi[j];
    for i:=1 to j-1 do
    begin {f=fi*U}
      kf:=kf+1;
      fj:=fj+fi[i]*offdiag[kf]
    end; {i}
    vj:=fj*diag[j]; {v=D*f}
    k[j]:=vj;
    ajlast:=alphaj;
    alphaj:=ajlast+vj*fj;
    diag[j]:=diag[j]*ajlas/alphaj/lambda;
    pj:=-fj/ajlast;
    for i:=1 to j-1 do
    begin
      {kj+1:=kj +vj*uj}
      {uj:=uj+pj*kj}
      ku:=ku+1;
      w:=offdiag[ku]+k[i]*pj;
      k[i]:=k[i]+offdiag[ku]*vj;
      offdiag[ku]:=w
    end; {i}
  end; {j}
end; {if n>1 then}
{Update parameter estimates}
for i:=1 to n do theta[i]:=theta[i]+perr*k[i]/alphaj;
{Updating of fi}
for i:=1 to n-1 do fi[n+1-i]:=fi[n-i];
fi[1]:=-y;
fi[na+1]:=u
end {with eststate do}
end; {LS}
```

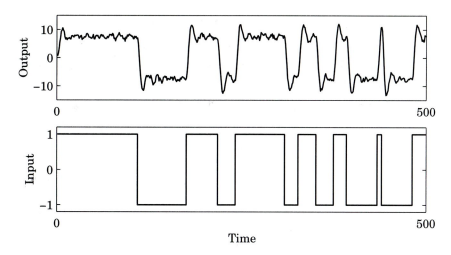

Figure 13.2 Input and output when the system of (13.17) is simulated. The input is a PRBS sequence.

13.6 Examples

Some examples show the use of identification methods. The first example shows the importance of using the correct model structure when estimating a model.

Example 13.3   Influence of model structure

Let the system be described by the model

$$
\begin{aligned}
y(k) - 1.5 y(k-1) &+ 0.7 y(k-2) \\
&= u(k-1) + 0.5 u(k-2) + e(k) - e(k-1) + 0.2 e(k-2)
\end{aligned}
\tag{13.17}
$$

where e has zero mean and standard deviation 0.5. This is a "standard" system that has been used often in the literature to test different identification methods. In (13.17), $C(q) \neq q^n$, which implies that the least-squares method will give biased estimates. However, the input-output relation of the process can be approximated by using the least-squares method for a higher-order model. Figure 13.2 shows a simulation of the system. The input is a Pseudo Random Binary Signal (PRBS) sequence with amplitude ±1. The data have been used to identify models of different orders using the least-squares and maximum-likelihood methods.

Figure 13.3 shows the step responses of the true system in (13.17) and of the estimated models when using the least-squares method with model orders $n = 1$, 2, and 4, and the maximum-likelihood method when the model order is 2. The least-squares method gives a poor description for a second-order model, and a good model is obtained when the model order is increased to 4. The maximum-likelihood method gives very good estimates of the dynamics and the noise characteristics for a second-order model. The estimated parameters for second-order models when using the least-squares method and the maximum-likelihood method are shown in Table 13.1. ∎

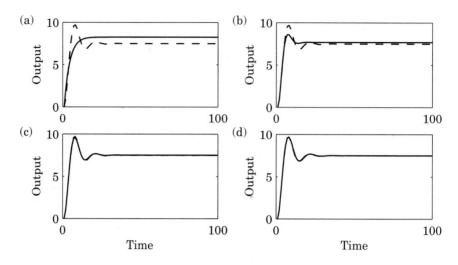

Figure 13.3 Step responses of the deterministic part of the system of (13.17) (dashed) and of the estimated models (solid) obtained when using the least-squares (LS) method with (a) $n = 1$, (b) $n = 2$, (c) $n = 4$, and (d) the maximum-likelihood (ML) method with $n = 2$.

Table 13.1 Estimated parameters and standard deviations for second-order models of the process in (13.17) when using the least-squares (LS) and the maximum-likelihood (ML) methods.

Parameter	True Value	LS $n = 2$	ML $n = 2$
a_1	-1.5	-1.265 ± 0.029	-1.513 ± 0.008
a_2	0.7	0.517 ± 0.023	0.704 ± 0.006
b_1	1.0	0.959 ± 0.101	1.041 ± 0.050
b_2	0.5	0.972 ± 0.131	0.394 ± 0.071
c_1	-1.0	—	-1.081 ± 0.045
c_2	0.2	—	0.215 ± 0.044

The second example illustrates recursive estimation using the least-squares method.

Example 13.4   Recursive estimation

Consider the process

$$y(k) + ay(k - 1) = bu(k - 1) + e(k)$$

with $a = -0.8$ and $b = 0.5$. The variance of the noise e is 0.25. The input signal is assumed to be a PRBS signal with amplitude ± 1. The input and the output data are shown in Fig. 13.4. The recursive equations (13.10) to (13.12) have been used

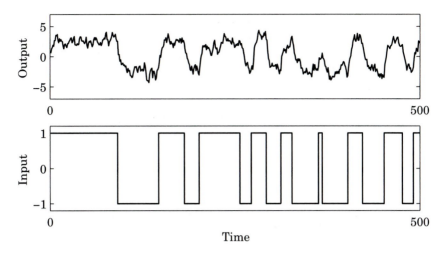

Figure 13.4   Input-output data for the process in Example 13.4.

to estimate a and b. The estimates are shown in Fig. 13.5 when the initial values of the parameters are zero and when $P(0)$ is 10 times the unit matrix. The estimates are after a few observations close to the true values. For the data in Fig. 13.4 we have

$$\begin{pmatrix} \hat{a}(500) \\ \hat{b}(500) \end{pmatrix} = \begin{pmatrix} -0.799 \\ 0.513 \end{pmatrix} \qquad P(500) = \begin{pmatrix} 0.880 & 1.560 \\ 1.560 & 4.771 \end{pmatrix} \cdot 10^{-3}$$

From Theorem 13.2 we get the following standard deviations for the estimates

$$\sigma_{\hat{a}} = 0.5\sqrt{8.80} \cdot 10^{-2} = 0.015$$
$$\sigma_{\hat{b}} = 0.5\sqrt{47.71} \cdot 10^{-2} = 0.035$$

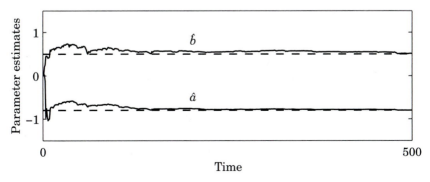

Figure 13.5   Recursive-parameter estimates when (13.10) to (13.12) are used on the data shown in Fig. 13.4. The true values are indicated by the dashed lines.

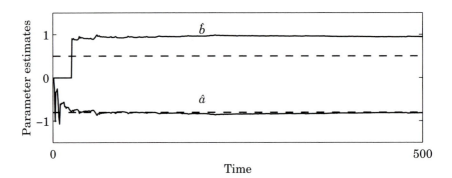

Figure 13.6   Recursive-parameter estimates for the system in Example 13.4 when the input is a unit pulse at time $t = 25$. The true values are indicated by the dashed lines.

The estimates are well within one standard deviation of their their true values. Now assume that the input signal is changed from a PRBS signal to a unit pulse at time $t = 25$. The estimates for the new experiment are shown in Fig. 13.6. The estimate of b is now very poor because the input signal is not sufficiently exciting. The estimate of a is, however, better because the output is excited by the noise and thus the output contains information about the parameter a.   ∎

The influence of feedback is illustrated in the next example.

Example 13.5   Influence of feedback

The system in Example 13.4 is simulated and the input is generated via feedback as

$$u(k) = -ky(k) = -0.2y(k) \tag{13.18}$$

The phase plane of the estimates is shown in Fig. 13.7. The identifiability of the parameters is lost due to the feedback. The estimates converge to a subspace that is determined by

$$\hat{b} = b + \frac{1}{k}(a - \hat{a}) = -3.5 - 5\hat{a}$$

The least-squares loss function has the same values for all parameters on this line.

The problem with the loss of identifiability disappears if a feedback law of sufficiently high complexity is used. In this example it is sufficient to introduce a delay in the controller and use

$$u(k) = -0.32y(k - 1) \tag{13.19}$$

The control laws (13.18) and (13.19) give approximately the same speed and output variance of the closed-loop system. The phase plane of the estimates are shown in Fig. 13.8. The estimates converge from all initial values to the correct value.   ∎

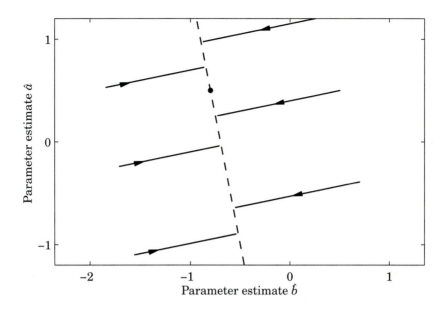

Figure 13.7 Recursive-parameter estimates for the system in Example 13.5 when the input is generated by the feedback (13.18). The dashed line shows the identifiable subspace. The dot shows the true values parameter values.

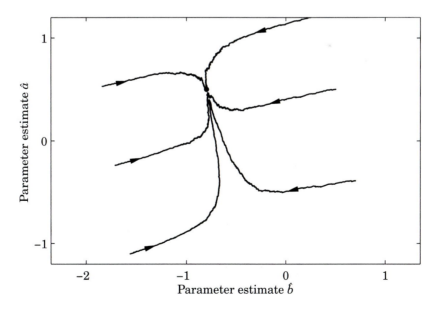

Figure 13.8 Recursive-parameter estimates for the system in Example 13.5 when the input is generated by the feedback (13.19). The dot shows the true values parameter values.

13.7 Summary

This chapter gives a short review of the identification problem. The presentation is concentrated on the least-squares method, because it is the basis for many other methods. On many occasions it is important to make the estimation in real time, and it is shown how the least-squares estimate can be obtained recursively. This is used, for instance, in adaptive controllers.

13.8 Problems

13.1 The following experiment has been made to determine the normal acceleration, g. A steel ball has been dropped without initial velocity from a high TV antenna. The position of the ball, l, has been determined at different times, giving the following measurements:

Time, s	Length of fall, in meters
1	8.49
2	20.05
3	50.65
4	72.19
5	129.85
6	171.56

The times of the measurements are exact, but there is an error in the measurement of the position. Determine the normal acceleration using the method of least squares from the model

$$l = \frac{gt^2}{2} + e$$

13.2 Derive recursive equations for increasing the number of parameters for the method of least squares. (*Hint:* Use the same idea as when making the observations recursively.)

13.3 Consider the process

$$y(k) + ay(k-1) = bu(k-1) + e(k) + ce(k-1)$$

where u and e are independent white-noise processes with zero mean and unit variance. Assume that the method of least squares is used to estimate a and b, as in Example 13.2. Determine the expected values of $\hat{a}$ and $\hat{b}$ as a function of a, b, and c.

13.4 The parameters b_1 and b_2 in the system

$$y(k) = b_1 u(k-1) + b_2 u(k-2) + e(k)$$

are determined using the method of least squares. Let the input be a step at time $k = 0$. Can the parameters b_1 and b_2 be determined with arbitrary accuracy when the number of observations increases? Will there be any changes if it is known that $b_2 = 0$?

13.5   Consider the system

$$y(k) = -ay(k-1) + bu(k-1) + e(k)$$

where e is zero-mean white noise. An experiment is done on the system to estimate a and b. The following data were calculated:

$$\Sigma y^2(k) = 30 \qquad \Sigma u^2(k) = 50$$
$$\Sigma y(k+1)y(k) = 1 \qquad \Sigma y(k)u(k) = 20$$
$$\Sigma y(k+1)u(k) = 36$$

All sums are from $k = 1$ to $k = 999$. Determine the least-squares estimate of a and b.

13.9 Notes and References

There are many books and papers dealing with identification methods. Some basic references in book form are Jenkins and Watts (1968), Eykhoff (1974), Goodwin and Payne (1977), Ljung and Söderström (1983), Norton (1986), Ljung (1987), Söderström and Stoica (1989), and Johansson (1993). An early survey of system identification is given in Åström and Eykhoff (1971).

Good sources for further references are Biermann (1977), Eykhoff (1981), Isermann (1981), Lawson and Hansson (1974), and the special issue on "Identification and system parameter estimation," *Automatica*, **17**, no. 1 (January 1981).

For someone interested in historical notes, see Gauss (1809) and Sorensen (1970).

A

Examples

Examples used in the book as "standard processes" are presented in this appendix.

Example A.1 Double integrator

The double integrator is used throughout the book as a main example to illustrate the theories presented. The process is described by the differential equation

$$\frac{d^2 y}{dt^2} = u \tag{A.1}$$

The transfer function is $G(s) = 1/s^2$. We introduce y and $\dot{y}$ as the states x_1 and x_2, respectively, of the system. The state-space representation is then

$$\frac{dx}{dt} = \begin{pmatrix} 0 & 1 \\ 0 & 0 \end{pmatrix} x + \begin{pmatrix} 0 \\ 1 \end{pmatrix} u$$

$$y = \begin{pmatrix} 1 & 0 \end{pmatrix} x \tag{A.2}$$

Sampling (A.2) using a zero-order hold with the sampling period h gives the discrete-time system (see Example 2.2)

$$x(kh + h) = \begin{pmatrix} 1 & h \\ 0 & 1 \end{pmatrix} x(kh) + \begin{pmatrix} h^2/2 \\ h \end{pmatrix} u(kh)$$

$$y(kh) = \begin{pmatrix} 1 & 0 \end{pmatrix} x(kh) \tag{A.3}$$

The pulse-transfer operator of (A.3) is given by

$$H(q) = \frac{h^2(q + 1)}{2(q - 1)^2} \tag{A.4}$$

There are several physical processes that can be described as double integrators. One such is the attitude of a satellite, which can be described by the equation

$$J \frac{d^2 \theta}{dt^2} = M_c + M_d$$

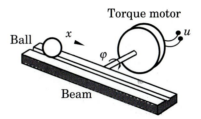

Figure A.1   Schematic illustration of the ball and beam.

where θ is the attitude angle, M_c is the control torque, M_d is the disturbing torque, and J is the moment of inertia.

Another example that can be described by the double integrator is a rolling ball on a tilting beam (see Fig. A.1). The equation of the ball and beam can be described by

$$J \frac{d^2\theta}{dt^2} = mgr \sin \varphi \approx mgr\varphi$$

$$x = r\theta$$

or

$$\frac{d^2x}{dt^2} = mgr^2 \varphi / J$$

where θ is the angle of the ball, g is the normal acceleration, x is the position of the ball, and φ is the tilting angle of the beam.   ∎

Example A.2   Motor

A DC motor can be described by a second-order model with one integrator and one time constant (see Fig. A.2). The input is the voltage to the motor and the output is the shaft position. The time constant is due to the mechanical parts of the system, and the dynamics due to the electrical parts are neglected. A normalized model of the process is then given by

$$Y(s) = \frac{1}{s(s+1)} U(s)$$

Introduce the velocity and the position of the motor shaft as states (see Fig. A.2). The state-space model of the motor is then given by

$$\frac{dx}{dt} = \begin{pmatrix} -1 & 0 \\ 1 & 0 \end{pmatrix} x + \begin{pmatrix} 1 \\ 0 \end{pmatrix} u$$

$$y = \begin{pmatrix} 0 & 1 \end{pmatrix} x \tag{A.5}$$

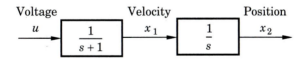

Figure A.2   Normalized model of a DC motor.

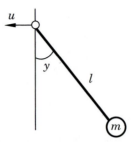

Figure A.3   Pendulum.

Sampling (A.5) using a zero-order hold gives the discrete-time model

$$x(kh + h) = \begin{pmatrix} e^{-h} & 0 \\ 1 - e^{-h} & 1 \end{pmatrix} x(kh) + \begin{pmatrix} 1 - e^{-h} \\ h - 1 + e^{-h} \end{pmatrix} u(kh)$$

$$y(kh) = \begin{pmatrix} 0 & 1 \end{pmatrix} x(kh)$$

(A.6)

(see Example 2.3). A current-controlled DC motor with the shaft velocity as output can also be described by the model of (A.5). Still another example that can be characterized by an integrator and a single pole is a ship. Let the input be the rudder angle and the output be the heading. The ship can then be described by the transfer function

$$G(s) = \frac{K}{s(1 + Ts)}$$

where the time constant may be positive or negative depending on the type of ship. For instance, large tankers are unstable.   ∎

Example A.3   Harmonic oscillator

Consider a pendulum (see Fig. A.3). The acceleration of the pivot point is the input and the angle y is the output. The system is then described by the normalized nonlinear equations

$$\frac{dx_1}{dt} = x_2$$

$$\frac{dx_2}{dt} = -\sin x_1 + u \cos x_1$$

$$y = x_1$$

where x_1 is the angle and x_2 is the angular velocity. Linearizing around $u = x_1 = 0$ gives

$$\frac{dx}{dt} = \begin{pmatrix} 0 & 1 \\ -1 & 0 \end{pmatrix} x + \begin{pmatrix} 0 \\ 1 \end{pmatrix} u$$

$$y = \begin{pmatrix} 1 & 0 \end{pmatrix} x$$

(A.7)

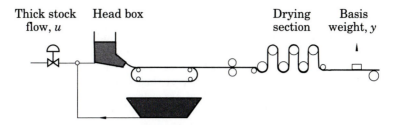

Figure A.4   Schematic diagram of a paper machine.

The transfer function of (A.7) is given by

$$G(s) = \frac{1}{s^2 + 1}$$

This transfer function can be generalized to

$$G(s) = \frac{\omega^2}{s^2 + \omega^2}$$

One state-space representation for this transfer function is

$$\frac{dx}{dt} = \begin{pmatrix} 0 & \omega \\ -\omega & 0 \end{pmatrix} x + \begin{pmatrix} 0 \\ \omega \end{pmatrix} u$$

$$y = \begin{pmatrix} 0 & 1 \end{pmatrix} x(t) \tag{A.8}$$

Sampling (A.8) using a zero-order hold gives the discrete-time system

$$x(kh + h) = \begin{pmatrix} \cos \omega h & \sin \omega h \\ -\sin \omega h & \cos \omega h \end{pmatrix} x(kh) + \begin{pmatrix} 1 - \cos \omega h \\ \sin \omega h \end{pmatrix} u(kh)$$

$$y(kh) = \begin{pmatrix} 1 & 0 \end{pmatrix} x(kh) \tag{A.9}$$

An overhead crane can also be modeled by (A.8).   ∎

Example A.4   Time-delay process

Many industrial processes can be approximated by first-order dynamics and a time delay. One example is a paper machine (see Fig. A.4). The input is the thick stock flow, that is, the amount of pulp. The output is the basic weight, that is, the thickness of the paper. The equations describing the system can be normalized to the transfer function

$$G(s) = \frac{1}{s + 1} e^{-s\tau} \tag{A.10}$$

Another physical process that can be described by (A.10) is a mixing system with long pipes. Example 2.8 gives the zero-order-hold sampling of (A.10).   ∎

Example A.5 An inventory model

An inventory is a typical example that can naturally be described as a discrete-time system. Orders and deliveries are obtained at regular intervals tied to the calendar—for example, each day or week.

Let $y(k)$ be the inventory at time k before any transaction is started. The deliveries to the inventory that are ordered at time k are $u(k)$. It is assumed that there is a delay of one period from the order until the goods start coming into the inventory. Finally, the delivery from the inventory is $v(k)$. Introduce the state variables $x_1(k) = y(k)$ and $x_2(k) = u(k-1)$. The inventory can be described by the following discrete-time state equations:

$$x_1(k+1) = x_1(k) + x_2(k) - v(k)$$
$$x_2(k+1) = u(k)$$

or

$$x(k+1) = \begin{pmatrix} 1 & 1 \\ 0 & 0 \end{pmatrix} x(k) + \begin{pmatrix} 0 \\ 1 \end{pmatrix} u(k) + \begin{pmatrix} -1 \\ 0 \end{pmatrix} v(k) \tag{A.11}$$
$$y(k) = \begin{pmatrix} 1 & 0 \end{pmatrix} x(k)$$

The input-output relation is given by

$$y(k) - y(k-1) = u(k-2) - v(k-1) \tag{A.12}$$

■

B

Matrices

B.1 Matrix Functions

In connection with sampled-data systems functions like $\exp A$ and $\ln A$, where A is a matrix, are of interest. The matrix exponential and matrix logarithm are both *matrix functions*. This section gives some properties of matrix functions and discusses some ways to compute them.

A useful property of a square matrix is given by Theorem B.1.

THEOREM B.1   THE CAYLEY-HAMILTON THEOREM   Let

$$a(\lambda) = \lambda^n + a_1 \lambda^{n-1} + \cdots + a_n = 0$$

be the characteristic equation of the square matrix A. Then A satisfies the following equation

$$a(A) = A^n + a_1 A^{n-1} + \cdots + a_n I = 0$$

That is, the matrix satisfies its own characteristic equation.   ∎

Let A be an $n \times n$ square matrix and $f(\lambda)$ a scalar function of a scalar argument λ. We now want to extend the function $f(\lambda)$ to a function with a matrix argument, that is, $f(A)$. If $f(\lambda)$ is a polynomical

$$f(\lambda) = \alpha_0 \lambda^m + \alpha_1 \lambda^{m-1} + \cdots + \alpha_m$$

then the matrix function $f(A)$ is defined as

$$f(A) = \alpha_0 A^m + \alpha_1 A^{m-1} + \cdots + a_m I$$

The eigenvalues of $f(A)$ can be found using the following theorem.

THEOREM B.2   EIGENVALUES OF A MATRIX FUNCTION   If $f(A)$ is a polynomial in A and e_i the eigenvectors of A associated with the eigenvalues λ_i, then

$$f(A)e_i = f(\lambda_i)e_i$$

so $f(\lambda_i)$ is an eigenvalue of $f(A)$ and e_i is the corresponding eigenvector.   ∎

Further, if $f(\lambda)$ can be defined by the power series

$$f(\lambda) = \sum_{i=0}^{\infty} c_i \lambda^i$$

which is assumed to be convergent for $|\lambda| < R$, then the matrix function

$$f(A) = \sum_{i=0}^{\infty} c_i A^i$$

is convergent if all the eigenvalues of A, λ_i satisfy $|\lambda_i| < R$.

By using the Cayley-Hamilton theorem, it can be shown that for every function f there is a polynomial p of degree less than n such that

$$f(A) = p(A) = \alpha_0 A^{n-1} + \alpha_1 A^{n-2} + \cdots + \alpha_{n-1} I \tag{B.1}$$

From Theorem B.2 we get

$$f(\lambda_i) = p(\lambda_i) \qquad i = 1, \ldots, n \tag{B.2}$$

If the eigenvalues are distinct, then these conditions are sufficient to determine α_i, $i = 0, \ldots, n - 1$. If there is a multiple eigenvalue with multiplicity m, then the additional conditions

$$f^{(1)}(\lambda_i) = p^{(1)}(\lambda_i)$$

$$\vdots \tag{B.3}$$

$$f^{(m-1)}(\lambda_i) = p^{(m-1)}(\lambda_i)$$

hold, where $f^{(i)}$ is the ith derivative with respect to λ.

By using (B.1) and the conditions in (B.2) and (B.3), it is possible to compute matrix functions. For low-order systems, this is a very convenient method for hand calculations.

Example B.1   Computation of matrix exponential

Let

$$A = \begin{pmatrix} 0 & 1 \\ -1 & 0 \end{pmatrix}$$

and determine

$$e^{Ah} = \alpha_0 Ah + \alpha_1 I$$

Ah has the eigenvalues $\pm ih$; the system of equations

$$e^{ih} = \alpha_0 ih + \alpha_1$$
$$e^{-ih} = -\alpha_0 ih + \alpha_1$$

holds, giving

$$\alpha_0 = \frac{1}{2ih}(e^{ih} - e^{-ih}) = \frac{\sin h}{h}$$
$$\alpha_1 = \frac{1}{2}(e^{ih} + e^{-ih}) = \cos h$$

Finally,

$$e^{Ah} = \sin h \begin{pmatrix} 0 & 1 \\ -1 & 0 \end{pmatrix} + \cos h \begin{pmatrix} 1 & 0 \\ 0 & 1 \end{pmatrix} = \begin{pmatrix} \cos h & \sin h \\ -\sin h & \cos h \end{pmatrix}$$

∎

Example B.2   Computation of matrix logarithm

Let

$$\Phi = \begin{pmatrix} 1 & h \\ 0 & 1 \end{pmatrix}$$

and compute $\ln \Phi$. The eigenvalues are given by $(\lambda - 1)^2 = 0$—that is, multiple eigenvalues exist. The matrix logarithm can now be written as

$$\ln \Phi = \alpha_0 \Phi + \alpha_1 I$$

where α_0 and α_1 are given by

$$\ln 1 = \alpha_0 + \alpha_1$$
$$\frac{\partial}{\partial \lambda}(\ln \lambda)\Big|_{\lambda = 1} = \alpha_0$$

which gives

$$0 = \alpha_0 + \alpha_1$$
$$1 = \alpha_0$$

Finally,

$$\ln \Phi = \begin{pmatrix} 1 & h \\ 0 & 1 \end{pmatrix} - \begin{pmatrix} 1 & 0 \\ 0 & 1 \end{pmatrix} = \begin{pmatrix} 0 & h \\ 0 & 0 \end{pmatrix}$$

∎

Remark. Instead of starting with the characteristic polynomial, it is possible to use the minimal polynomial of the matrix. The degree of the series in (B.1) will then be the degree of the minimal polynomial minus one. In general, this will not reduce the computing time because the minimal polynomial—or, alternatively, the Jordan form—has to be computed.

B.2 Matrix-Inversion Lemma

The following lemma is used in Sec. 13.5 to invert a matrix.

LEMMA B.1   MATRIX-INVERSION LEMMA   Let A, C, and $C^{-1} + DA^{-1}B$ be nonsingular square matrices; then

$$(A + BCD)^{-1} = A^{-1} - A^{-1}B(C^{-1} + DA^{-1}B)^{-1}DA^{-1}$$

Proof.   By direct substitution,

$$(A+BCD)\left(A^{-1} - A^{-1}B(C^{-1} + DA^{-1}B)^{-1}DA^{-1}\right)$$
$$= I + BCDA^{-1} - B(C^{-1} + DA^{-1}B)^{-1}DA^{-1}$$
$$\quad - BCDA^{-1}B(C^{-1} + DA^{-1}B)^{-1}DA^{-1}$$
$$= I + BCDA^{-1} - BC(C^{-1} + DA^{-1}B)(C^{-1} + DA^{-1}B)^{-1}DA^{-1}$$
$$= I + BCDA^{-1} - BCDA^{-1} = I$$

B.3 Notes and References

Further properties of matrices can be found in Gantmacher (1960), Bellman (1970), Barnett (1971), and Golub and Van Loan (1989).

Bibliography

ACKERMANN, J. (1972). *Abtastregelung*. Springer-Verlag, Berlin.

— (1996). *Sampled-Data Control Systems—Analysis and Synthesis, Robust Design*. Springer-Verlag, Berlin.

ANDERSON, B. D. O., and J. B. MOORE (1971). *Linear Optimal Control*. Prentice Hall, Englewood Cliffs, N.J.

— (1979). *Optimal Filtering*. Prentice Hall, Englewood Cliffs, N.J.

— (1990). *Optimal Control—Linear Quadratic Methods*. Prentice Hall, Englewood Cliffs, N.J.

ANTONIOU, A. (1979). *Digital Filters: Analysis and Design*. McGraw-Hill, New York.

ARAKI, M., and Y. ITO (1993). "Frequency-response of sampled-data systems I: Open-loop consideration." In *Preprints of the IFAC 12th World Congress*, vol. 7, pp. 289–292. Sydney.

ARNOLD III, A. F., and A. J. LAUB (1984). "Generalized eigenproblem algorithms and software for algebraic Riccati equations." *Proc. IEEE*, **72**, pp. 1746–1754.

ÅSTRÖM, K. J. (1963). "On the choice of sampling rates in optimal linear systems." Technical Report. IBM San José Research Laboratory.

— (1965). "Notes on the regulation problem." Technical Report CT211. IBM Nordic Laboratory, Lidingö.

— (1967). "Computer control of a paper machine: An application of linear stochastic control theory." *IBM J. Res. Dev.*, **II**, pp. 389–405.

— (1970). *Introduction to Stochastic Control Theory*. Academic Press, New York.

— (1980). "Piece-wise deterministic signals." In O. D. ANDERSON, Ed., *Time Series*. North Holland, Amsterdam.

— (1983a). "Computer-aided modeling, analysis and design of control systems: A perspective." *IEEE Control Syst. Mag.*, **3:2**, pp. 4–16.

— (1983b). "Theory and application of adaptive control." *Automatica*, **19**, pp. 471–486.

— (1987). "Adaptive feedback control." *Proc. IEEE*, **75**, pp. 185–217.

ÅSTRÖM, K. J., and P. E. EYKHOFF (1971). "System identification: A survey." *Automatica*, **7**, pp. 123–162.

ÅSTRÖM, K. J., P. HAGANDER, and J. STERNBY (1984). "Zeros of sampled systems." *Automatica*, **20**, pp. 31–38.

ÅSTRÖM, K. J., and T. HÄGGLUND (1995). *PID Controllers: Theory, Design, and Tuning*, 2nd ed. Instrument Society of America, Research Triangle Park, N.C.

ÅSTRÖM, K. J., and J. KANNIAH (1994). "A fast adaptive controller for motion control." *J. Syst. Eng.*, **4**, pp. 70–75.

ÅSTRÖM, K. J., and B. WITTENMARK (1973). "On self-tuning regulators." *Automatica*, **9**, pp. 185–199.

— (1980). "Self-tuning controllers based on pole-zero placement." *Proc. IEE, pt. D*, **127**, pp. 120–130.

— (1995). *Adaptive Control*, 2nd ed. Addison-Wesley, Reading, Mass.

ATHANS, M., and P. L. FALB (1966). *Optimal Control*. McGraw-Hill, New York.

ATHERTON, D. P. (1975). *Nonlinear Control Engineering—Describing Function Analysis and Design*. Van Nostrand Reinhold, London.

— (1982). "Limit cycles in relay systems." *Electron. Lett.*, **18**, pp. 922–923.

BALCHEN, J. G., and K. I. MUMMÉ (1988). *Process Control*. Van Nostrand Reinhold, New York.

BARKER, R. H. (1952). "The pulse transfer function and its applications to sampling servosystems." *Proc. IEE*, **99**, pp. 302–317.

BARNES, J. G. P. (1982). *Programming in Ada*. Addison-Wesley, New York.

BARNETT, S. (1971). *Matrices in Control Theory*. Van Nostrand Reinhold, New York.

— (1983). *Polynomials and Linear Control Systems*. Marcel Dekker, New York.

BELLMAN, R. (1957). *Dynamic Programming*. Princeton University Press, Princeton, N.J.

— (1961). *Adaptive Control: A Guided Tour*. Princeton University Press, Princeton, N.J.

— (1970). *Introduction to Matrix Analysis*. McGraw-Hill, New York.

BELLMAN, R., I. GLICKSBERG, and O. A. GROSS (1958). "Some aspects of the mathematical theory of control processes." Technical Report R-313. The RAND Corporation, Santa Monica, Calif.

BENNER, P., A. L. LAUB, and V. MEHRMANN (1995). "A collection of benchmark examples for the numerical solution of algebraic Riccati equations II: Discrete-time case." Technical Report SPC 95_23. Fak. f. Mathematik, TU Chemnitz-Zwickau, Chemnitz, FRG.

BERNHARDSSON, B. (1990). "The predictive first order hold circuit." In *Proceedings of the 29th IEEE Conference on Decision and Control*, pp. 1890–1891. Honolulu.

— (1993). "Sampling of state space systems with several time delays." In *Preprints of the IFAC 12th World Congress*, pp. 361–364. Sydney, Australia.

BERTRAM, J. E. (1958). "The effect of quantization in sampled-feedback systems." *Trans. AIEE*, **77**, pp. 177–182.

BIERMANN, G. (1977). *Factorization Methods for Discrete Estimation*. Academic Press, New York.

BITTANTI, S., A. J. LAUB, and J. C. WILLEMS (1991). *The Riccati Equation*. Springer-Verlag, Berlin.

BJÖRK, G., Å. DAHLQVIST, and N. ANDERSSON (1974). *Numerical Methods*. Prentice Hall, Englewood Cliffs, N.J.

BLANKENSHIP, W. A. (1963). "A new version of the euclidean algorithm." *Amer. Math. Monthly*, **70**, pp. 742–745.

BLOMBERG, H., and R. YLINEN (1983). *Algebraic Theory for Multivariable Linear Systems*. Academic Press, New York.

BOUDREAU, J. A. (1976). "Integrated flight control system design for CCV." In *Proceedings of the AIAA Conference on Flight Mechanics, Guidance and Control*.

BOX, G. E. P., and G. M. JENKINS (1970). *Time Series Analysis, Forecasting, and Control*. Holden-Day, San Francisco.

— (1976). *Time Series Analysis and Control*. Holden-Day, San Francisco.

BRINCH-HANSEN, P. (1973). *Operating System Principles*. Prentice Hall, Englewood Cliffs, N.J.

BRISTOL, E. H. (1977). "Design and programming control algorithms for DDC systems." *Control Eng.*, January, pp. 24–26.

— (1980). "Strategic design: A practical chapter in a textbook on control." In *Preprints of the JACC*. Paper WA4-A. San Francisco.

BROWN, G. S., and D. P. CAMPBELL (1948). *Principles of Servomechanisms*. John Wiley, New York.

BRYSON, A. E., and Y.-C. HO (1969). *Applied Optimal Control—Optimization, Estimation, and Control*. Ginn, Waltham, Mass.

BUCKLEY, P. S. (1964). *Techniques of Process Control*. John Wiley, New York.

— (1978). "Distillation column design using variable control, Part 1: Process and control design; Part 2: Economics, energy, and equipment." *Instrumentation Tech.* September, 115–122; October, 49–53.

BUCY, R. S. (1959). "Optimum finite time filters for a special nonstationary class of inputs." Technical Report. Applied Physics Laboratory, Johns Hopkins University, Baltimore.

BURNS, A., and G. DAVIES (1993). *Concurrent Programming*. International Computer Science Series. Addison–Wesley, Reading, Mass.

BURNS, A., and A. WELLINGS (1990). *Real-time systems and their programming languages*. International computer science series. Addison-Wesley, Reading, Mass.

BURNS, B. R. A. (1976). "Fly-by-wire and control configured vehicles–Rewards and risks." *Aeronaut. J.*, February.

BUTTNER, M. (1977). "Elimination of limit cycles in digital filters with very low increase in quantization noise." *IEEE Trans. Circ. Syst.*, **CAS-24**, pp. 300–304.

CAINES, P. E. (1988). *Linear Stochastic Systems.* John Wiley, New York.

CATTHOOR, F., J. RABAEY, G. GOOSSENS, J. L. V. MEERBERGER, R. JAIN, H. J. D. MAN, and J. VANDEWALLE (1988). "Architectural strategies for an application-specific synchronous multiprocessor environment." *IEEE Trans. Acoust.,* **36**, pp. 265–284.

CELLIER, F. E. (1991). *Continuous System Modeling.* Springer-Verlag, New York.

CHAR, B. W. (1992). *First Leaves: A Tutorial Introduction to Maple V.* Springer-Verlag, New York.

CHESTNUT, H., and R. W. MAYER (1959). *Servomechanisms and Regulating System Design,* vol. 1. John Wiley, New York.

CHUNG, K. L. (1974). *A Course in Probability Theory.* Academic Press, New York.

COHN, A. (1922). "Über die Anzahl der Wurzeln einer Algebraischen Gleichung in einem Kreise." *Matematische Zeitschrift,* **14**, pp. 110–148.

CROCHIEVE, R. E., and L. R. RABINER (1983). *Multirate Digital Signal Processing.* Prentice Hall, Englewood Cliffs, N.J.

CURRY, E. E. (1967). "The analysis of round-off and truncation errors in a hybrid control system." *IEEE Trans. Automat. Control,* **AC-12**, pp. 601–604.

DAHLIN, E. B. (1968). "Designing and tuning digital controllers." *Instrum. Control Syst.,* **41:6**, pp. 77–83.

DE SOUZA, C. E., and G. C. GOODWIN (1984). "Intersample variances in discrete minimum variance control." *IEEE Trans. Automat. Control,* **AC-29**, pp. 759–761.

DOETSCH, G. (1971). *Guide to the Applications of the Laplace and Z-Transforms.* Van Nostrand Reinhold, New York.

DORF, R. C., and R. H. BISHOP (1995). *Modern Control Systems,* 7th ed. Addison-Wesley, Reading, Mass.

DOYLE, J. C., and G. STEIN (1981). "Multivariable feedback design: Concepts for a classical/modern synthesis." *IEEE Trans. Automat. Control,* **AC-26**, pp. 4–16.

EMAMI-NAEINI, A., and G. F. FRANKLIN (1980). "Comments on 'The numerical solution of the discrete time algebraic Riccati equation'." *IEEE Trans. Automat. Control,* **AC-25**, pp. 1015–1016.

EYKHOFF, P. (1974). *System Identification: Parameter and State Estimation.* John Wiley, London.

— Ed. (1981). *Trends and Progress in System Identification.* Pergamon Press, Oxford.

FATEMAN, R. J. (1982). "High-level language implications of the proposed IEEE floating-point standard." *ACM Trans. Prog. Lang. Syst.,* **4:2**, pp. 239–257.

FLOWER, J. O., G. P. WINDETT, and S. C. FORGE (1971). "Aspects of the Frequency Response Testing of Simple Sampled Systems." *Int. J. Control,* **14**, pp. 881–896.

FOSS, A. S. (1973). "Critique of chemical process control theory." *IEEE Trans. Automat. Control,* **AC-18**, pp. 646–652.

FRANKLIN, G. F., and J. D. POWELL (1989). *Digital Control of Dynamic Systems,* 2nd ed. Addison-Wesley, Reading, Mass.

FRANKLIN, G. F., J. D. POWELL, and A. EMAMI-NAEINI (1994). *Feedback Control of Dynamic Systems*, 3rd ed. Addison-Wesley, Reading, Mass.

GANTMACHER, F. R. (1960). *The Theory of Matrices*, vol. I and II. Chelsea, New York.

GARDENHIRE, L. W. (1964). "Selection of sample rates." *ISA J.*, April, pp. 59–64.

GAUSS, K. F. (1809). *Theoria Motus Corporum Coelestium* (Theory of motion of the heavenly bodies). (English trans., 1963.) Dover, New York.

GAWTHROP, P. J. (1980). "Hybrid self-tuning control." *Proc. IEE*, **127**, pp. 229–236.

GEVERS, M., and G. LI (1993). *Parametrizations in Control, Estimation, and Filtering Problems: Accuracy Aspects*. Springer-Verlag, London.

GILLE, J. C., M. J. PELEGRIN, and P. DECAULNE (1959). *Feedback Control Systems*. McGraw-Hill, New York.

GOFF, K. W. (1966). "A systemic approach to DDC design." *ISA J.*, December, pp. 44–54.

GOLUB, G. H., and C. F. VAN LOAN (1989). *Matrix Computations*, 2nd ed. John Hopkins University Press, Baltimore.

GOODWIN, G. C., and R. L. PAYNE (1977). *Dynamic System Identification: Experiment Design and Data Analysis*. Academic Press, New York.

GOODWIN, G. C., and K. S. SIN (1984). *Adaptive Filtering, Prediction and Control*. Prentice Hall, Englewood Cliffs, N.J.

GORDON, G. (1969). *System Simulation*. Prentice Hall, Englewood Cliffs, N.J.

GRÆBE, S. F., and A. L. B. AHLÉN (1996). "Dynamic transfer among alternative controllers and its relation to antiwindup controller design." *IEEE Trans. Control Syst. Tech.*, **4**, pp. 92–99.

GUNKEL, III, T. L., and G. F. FRANKLIN (1963). "A general solution for linear sampled data control." *Trans. ASME J. Basic Eng.*, **85-D**, pp. 197–201.

GUPTA, M. M., Ed. (1986). *Adaptive Methods for Control System Design*. IEEE Press, New York.

GUSTAFSSON, K., and P. HAGANDER (1991). "Discrete-time LQG with cross-terms in the loss function and noise description." Technical Report TFRT-7475. Department of Automatic Control, Lund Institute of Technology.

GUSTAFSSON, K., M. LUNDH, and G. SÖDERLIND (1988). "A PI stepsize control for the numerical solution of ordinary differential equations." *BIT (Nordisk Tidskrift för Informationsbehandling)*, **28:2**, pp. 270–287.

HAGANDER, P., and A. HANSSON (1996). "How to solve singular discrete-time Riccati-equations." In *Preprints of the IFAC World Congress*, vol. C, pp. 313–318. San Francisco.

HAIRER, E., and G. WANNER (1991). *Solving Ordinary Differential Equations II—Stiff- and Differential-Algebraic Problems*. Springer-Verlag, New York.

HANUS, R. (1988). "Antiwindup and bumpless transfer: A survey." In *Proceedings of the 12th World Congress on Scientific Computation, IMACS*, vol. 2, pp. 59–65. Paris.

HIGHAM, J. D. (1968). "Single-term' control of first- and second-order processes with dead time." *Control*, February, pp. 136–140.

HUREWICZ, W. (1947). "Filters and servo systems with pulsed data." In H. M. JAMES et al., Eds., *Theory of Servomechanism.* McGraw-Hill, New York.

ISERMANN, R. (1989). *Digital Control Systems, Vol. 1: Fundamentals, Deterministic Control,* 2nd ed. Springer-Verlag, Berlin.

— (1991). *Digital Control Systems, Vol. 2: Stochastic Control, Multivariable Control, Adaptive Control, Applications,* 2nd ed. Springer-Verlag, Berlin.

— Ed. (1981). *System Identification.* Tutorial presented at the 5th IFAC Symposium on Identification and System Parameter Estimation, Darmstadt. Pergamon Press, Oxford.

JACKSON, L. B. (1970a). "On the interaction of roundoff noise and dynamic range in digital filters." *Bell Syst. Tech. J.,* **49**, pp. 159–184.

— (1970b). "Roundoff noise analysis for fixed-point digital filters realized in cascade of parallel form." *IEEE Trans. Audio Electroacoust.,* **AU-18**, pp. 107–122.

— (1979). "Limit cycles in state-space structures for digital filters." *IEEE Trans. Circ. Syst.,* **CAS-26**, pp. 67–68.

JAMES, H. M., N. B. NICHOLS, and R. S. PHILIPS (1947). *Theory of Servomechanisms.* McGraw-Hill, New York.

JENKINS, G. M., and D. G. WATTS (1968). *Spectral Analysis and Its Applications.* Holden-Day, San Francisco.

JERRI, A. J. (1977). "The Shannon sampling theorem—Its various extensions and applications: A tutorial review." *Proc. IEEE,* **65**, pp. 1565–1595.

JEŽEK, J. (1982). "New algorithm for minimal solution of linear polynomial equations." *Kybernetica,* **18**, pp. 505–516.

JOHANSSON, R. (1993). *System Modeling and Identification.* Prentice Hall, Englewood Cliffs, N. J.

JURY, E. I. (1956). "Synthesis and critical study of sampled-data control systems." *AIEE Trans.,* **75, pt. II**, pp. 141–151.

— (1957). "Hidden oscillations in sampled-data control systems." *AIEE Trans.,* **75**, pp. 391–395.

— (1958). *Sampled-Data Control Systems.* John Wiley, New York.

— (1961). "Sampling schemes in sampled-data control systems." *IRE Trans. Automat. Control,* **AC-6**, pp. 88–90.

— (1967a). "A general z-transform formula for sampled-data systems." *IEEE Trans. Automat. Control,* **AC-12**, pp. 606–608.

— (1967b). "A note on multirate sampled-data systems." *IEEE Trans. Automat. Control,* **AC-12**, pp. 319–320.

— (1980). "Sampled-data systems, revisited: Reflections, recollections, and reassessments." *Trans. ASME, J. Dyn. Syst., Measure., Control,* **102**, pp. 208–216.

— (1982). *Theory and Application of the z-Transform Method.* Krieger, Malabar, Fla.

JURY, E. I., and J. BLANCHARD (1961). "A stability test for linear discrete time systems in table form." *Proc. IRE,* **49**, pp. 1947–1948.

JURY, E. I., and Y. Z. TSYPKIN (1971). "On the theory of discrete systems." *Automatica*, **7**, pp. 89–107.

KAILATH, T. (1980). *Linear Systems*. Prentice Hall, Englewood Cliffs, N.J.

KALMAN, R. E. (1960a). "Contributions to the theory of optimal control." *Boletin de la Sociedad Matematica Mexicana*, **5**, pp. 102–119.

— (1960b). "A new approach to linear filtering and prediction problems." *Trans. ASME, Ser. D. J. Basic Eng.*, **82**, pp. 34–45.

— (1961). "On the general theory of control systems." In *Proceedings of the First IFAC Congress*, pp. 481–492.

KALMAN, R. E., and J. E. BERTRAM (1958). "General synthesis procedure for computer control of single and multiloop linear systems." *AIEE Trans*, **77**, pp. 602–609.

— (1960). "Control system analysis and design via the second method of Lyapunov: II. Discrete-time systems." *Trans. ASME. Ser. D. J. Basic Eng.*, **82:3**, pp. 394–400.

KALMAN, R. E., and R. S. BUCY (1961). "New results in linear filtering and prediction theory." *Trans. ASME, Ser. D., J. Basic Eng.*, **83**, pp. 95–107.

KALMAN, R. E., P. L. FALB, and M. A. ARBIB (1969). *Topics in Mathematical System Theory*. McGraw-Hill, New York.

KALMAN, R. E., Y. C. HO, and K. S. NARENDRA (1963). "Controllability of linear dynamical systems." In *Contributions to Different Equations*, vol. 1, pp. 189–213. John Wiley, New York.

KARLIN, S. (1966). *A First Course in Stochastic Processes*. Academic Press, New York.

KHEIR, N. A. (1988). *Systems Modeling and Computer Simulation*. Marcel Dekker, New York.

KLEINMAN, D. L. (1968). "On an iterative technique for Riccati equation computations." *IEEE Trans. Automat. Control*, **AC-13**, pp. 114–115.

KNOWLES, J. B., and R. EDWARDS (1965). "Effect of a finite-word-length computer in a sampled-data feedback system." *Proc. IEE*, **112**, pp. 1197–1207.

KOLMOGOROV, A. N. (1941). "Interpolation and extrapolation of stationary random sequences." Technical Report. Ser. Math. 5. Moscow University.

KONAR, A. F., and J. K. MAHESH (1978). "Analysis methods for multirate digital control systems." Honeywell Report No F0636-TRI. Honeywell Systems and Research Center, Minneapolis, Minn.

KOTELNIKOV, V. A. (1933). "On the transmission capacity of 'Ether' and wire in electrocommunication." In *Proceedings First All-union Conference on Questions of Communication*. Moscow.

KRANC, G. M. (1957). "Input-output analysis of multirate feedback systems." *IRE Trans. Automat. Control*, **AC-3**, pp. 21–28.

KUČERA, V. (1979). *Discrete Linear Control*. Academia, Prague.

— (1984). "The LQG problem: A study of common factors." *Probl. Control Inform. Theory*, **13**, pp. 239–251.

— (1991). *Analysis and Design of Discrete Linear Control Systems*. Prentice-Hall International, London.

— (1993). "Diophantine equations in control: A survey." *Automatica*, **29**, pp. 1361–1375.

KUMAR, P. R., and P. VARAIYA (1986). *Stochastic Systems: Estimation, Identification and Adaptive Control*. Prentice Hall, Englewood Cliffs, N.J.

KUO, B. C. (1980). *Digital Control Systems*. Holt-Saunders, Tokyo.

KWAKERNAAK, H., and R. SIVAN (1972). *Linear Optimal Control Systems*. John Wiley, New York.

KWONG, R. H. (1991). "On the linear quadratic Gaussian problem with correlated noise and its relation to minimum variance control." *SIAM J. of Control Optim.*, **29**, pp. 139–152.

LANING, J. H., and R. H. BATTIN (1956). *Random Processes in Automatic Control*. McGraw-Hill, New York.

LAWDEN, D. F. (1951). "A general theory of sampling servomechanisms." *Proc. IEE*, **98**, pp. 31–36.

LAWSON, C. L., and R. J. HANSSON (1974). *Solving Least Squares Problems*. Prentice Hall, Englewood Cliffs, N.J.

LEE, E. A. (1988). "Programmable DSP architectures: Part 1." *IEEE ASSP Mag.*, **5:4**, October, pp. 4–19.

LENNARTSON, B. (1987). "On the choice of controller and sampling period for linear stochastic control." *Preprints of the 10th IFAC World Congress*, **9**, pp. 241–246.

LENNARTSON, B., and T. SÖDERSTRÖM (1986). "An investigation of the intersample variance for linear stochastic control." *Preprints of the 25th IEEE Conference on Decision and Control*, pp. 1770–1775.

LI, Y. T., J. L. MEIRY, and R. E. CURRY (1972). "On the ideal-sampler approximation." *IEEE Trans. Automat. Control*, **AC-17**, pp. 167–168.

LINDORFF, D. P. (1965). *Theory of Sampled-Data Control Systems*. John Wiley, New York.

LINVILL, W. K. (1951). "Sampled-data control systems studied through comparison sampling with amplitude modulation." *AIEE Trans.*, **70, pt. II**, pp. 1778–1788.

LJUNG, L. (1987). *System Identification: Theory for the User*. Prentice Hall, Englewood Cliffs, N.J.

LJUNG, L., and T. SÖDERSTRÖM (1983). *Theory and Practice of Recursive Identification*. The MIT Press, Cambridge, Mass.

LUCAS, M. P. (1986). *Distributed Control Systems—Their Evaluation and Design*. Van Nostrand Reinhold, New York.

LUENBERGER, D. G. (1964). "Observing the state of a linear system." *IEEE Trans. Mil. Electron.*, **MIL-8**, pp. 74–80.

— (1971). "An introduction to observers." *IEEE Trans. Automat. Control*, **AC-16**, pp. 596–603.

MACCOLL, L. A. (1945). *Fundamental Theory of Servomechanisms*. D. Van Nostrand, New York.

MACGREGOR, J. F. (1976). "Optimal choice of the sampling interval for discrete process control." *Technometrics*, **18:2**, pp. 151–160.

MATTSSON, S.-E., M. ANDERSSON, and K. J. ÅSTRÖM (1993). "Object-oriented modeling and simulation." In D. A. LINKENS, Ed., *CAD for Control Systems*, pp. 31–69. Marcel Dekker, New York.

MCGARTY, T. P. (1974). *Stochastic Systems and State Estimation*. John Wiley, New York.

MELZER, S. M., and B. C. KUO (1971). "Sampling period sensitivity of the optimal sampled data linear regulator." *Automatica*, **7**, pp. 367–370.

MIDDLETON, R. H., and G. C. GOODWIN (1987). "Improved finite word length characteristics in digital control using delta operators." *IEEE Trans. Automat. Control*, **AC-31**, pp. 1015–1021.

— (1989). *Digital Control and Estimation: A Unified Approach*. Prentice Hall, Englewood Cliffs, N.J.

MIMINIS, G. S., and C. C. PAIGE (1982). "An algorithm for pole assignment of time invariant linear systems." *Int. J. Control*, **35:2**, pp. 341–354.

— (1988). "A direct algorithm for pole assignment of time-invariant multi-input systems using state feedback." *Automatica*, **24**, pp. 343–356.

MORARI, M., and J. H. LEE (1991). "Model predictive control: The good, the bad, and the ugly." In *Chemical Process Control, CPCIV*, pp. 419–442. Padre Island, Texas.

MORARI, M., and E. ZAFIRIOU (1989). *Robust Process Control*. Prentice Hall, Englewood Cliffs, N.J.

MORONEY, P. (1983). *Issues in the Implementation of Digital Feedback Compensators*. The MIT Press, Cambridge, Mass.

MOSCA, E., L. GIARRE, and A. CASAVOLA (1990). "On the polynomial equations for the MIMO LQ stochastic regulator." *IEEE Trans. Automat. Control*, **AC-35**, pp. 320–322.

NEUMAN, C. P., and C. S. BARADELLO (1979). "Digital transfer functions for microcomputer control." *IEEE Trans. Syst., Man, Cybern.*, **SMC-9**, pp. 856–860.

NEWTON, JR, G. C., L. A. GOULD, and J. F. KAISER (1957). *Analytical Design of Linear Feedback Controls*. John Wiley, New York.

NORTON, J. P. (1986). *An Introduction to Identification*. Academic Press, London.

NYQUIST, H. (1928). "Certain topics in telegraph transmission theory." *AIEE Trans.*, **47**, pp. 617–644.

OLDENBURG, R. C., and H. SARTORIUS (1948). *The Dynamics of Automatic Control*. ASME, New York.

OPPENHEIM, A. V., and R. W. SCHAFER (1989). *Discrete-Time Signal Processing*. Prentice Hall, Englewood Cliffs, N.J.

PAPOULIS, A. (1965). *Probability, Random Variables, and Stochastic Processes*. McGraw-Hill, New York.

PAPPAS, T., A. J. LAUB, and N. R. SANDELL, JR (1980). "On the numerical solution of the discrete time Riccate equation." *IEEE Trans. Automat. Control*, **AC-25**, pp. 631–641.

PARKER, S. R., and S. F. HESS (1971). "Limit cycle oscillations in digital filters." *IEEE Trans. Circ. Theory.*, **CT-18**, pp. 687–697.

PARZEN, E. (1962). *Stochastic Processes*. Holden-Day, San Francisco.

PERNEBO, L. (1981). "An algebraic theory for the design of controllers for multivariable systems—Part I: Structure matrices and feedforward design; Part II: Feedback realizations and feedback design." *IEEE Trans. Automat. Control*, **AC-26**, pp. 171–182 and 183–194.

PETERKA, V. (1972). "On steady-state minimum variance control strategy." *Kybernetika*, **8**, pp. 219–232.

PETKOV, P. H., N. D. CHRISTOV, and M. M. KONSTANTINOV (1984). "A computational algorithm for pole assignment of linear input systems." In *Preprints of the 23rd IEEE Conference on Decision and Control*, pp. 1770–1773.

POLYA, G. (1945). *How to Solve It*. Princeton University Press, Princeton, N.J.

PONTRYAGIN, L. S., V. G. BOLTYANSKII, R. V. GAMKRELIDZE, and E. F. MISCHENKO (1962). *The Mathematical Theory of Optimal Processes*. John Wiley, New York.

RABINER, L. R., and B. GOLD (1975). *Theory and Application of Digital Signal Processing*. Prentice Hall, Englewood Cliffs, N.J.

RAGAZZINI, J. R., and G. F. FRANKLIN (1958). *Sampled-Data Control Systems*. McGraw-Hill, New York.

RAGAZZINI, J. R., and L. A. ZADEH (1952). "The analysis of sampled-data systems." *AIEE Trans.*, **71, pt. II**, pp. 225–234.

RICHALET, J., A. RAULT, J. L. TESTUD, and J. PAPON (1978). "Model predictive heuristic control: Applications to industrial processes." *Automatica*, **14**, pp. 413–428.

RINK, R. E., and H. Y. CHONG (1979). "Performance of state regulator systems with floating-point computation." *IEEE Trans. Automat. Control*, **AC-24**, pp. 411–421.

RISSANEN, J. (1960). "Control system synthesis by analogue computer based on 'Generalized linear feedback' concept." In *Proceedings of the Symposium on Analog Computation Applied to the Study of Chemical Processes*, pp. 1–13. Brussels.

RÖNNBÄCK, S., K. S. WALGAMA, and J. STERNBY (1992). "An extension to the generalized anti-windup compensator." In P. BORNE et al., Eds., *Mathematics of the Analysis and Design of Process Control*, pp. 275–285. Elsevier, Amsterdam.

ROSENBROCK, H. H. (1970). *State-Space and Multivariable Theory*. Nelson, London.

SAFONOV, M. G. (1980). *Stability and Robustness of Multivariable Feedback Systems*. The MIT Press, Cambridge, Mass.

SANCHIS, R., and P. ALBERTOS (1995). "Design of ripple-free controllers." In *Proceedings of the 3rd European Control Conference*, pp. 3660–3664. Rome.

SCHUR, J. (1918). "Über Potenzreihen, die im inneren des Einheitskreises beschänkt sind, II." *Zeitschrift für die reine und angewandte Mathematik*, **148**, pp. 122–145.

SHANNON, C. E. (1949). "Communication in presence of noise." *Proc. IRE*, **37**, pp. 10–21.

SHINSKEY, F. G. (1988). *Process Control Systems*. McGraw-Hill, New York.

SIMON, H. A. (1956). "Dynamic programming under uncertainty with quadratic criterion function." *Econometrica*, **24**, p. 74.

SLAUGHTER, J. B. (1964). "Quantization errors digital control systems." *IEEE Trans. Automat. Control*, **AC-9**, pp. 70–74.

SMITH, O. J. M. (1957). "Closer control of loops with deadtime." *Chem. Eng. Prog.*, **53**, pp. 217–219.

SÖDERSTRÖM, T., and P. STOICA (1989). *System Identification*. Prentice-Hall International, Hemel Hempstead, U.K.

SORENSEN, H. W. (1970). "Least-squares estimation: From Gauss to Kalman." *IEEE Spectrum*, **7:7**, pp. 63–68.

TAKAHASHI, Y., C. S. CHAN, and D. M. AUSLANDER (1971). "Parametereinstellung bei linearen DDC-Algorithmen." *Regelungstechnik und Process-Datenverarbeitung*, **19**, pp. 237–244.

TOU, J. T. (1959). *Digital and Sampled-Data Systems*. McGraw-Hill, New York.

TSCHAUNER, J. (1963a). "A general formulation of the stability constraints for sampled-data controller systems." *Proc. IEEE*, **51**, pp. 619–620.

— (1963b). "Stability of sampled-data systems." *Proc. IEEE*, **51**, pp. 621–622.

TSIEN, H, S. (1955). *Engineering Cybernetics*. McGraw-Hill, New York.

TSYPKIN, Y. Z. (1949). "Theory of discontinuous control I and II." *Avtomat i Telemekh*, **10**, pp. 189–224 and 342–361.

— (1950). "Theory of discontinuous control III." *Avtomat i Telemekh*, **11**, pp. 300–319.

— (1958). *Theory of Impulse Systems*. State Publisher for Physical Mathematical Literature, Moscow.

— (1984). *Relay Control Systems*. Cambridge University Press, Cambridge, Mass.

VAN DOOREN, P. (1981). "A generalized eigenvalue approach for solving Riccati equations." *SIAM J. Sci. Statist. Comput.*, **2**, pp. 121–135.

WHITBECK, R. F. (1980). "Multirate digital control systems with simulation applications." Report AFWAL-TR-80-3101, vols. I, II, and III. Flight Dynamics Laboratory, Air Force Wright Aeronautical Laboratory, Wright-Patterson Air Force Base, Ohio.

WHITTLE, P. (1963). *Prediction and Regulation by Linear Least-Squares Methods*. English Universities Press, London.

WIENER, N. (1949). *Extrapolation, Interpolation & Smoothing of Stationary Time Series*. The MIT Press, Cambridge, Mass.

WILLEMS, J. L., and H. VAN DE VOORDE (1978). "The return difference for discrete-time optimal feedback systems." *Automatica*, **14**, pp. 511–513.

WILLIAMS, A. B. (1981). *Network Analysis and Synthesis*. McGraw-Hill, New York.

WILLIAMSON, D. (1991). *Digital Control and Implementation: Finite Wordlength Considerations*. Prentice Hall, Englewood Cliffs, N.J.

WILLSKY, A. S. (1979). *Digital Signal Processing and Control and Estimation Theory*. The MIT Press, Cambridge, Mass.

WILLSON, JR, A. N. (1972a). "Limit cycles due to adder overflow in digital filters." *IEEE Trans. Circ. Theory*, **CT-19**, pp. 342–346.

— (1972b). "Some effects of quantization and adder overflow on the forced response of digital filters." *Bell Syst. Tech. J.*, **51**, pp. 863–887.

WIRTH, N. (1979). *Algorithms + Data Structures = Programs*. Prentice Hall, Englewood Cliffs, N.J.

WITTENMARK, B. (1985a). "Design of digital controllers—The servo problem." In S. TZAFESTAS, Ed., *Applied Digital Control*. Elsevier, Amsterdam.

— (1985b). "Sampling of a system with a time-delay." *IEEE Trans. Automat. Control*, **AC-30**, pp. 507–510.

WOLFRAM, S. (1988). *Mathematica: A System for Doing Mathematics*. Addison-Wesley, Reading, Mass.

WOLOWICH, W. A. (1974). *Linear Multivariable Systems*. Springer-Verlag, New York.

WONHAM, W. M. (1974). *Linear Multivariable Control: A Geometric Approach*. Springer-Verlag, New York.

YAMAMOTO, Y. (1994). "A function space approach to sampled-data control systems and tracking problems." *IEEE Trans. Automat. Control*, **AC-39**, pp. 703–712.

YAMAMOTO, Y., and M. ARAKI (1994). "Frequency responses for sampled-data systems—their equivalence and relationships." *Linear Algebra Appl.*, **205-206**, pp. 1319–1339.

YAMAMOTO, Y., and P. P. KHARGONEKAR (1996). "Frequency response of sampled-data systems." *IEEE Trans. Automat. Control*, **AC-41**, pp. 166–176.

YOULA, D. C., J. J. BONGIORNO, and H. A. JABR (1976). "Modern Wiener-Hopf design of optimal controllers. Part I: The single-input-single-output case; Part II: The multivariable case." *IEEE Trans. Automat. Control*, **AC-21**, pp. 319–338.

ZIEGLER, J. G., and N. B. NICHOLS (1942). "Optimum settings for automatic controllers." *Trans. ASME*, **64**, pp. 759–768.

Index